STRUCTURAL MECHANICS

STRUCTURAL MECHANICS
MODELLING AND ANALYSIS OF FRAMES AND TRUSSES

Karl-Gunnar Olsson

Chalmers University of Technology, Sweden

Ola Dahlblom

Lund University, Sweden

Library of Congress Cataloging-in-Publication Data

Olsson, Karl-Gunnar, author.
    Structural mechanics : modelling and analysis of frames and trusses / Karl-Gunnar Olsson, Ola Dahlblom.
        pages cm
    Includes index.
    ISBN 978-1-119-15933-9 (pbk.)
1. Trusses–Design and construction. 2. Structural frames–Design and construction. 3. Trusses–Mathematical models. 4. Structural frames–Mathematical models. 5. Structural analysis (Engineering)–Mathematics. I. Dahlblom, Ola, author. II. Title.
    TA660.T8O47 2016
    624.1′773–dc23
                                                                                    2015028041

A catalogue record for this book is available from the British Library.

Typeset in 10/12pt TimesLTStd by SPi Global, Chennai, India

Printed and bound in Singapore by Markono Print Media Pte Ltd

1   2016

Contents

Preface

The autumn sun shines on Sunnibergbrücke at Klosters in the canton of Graubünden in south-western Switzerland. On the cover picture one can sense how the bridge elegantly migrates through the landscape. The steel and concrete structure and the architecture merge into one of the most elegant buildings of our time. The engineer who designed the bridge is named Christian Menn. It is late in October 2009, and a group of Swedish students sketch, photograph and enthusiastically discuss the shape and the structural behaviour of the bridge. In a week they will start a course in structural mechanics.

Structural mechanics is the branch of physics that describes how different materials, which have been shaped and joined together to structures, carry their loads. Knowledge on the modes of action of these structures can be used in different contexts and for different purposes. The Roman architect and engineer Vitruvius, who lived during the first century BC summarises in the work *De architectura libri decem* ('Ten books on architecture') the art of building with the three classical notions of *firmitas*, *utilitas* and *venustas* (strength, functionality and beauty). Engineering of our time has basically the same goal. It is about utilising the knowledge and practices of our time in a creative process where sustainable and efficient, functional and expressive buildings are designed.

At an early design stage a structural engineer needs to be trained to see how to efficiently use material and shape to provide the construction with stability, stiffness and strength. Using simple models, structural behaviour can be evaluated and cross-section sizes estimated. As the design develops the need for precision of the analyses increases. In all this, the ability to formulate computational models and to carry out simulations is of crucial importance.

A useful computational model should be simple enough to be easily manageable and, simultaneously, sufficiently complex to provide an adequate accuracy. In recent years, the finite element method has become the dominant method for formulating computational models and conducting analyses. The FE method is based on expressing forces and deformations as discrete entities in a chosen and representative set of degrees of freedom. Between the degrees of freedom simple bodies (elements) are placed and together they constitute the structure to be modelled. Each element may describe a unique mode of action and can be given a specific geometry. In all this, FEM provides opportunities for both accurate analyses of structures with complex geometry and material behaviour, and for quick estimates in early design stages.

Here, we present a new textbook in structural mechanics, dealing with the modelling and analysis of trusses and frames. The textbook is based on the finite element method. Gradually, an understanding of basic elements of structural mechanics – springs, bars, beams, foundations and so on is built up. Methods for assembling them into complex load-bearing structures are presented, and tools for analysis and simulation are provided. The book has been limited to treating trusses and frames in two and three dimensions. To demonstrate the generality of the methodology the book also has a chapter, 'Flows in Networks', that addresses other areas of applied mechanics, including thermal conduction and electrical flow.

The textbook supports three kinds of learning outcome:

- *Knowledge of basic theory of structural mechanics*. The textbook has a structure that highlights the theory as a whole. Different modes of action in structural mechanics are described in a common format where basic concepts and relationships recur at different scale levels. One aim is to highlight the mechanisms that determine how structures carry their loads and how we by this knowledge can manipulate the distribution of internal forces as well as patterns of deformations.
- *Skills in modelling and analysis of structures*. Being able to describe a structure by a mathematical model and perform computations is one of the most important engineering skills. The matrix-based presentation of the textbook practices a computation methodology that is general and can be applied for phenomena and geometries of structural mechanics as well as for simulations in a variety of engineering areas far beyond the textbook limitations. Through exercises and with support from the computer program Matlab/CALFEM students in a course formulate about 30 computer algorithms of their own, each with increasing complexity.
- *Ability to evaluate and optimise designs proposed*. Having an eye trained for patterns of forces and deformations helps to evaluate and improve the efficiency of structural designs. This facilitates modification of the design of a structure in the desired direction, thus creating an efficient structural behaviour, for example by reducing bending in the favour of axial only forces – compression and tension.

The textbook is intended for engineering students at the bachelor level. The presentation assumes knowledge of calculus in one variable, linear algebra, classical mechanics and basic solid/structural mechanics. Chapters 1–5 are a unit and should be read in the order they appear, while Chapters 6–10 are independent of each other and can be read in any order. For a limited course, we recommend primarily Chapters 1–6.

The Division of Structural Mechanics at Lund University has a long tradition in the development of teaching materials in structural mechanics and the finite element method. A key person behind this development is Hans Petersson who came to the division as a professor in 1977. Within a few years, a group of young Ph.D. students and teachers gathered around Hans, taking note of his knowledge and absorbed his enthusiasm about teaching and its tools. We were two of them. Earlier, the framework of the computer program CALFEM (Computer Aided Learning of the Finite Element Method) was developed, and based on his concept the textbook 'Konstruktionsberäkningar med dator' (Design calculations using a computer) was written with Sven Thelandersson as author. In this spirit, the division has continued to develop teaching materials, and approaches. In more than 30 years time, both ideas and collaborators spread. CALFEM is today a toolbox to the computer program Matlab and is used worldwide. In Sweden, collaboration between Lund University, Chalmers and KTH Royal Institute of Technology has been established, and from the site www.structarch.org, CALFEM as well as other software for structural mechanics analysis and conceptual design can be downloaded free of charge.

The contents of this textbook have been developed over many years and there are many students and colleagues at Lund University, Chalmers and Linnæus University, who contributed with ideas, suggestions, corrections and translations during the creation of the book. We would particularly like to mention Professor Per-Erik Austrell, Dr. Henrik Danielsson, Dr. Susanne Heyden and Professor Kent Persson at Structural Mechanics in Lund, Dr. Mats Ander and Dr. Peter Möller at Applied Mechanics at Chalmers and Ms. Louise Blyberg and Professor Anders Olsson at Linnæus University in Växjö. Professor Emeritus Bengt Åkesson at Chalmers has with great precision and sharpness examined facts of the manuscript and given us reason to examine and modify the conceptual choices and formulations. Dr. Samar Malek has thoroughly proofread the English version of the text. Mr. Bo Zadig at Structural Mechanics in Lund has skilfully drawn the figures. Sincere thanks to all of you for your commitment and wise observations. And to Professor Göran Sandberg who with his character, his knowledge and in his role as head of the department has built and continues to build a creative environment for the teaching and development of teaching concepts and tools. We want to thank people at John Wiley & Sons and their partners for cooperation and guidance. In particular we are grateful to Eric Willner, Anne Hunt, Clive Lawson and Lincy Priya.

The textbook is also available in Swedish, with the reverse order of authors.

Karl-Gunnar Olsson and Ola Dahlblom
Gothenburg and Lund in October 2015

1

Matrix Algebra

The method used in this textbook to formulate computational models is characterised by the use of matrices. The different quantities – load, section force, stiffness and displacement – are separated and gathered into groups of numbers. All load values are gathered in a load matrix and all stiffnesses in a stiffness matrix. This is one of the primary strengths of the method. With a matrix formulation, the formulae describing the relations between quantities are compact and easy to view. Physical mechanisms and underlying principles become clear. We begin with a short summary of the matrix algebra and the notations that are used.

1.1 Definitions

A matrix consists of a set of *matrix elements* ordered in *rows* and *columns*. If the matrix consists of only one column it is referred to as a *column matrix* and if it has only one row it is referred to as a *row matrix*. Such matrices are *one-dimensional* and may also be referred to as *vectors*. A vector is denoted by a lower case letter set in bold:

$$\mathbf{a} = \begin{bmatrix} a_1 \\ a_2 \\ a_3 \end{bmatrix} \tag{1.1}$$

where a_1, a_2 and a_3 are the components of the vector. A *two-dimensional* matrix is denoted by a capital letter set in bold:

$$\mathbf{A} = \begin{bmatrix} A_{11} & A_{12} & A_{13} \\ A_{21} & A_{22} & A_{23} \\ A_{31} & A_{32} & A_{33} \\ A_{41} & A_{42} & A_{43} \end{bmatrix}; \quad \mathbf{B} = \begin{bmatrix} B_{11} & B_{12} & B_{13} \\ B_{21} & B_{22} & B_{23} \\ B_{31} & B_{32} & B_{33} \end{bmatrix} \tag{1.2}$$

where A_{11}, A_{12} and so on are elements of the matrix $\mathbf{A}$. An arbitrary component of a matrix is denoted A_{ij}, where the first index refers to the row number and the second index to the column number. The matrix $\mathbf{A}$ in (1.2) has the *dimensions* 4×3 and the matrix $\mathbf{B}$ has the dimensions 3×3.

Structural Mechanics: Modelling and Analysis of Frames and Trusses, First Edition.
Karl-Gunnar Olsson and Ola Dahlblom.
© 2016 John Wiley & Sons, Ltd. Published 2016 by John Wiley & Sons, Ltd.

Since the number of rows and columns in **B** are equal, it is a *square matrix*. If it is only the *diagonal elements* B_{ii} that are different from 0, the matrix is a *diagonal matrix*. A diagonal matrix where all the diagonal elements are equal to 1 is an *identity matrix* and is usually denoted **I**. The *transposed matrix* $\mathbf{A}^T$ of a matrix **A** is formed by letting the rows of **A** become columns of $\mathbf{A}^T$, that is the *transpose* of **A** in (1.2) is

$$\mathbf{A}^T = \begin{bmatrix} A_{11} & A_{21} & A_{31} & A_{41} \\ A_{12} & A_{22} & A_{32} & A_{42} \\ A_{13} & A_{23} & A_{33} & A_{43} \end{bmatrix} \tag{1.3}$$

A matrix **A** is *symmetric* if $\mathbf{A} = \mathbf{A}^T$. Only square matrices can be symmetric. A matrix with all elements equal to 0 is referred to as a *zero matrix* and is usually denoted **0**.

1.2   Addition and Subtraction

Matrices of equal dimensions can be added and subtracted. The result is a new matrix of the same dimensions, where each element is the sum of or the difference between the corresponding elements of the two matrices. If

$$\mathbf{A} = \begin{bmatrix} A_{11} & A_{12} & A_{13} \\ A_{21} & A_{22} & A_{23} \\ A_{31} & A_{32} & A_{33} \end{bmatrix}; \quad \mathbf{B} = \begin{bmatrix} B_{11} & B_{12} & B_{13} \\ B_{21} & B_{22} & B_{23} \\ B_{31} & B_{32} & B_{33} \end{bmatrix} \tag{1.4}$$

the sum of **A** and **B** is given by

$$\mathbf{C} = \mathbf{A} + \mathbf{B} \tag{1.5}$$

where

$$\mathbf{C} = \begin{bmatrix} A_{11} + B_{11} & A_{12} + B_{12} & A_{13} + B_{13} \\ A_{21} + B_{21} & A_{22} + B_{22} & A_{23} + B_{23} \\ A_{31} + B_{31} & A_{32} + B_{32} & A_{33} + B_{33} \end{bmatrix} \tag{1.6}$$

and the difference between **A** and **B** is given by

$$\mathbf{D} = \mathbf{A} - \mathbf{B} \tag{1.7}$$

where

$$\mathbf{D} = \begin{bmatrix} A_{11} - B_{11} & A_{12} - B_{12} & A_{13} - B_{13} \\ A_{21} - B_{21} & A_{22} - B_{22} & A_{23} - B_{23} \\ A_{31} - B_{31} & A_{32} - B_{32} & A_{33} - B_{33} \end{bmatrix} \tag{1.8}$$

1.3   Multiplication

Multiplying a matrix **A** with a scalar c results in a matrix with the same dimensions as **A** and where each element is the corresponding element of **A** multiplied by c, that is

$$c\mathbf{A} = \begin{bmatrix} cA_{11} & cA_{12} & cA_{13} \\ cA_{21} & cA_{22} & cA_{23} \\ cA_{31} & cA_{32} & cA_{33} \end{bmatrix} \tag{1.9}$$

Multiplication between two matrices

$$\mathbf{C} = \mathbf{AB} \tag{1.10}$$

can be performed only if the number of columns in $\mathbf{A}$ equals the number of rows in $\mathbf{B}$. The element C_{ij} is then computed according to

$$C_{ij} = \sum_{k=1}^{n} A_{ik} B_{kj} \tag{1.11}$$

For

$$\mathbf{A} = \begin{bmatrix} A_{11} & A_{12} \\ A_{21} & A_{22} \end{bmatrix}; \quad \mathbf{B} = \begin{bmatrix} B_{11} & B_{12} \\ B_{21} & B_{22} \end{bmatrix} \tag{1.12}$$

the product of the matrices, $\mathbf{C} = \mathbf{AB}$, is obtained from

$$\mathbf{C} = \begin{bmatrix} A_{11} & A_{12} \\ A_{21} & A_{22} \end{bmatrix} \begin{bmatrix} B_{11} & B_{12} \\ B_{21} & B_{22} \end{bmatrix} = \begin{bmatrix} A_{11}B_{11} + A_{12}B_{21} & A_{11}B_{12} + A_{12}B_{22} \\ A_{21}B_{11} + A_{22}B_{21} & A_{21}B_{12} + A_{22}B_{22} \end{bmatrix} \tag{1.13}$$

In general,

$$\mathbf{BA} \neq \mathbf{AB} \tag{1.14}$$

1.4 Determinant

For every quadratic matrix $\mathbf{A}$ $(n \times n)$, it is possible to compute a scalar value called a *determinant*. For $n = 1$,

$$\det \mathbf{A} = A_{11} \tag{1.15}$$

For $n > 1$, the determinant $\det \mathbf{A}$ is computed according to the expression

$$\det \mathbf{A} = \sum_{k=1}^{n} (-1)^{i+k} A_{ik} \det M_{ik} \tag{1.16}$$

where i is an arbitrary row number and $\det M_{ik}$ is the determinant of the matrix obtained when the ith row and the kth column is deleted from the matrix $\mathbf{A}$. For $n = 2$, this results in

$$\det \mathbf{A} = A_{11}A_{22} - A_{12}A_{21} \tag{1.17}$$

and for $n = 3$

$$\det \mathbf{A} = A_{11}A_{22}A_{33} + A_{12}A_{23}A_{31} + A_{13}A_{21}A_{32} - A_{11}A_{23}A_{32} - A_{12}A_{21}A_{33} - A_{13}A_{22}A_{31} \tag{1.18}$$

1.5 Inverse Matrix

The quadratic matrix $\mathbf{A}$ is *invertible* if there exists a matrix $\mathbf{A}^{-1}$ such that

$$\mathbf{A}^{-1}\mathbf{A} = \mathbf{I} \tag{1.19}$$

The matrix $\mathbf{A}^{-1}$ is then the *inverse* of $\mathbf{A}$. For the inverse $\mathbf{A}^{-1}$ to exist, it is necessary that $\det \mathbf{A} \neq 0$. If

$$\mathbf{A}^{-1} = \mathbf{A}^T \tag{1.20}$$

the matrix $\mathbf{A}$ is *orthogonal* and then

$$\mathbf{A}^T \mathbf{A} = \mathbf{A} \mathbf{A}^T = \mathbf{I} \tag{1.21}$$

1.6 Counting Rules

The following counting rules apply to matrices (under the condition that the dimensions of the matrices included are such that the operations are defined).

$$\mathbf{A} + \mathbf{B} = \mathbf{B} + \mathbf{A} \tag{1.22}$$

$$\mathbf{A} + (\mathbf{B} + \mathbf{C}) = (\mathbf{A} + \mathbf{B}) + \mathbf{C} \tag{1.23}$$

$$(\mathbf{A} + \mathbf{B})^T = \mathbf{A}^T + \mathbf{B}^T \tag{1.24}$$

$$(\mathbf{AB})^T = \mathbf{B}^T \mathbf{A}^T \tag{1.25}$$

$$\mathbf{IA} = \mathbf{A} \tag{1.26}$$

$$c(\mathbf{AB}) = (c\mathbf{A})\mathbf{B} = \mathbf{A}(c\mathbf{B}) \tag{1.27}$$

$$(c + d)\mathbf{A} = c\mathbf{A} + d\mathbf{A} \tag{1.28}$$

$$c(\mathbf{A} + \mathbf{B}) = c\mathbf{A} + c\mathbf{B} \tag{1.29}$$

$$(\mathbf{AB})\mathbf{C} = \mathbf{A}(\mathbf{BC}) \tag{1.30}$$

$$(\mathbf{A} + \mathbf{B})\mathbf{C} = \mathbf{AC} + \mathbf{BC} \tag{1.31}$$

$$\mathbf{A}(\mathbf{B} + \mathbf{C}) = \mathbf{AB} + \mathbf{AC} \tag{1.32}$$

$$\det \mathbf{AB} = \det \mathbf{A} \det \mathbf{B} \tag{1.33}$$

$$\det \mathbf{A}^{-1} = 1 / \det \mathbf{A} \tag{1.34}$$

$$\det c\mathbf{A} = c^n \det \mathbf{A} \tag{1.35}$$

$$(\mathbf{A}^{-1})^T = (\mathbf{A}^T)^{-1} \tag{1.36}$$

$$(\mathbf{AB})^{-1} = \mathbf{B}^{-1} \mathbf{A}^{-1} \tag{1.37}$$

1.7 Systems of Equations

A *linear system of equations* with n equations and p unknowns can be written in matrix form as

$$\mathbf{K}\,\mathbf{a} = \mathbf{f} \tag{1.38}$$

where $\mathbf{K}$ has the dimensions $n \times p$, $\mathbf{a}$ the dimensions $p \times 1$ and $\mathbf{f}$ the dimensions $n \times 1$. Usually, the coefficients in $\mathbf{K}$ are known, while the coefficients in $\mathbf{a}$ and $\mathbf{f}$ can be known as well as unknown. For the case when all the components of **a** are unknown and all the components of **f** are known, there are three types of systems of equations:

- $n = p$, the number of equations equals the number of unknowns. The matrix **K** is quadratic. Depending on the contents of **K** and **f**, four different characteristic cases can be recognised. These are often indications of different states or behaviours that may be important to notice: If $\det \mathbf{K} \neq 0$, there is a *unique solution*.
  - For $\mathbf{f} = \mathbf{0}$, this solution is the trivial one, $\mathbf{a} = \mathbf{0}$.
  - For $\mathbf{f} \neq \mathbf{0}$, there is a unique solution, $\mathbf{a} \neq \mathbf{0}$. In general, this is an indication of a functioning physical model.

  If $\det \mathbf{K} = 0$, there is no unique solution. This may be an indication of an, in some way, unstable physical model.
  - For $\mathbf{f} = \mathbf{0}$, there are infinitely many solutions. This is the case for eigenvalue problems, which, for example, can be a method to gain knowledge about unstable states of the model.
  - For $\mathbf{f} \neq \mathbf{0}$, there is either none or infinitely many solutions; there may be elements missing in the model or the set of boundary conditions may be incomplete.
- $n < p$, the number of equations is less than the number of unknowns. The system is underdetermined. There are infinitely many solutions.
- $n > p$, the number of equations exceeds the number of unknowns. The system is overdetermined. In general, there is no solution.

In the following symmetric matrices, **K** and **A** are considered which are common in the forthcoming applications.

1.7.1 Systems of Equations with Only Unknown Components in the Vector **a**

For the case when $\det \mathbf{K} \neq 0$ and $\mathbf{f} \neq \mathbf{0}$, the unknowns in the vector **a** can be determined by Gaussian elimination. This is shown in the following example.

Example 1.1 Solving a system of equations with only unknown components in the vector a

We are looking for a solution to the system of equations

$$\begin{bmatrix} 8 & -4 & -2 \\ -4 & 10 & -4 \\ -2 & -4 & 10 \end{bmatrix} \begin{bmatrix} a_1 \\ a_2 \\ a_3 \end{bmatrix} = \begin{bmatrix} -8 \\ 18 \\ 6 \end{bmatrix} \tag{1}$$

The unknowns are determined by Gaussian elimination. In this procedure, all elements different from 0 are eliminated below the diagonal: let the first row remain unchanged. From row 2 we subtract row 1 multiplied by the quotient $K_{21}/K_{11} = -4/8 = -0.5$. From row 3 we subtract row 1 multiplied by the quotient $K_{31}/K_{11} = -2/8 = -0.25$. In this way, we obtain

$$\begin{bmatrix} 8 & -4 & -2 \\ 0 & 8 & -5 \\ 0 & -5 & 9.5 \end{bmatrix} \begin{bmatrix} a_1 \\ a_2 \\ a_3 \end{bmatrix} = \begin{bmatrix} -8 \\ 14 \\ 4 \end{bmatrix} \tag{2}$$

In the next step, we let the rows 1 and 2 remain. From row 3 we subtract row 2 multiplied by the quotient $K_{32}/K_{22} = -5/8 = -0.625$. We have triangularised the coefficient matrix $\mathbf{K}$ and obtain

$$\begin{bmatrix} 8 & -4 & -2 \\ 0 & 8 & -5 \\ 0 & 0 & 6.375 \end{bmatrix} \begin{bmatrix} a_1 \\ a_2 \\ a_3 \end{bmatrix} = \begin{bmatrix} -8 \\ 14 \\ 12.75 \end{bmatrix} \tag{3}$$

With the system of equations in this form, we can determine a_3, a_2 and a_1 by back-substitution

$$a_3 = \frac{12.75}{6.375} = 2; \quad a_2 = \frac{14 - (-5)a_3}{8} = 3;$$

$$a_1 = \frac{-8 - (-4)a_2 - (-2)a_3}{8} = 1 \tag{4}$$

and with that, we have the solution

$$\begin{bmatrix} a_1 \\ a_2 \\ a_3 \end{bmatrix} = \begin{bmatrix} 1 \\ 3 \\ 2 \end{bmatrix} \tag{5}$$

To check the results, we can substitute the solution into the original system of equations and carry out the matrix multiplication

$$\begin{bmatrix} 8 & -4 & -2 \\ -4 & 10 & -4 \\ -2 & -4 & 10 \end{bmatrix} \begin{bmatrix} 1 \\ 3 \\ 2 \end{bmatrix} \quad \text{which gives} \quad \begin{bmatrix} -8 \\ 18 \\ 6 \end{bmatrix} \tag{6}$$

This is equal to the original right-hand side of the system of equations, that is the solution found is correct.

1.7.2 Systems of Equations with Known and Unknown Components in the Vector **a**

The systems of equations that we consider, in general, has a square matrix $\mathbf{K}$, initially with $\det \mathbf{K} = 0$, and a vector $\mathbf{f} \neq \mathbf{0}$. Moreover, it is usually the case that some components of $\mathbf{a}$ are known and the corresponding components of $\mathbf{f}$ are unknown. One systematic way to solve such a system of equations begins with a *partition* of the matrices, which means that they are divided into *submatrices*

$$\mathbf{K} = \begin{bmatrix} \mathbf{A}_1 & \mathbf{A}_2 \\ \mathbf{A}_3 & \tilde{\mathbf{K}} \end{bmatrix}; \quad \mathbf{a} = \begin{bmatrix} \mathbf{g} \\ \tilde{\mathbf{a}} \end{bmatrix}; \quad \mathbf{f} = \begin{bmatrix} \mathbf{r} \\ \tilde{\mathbf{f}} \end{bmatrix} \tag{1.39}$$

where the matrices $\mathbf{A}_1, \mathbf{A}_2, \mathbf{A}_3, \tilde{\mathbf{K}}, \mathbf{g}$ and $\tilde{\mathbf{f}}$ contain known quantities, while $\tilde{\mathbf{a}}$ and $\mathbf{r}$ are unknown. With use of these submatrices, the system of equations (1.38) can be expressed as

$$\begin{bmatrix} \mathbf{A}_1 & \mathbf{A}_2 \\ \mathbf{A}_3 & \tilde{\mathbf{K}} \end{bmatrix} \begin{bmatrix} \mathbf{g} \\ \tilde{\mathbf{a}} \end{bmatrix} = \begin{bmatrix} \mathbf{r} \\ \tilde{\mathbf{f}} \end{bmatrix} \tag{1.40}$$

The system of equations can be divided into two parts and then be written as

$$\mathbf{A}_1 \mathbf{g} + \mathbf{A}_2 \tilde{\mathbf{a}} = \mathbf{r} \tag{1.41}$$

$$\mathbf{A}_3 \mathbf{g} + \tilde{\mathbf{K}} \tilde{\mathbf{a}} = \tilde{\mathbf{f}} \tag{1.42}$$

or

$$\tilde{\mathbf{K}} \tilde{\mathbf{a}} = \tilde{\mathbf{f}} - \mathbf{A}_3 \mathbf{g} \tag{1.43}$$

$$\mathbf{r} = \mathbf{A}_1 \mathbf{g} + \mathbf{A}_2 \tilde{\mathbf{a}} \tag{1.44}$$

where the right-hand side of the equation (1.43) consists of known quantities. The purpose of the partition of the system of equations is to, within the original system of equations, find a sub-system with det $\tilde{\mathbf{K}} \neq 0$, that is a system with a unique solution. The unknowns in $\tilde{\mathbf{a}}$ can then be computed from (1.43). One way to perform this computation is to use Gaussian elimination. Once $\tilde{\mathbf{a}}$ has been determined, $\mathbf{r}$ can be computed from (1.44).

Example 1.2 Solving a system of equations with both known and unknown components in the vector a

In the system of equations

$$
\left[
\begin{array}{cccc|cc}
20 & 0 & 0 & 0 & -20 & 0 \\
0 & 15 & 0 & -15 & 0 & 0 \\
0 & 0 & 16 & 12 & -16 & -12 \\
0 & -15 & 12 & 24 & -12 & -9 \\
\hline
-20 & 0 & -16 & -12 & 36 & 12 \\
0 & 0 & -12 & -9 & 12 & 9
\end{array}
\right]
\left[
\begin{array}{c}
0 \\
0 \\
-3 \\
0 \\
a_5 \\
a_6
\end{array}
\right]
=
\left[
\begin{array}{c}
f_1 \\
f_2 \\
f_3 \\
f_4 \\
0 \\
-15
\end{array}
\right]
\tag{1}
$$

the vector $\mathbf{a}$ has known and unknown components. The solution can then be systematised using partitioning (1.40). The auxiliary lines show this partition. The system of equations is partitioned into two parts according to (1.41) and (1.42):

$$
\left[
\begin{array}{cccc}
20 & 0 & 0 & 0 \\
0 & 15 & 0 & -15 \\
0 & 0 & 16 & 12 \\
0 & -15 & 12 & 24
\end{array}
\right]
\left[
\begin{array}{c}
0 \\
0 \\
-3 \\
0
\end{array}
\right]
+
\left[
\begin{array}{cc}
-20 & 0 \\
0 & 0 \\
-16 & -12 \\
-12 & -9
\end{array}
\right]
\left[
\begin{array}{c}
a_5 \\
a_6
\end{array}
\right]
=
\left[
\begin{array}{c}
f_1 \\
f_2 \\
f_3 \\
f_4
\end{array}
\right]
\tag{2}
$$

$$
\left[
\begin{array}{cccc}
-20 & 0 & -16 & -12 \\
0 & 0 & -12 & -9
\end{array}
\right]
\left[
\begin{array}{c}
0 \\
0 \\
-3 \\
0
\end{array}
\right]
+
\left[
\begin{array}{cc}
36 & 12 \\
12 & 9
\end{array}
\right]
\left[
\begin{array}{c}
a_5 \\
a_6
\end{array}
\right]
=
\left[
\begin{array}{c}
0 \\
-15
\end{array}
\right]
\tag{3}
$$

In the lower system of equations, there are two equations and two unknowns. If the known terms of the system are gathered on the right-hand side of the equal sign, cf. (1.43), we obtain

$$
\left[
\begin{array}{cc}
36 & 12 \\
12 & 9
\end{array}
\right]
\left[
\begin{array}{c}
a_5 \\
a_6
\end{array}
\right]
=
\left[
\begin{array}{c}
0 \\
-15
\end{array}
\right]
-
\left[
\begin{array}{cccc}
-20 & 0 & -16 & -12 \\
0 & 0 & -12 & -9
\end{array}
\right]
\left[
\begin{array}{c}
0 \\
0 \\
-3 \\
0
\end{array}
\right]
\tag{4}
$$

or

$$\begin{bmatrix} 36 & 12 \\ 12 & 9 \end{bmatrix}\begin{bmatrix} a_5 \\ a_6 \end{bmatrix} = \begin{bmatrix} -48 \\ -51 \end{bmatrix} \tag{5}$$

From this system of equations, the unknown elements can be determined by Gaussian elimination: the first row remains unchanged. From row 2 we subtract row 1 multiplied by the quotient $K_{21}/K_{11} = 12/36 = 0.33333$. In this way, we obtain

$$\begin{bmatrix} 36 & 12 \\ 0 & 5 \end{bmatrix}\begin{bmatrix} a_5 \\ a_6 \end{bmatrix} = \begin{bmatrix} -48 \\ -35 \end{bmatrix} \tag{6}$$

and the unknown a_5 and a_6 can be determined by back-substitution

$$a_6 = \frac{-35}{5} = -7; \quad a_5 = \frac{-48 - 12a_6}{36} = 1 \tag{7}$$

$$\begin{bmatrix} a_5 \\ a_6 \end{bmatrix} = \begin{bmatrix} 1 \\ -7 \end{bmatrix} \tag{8}$$

With a_5 and a_6 being known, the unknown coefficients in $\mathbf{f}$ can be determined using the upper system of equations obtained from the partition, cf. (1.44),

$$\begin{bmatrix} f_1 \\ f_2 \\ f_3 \\ f_4 \end{bmatrix} = \begin{bmatrix} 20 & 0 & 0 & 0 \\ 0 & 15 & 0 & -15 \\ 0 & 0 & 16 & 12 \\ 0 & -15 & 12 & 24 \end{bmatrix}\begin{bmatrix} 0 \\ 0 \\ -3 \\ 0 \end{bmatrix} + \begin{bmatrix} -20 & 0 \\ 0 & 0 \\ -16 & -12 \\ -12 & -9 \end{bmatrix}\begin{bmatrix} 1 \\ -7 \end{bmatrix} = \begin{bmatrix} -20 \\ 0 \\ 20 \\ 15 \end{bmatrix} \tag{9}$$

and with that, all the unknowns are determined.

1.7.3   Eigenvalue Problems

At times it is of interest to study the case when $\det \mathbf{K} = 0$ and $\mathbf{f} = \mathbf{0}$. Mainly, two different types of problems appear. A system of equations in the form

$$(\mathbf{A} - \lambda\mathbf{I})\mathbf{a} = \mathbf{0} \tag{1.45}$$

is referred to as an *eigenvalue problem* or sometimes *standard eigenvalue problem*. For a solution to exist, it is required that

$$\det(\mathbf{A} - \lambda\mathbf{I}) = 0 \tag{1.46}$$

A system of equations in the form

$$(\mathbf{A} - \lambda\mathbf{B})\mathbf{a} = \mathbf{0} \tag{1.47}$$

is referred to as a *generalised eigenvalue problem* and for a solution to exist it is required that

$$\det(\mathbf{A} - \lambda\mathbf{B}) = 0 \tag{1.48}$$

Solving an eigenvalue problem means that the values of λ, which fulfil Equations (1.46) and (1.48) are determined, that is the eigenvalues λ_i are computed. The number of eigenvalues

λ_i is equal to the number of unknowns in the system of equations. Two or more eigenvalues may coincide. A symmetric matrix K with real elements has only real eigenvalues. For each eigenvalue λ_i there is an eigenvector a_i. The unknowns in the eigenvector a_i cannot be uniquely determined, but their relative magnitude can be computed.

If the product of two vectors $b^T c = 0$, then the vectors b and c are orthogonal. For eigenvectors, we have $a_i^T a_j = 0$ for $i \neq j$, that is any two eigenvectors are always orthogonal.

The following example shows how an eigenvalue problem is solved:

Example 1.3 Solving an eigenvalue problem
We want to find a solution to the eigenvalue problem

$$(A - \lambda I)a = 0 \tag{1}$$

where

$$A = \begin{bmatrix} 5 & -2 \\ -2 & 8 \end{bmatrix} \tag{2}$$

The determinant of $(A - \lambda I)$ can be computed as

$$\det(A - \lambda I) = \det \begin{bmatrix} 5 - \lambda & -2 \\ -2 & 8 - \lambda \end{bmatrix} = (5 - \lambda)(8 - \lambda) - 4 = \lambda^2 - 13\lambda + 36 \tag{3}$$

When this expression is set to zero, the equation

$$\lambda^2 - 13\lambda + 36 = 0 \tag{4}$$

is obtained. The solutions to this equation are the eigenvalues

$$\lambda_1 = 4; \quad \lambda_2 = 9 \tag{5}$$

By substituting the computed eigenvalues into the first equation in the original system of equations we obtain

$$(5 - 4)a_1 - 2a_2 = 0; \quad a_1 = t_1 \begin{bmatrix} 2 \\ 1 \end{bmatrix} \tag{6}$$

and

$$(5 - 9)a_1 - 2a_2 = 0; \quad a_2 = t_2 \begin{bmatrix} 1 \\ -2 \end{bmatrix} \tag{7}$$

where t_1 and t_2 are arbitrary scalar multipliers, $t_1 \neq 0$, $t_2 \neq 0$. Had we substituted the eigenvalues into the second equation instead, the results would be the same. Computation of the product of the two eigenvectors yields

$$a_1^T a_2 = t_1 t_2 \begin{bmatrix} 2 & 1 \end{bmatrix} \begin{bmatrix} 1 \\ -2 \end{bmatrix} = 0 \tag{8}$$

The fact that the product is 0 means that the eigenvectors a_1 and a_2 are orthogonal.

Exercises

1.1 Begin with the matrices

$$\mathbf{A} = \begin{bmatrix} 2 & 3 & -1 \\ 4 & 8 & 0 \end{bmatrix}; \quad \mathbf{B} = \begin{bmatrix} 0 & -2 & 4 \\ 1 & 0 & 2 \end{bmatrix}; \quad \mathbf{C} = \begin{bmatrix} 1 & 0 & 3 \\ 4 & 2 & 1 \\ 3 & 4 & 1 \end{bmatrix}$$

and perform the following matrix operations manually.
(a) $\mathbf{A} + \mathbf{B}$
(b) $\mathbf{A}\mathbf{B}^T$
(c) $\mathbf{B}^T\mathbf{A}$
(d) $\mathbf{A}\mathbf{C}$
(e) $\det \mathbf{C}$

1.2 Introduce the matrices

$\mathbf{A}$ with dimensions 4×3

$\mathbf{B}$ with dimensions 3×6

$\mathbf{C}$ with dimensions 1×8

$\mathbf{D}$ with dimensions 6×1

Which of the following operations are possible to perform? For the possible operations, give the dimensions of $\mathbf{E}$
(a) $\mathbf{E} = \mathbf{AB}$
(b) $\mathbf{E} = \mathbf{BD}$
(c) $\mathbf{E} = \mathbf{ABCD}$
(d) $\mathbf{E} = \mathbf{ABDC}$
(e) $\mathbf{E} = \mathbf{B}^T\mathbf{A}^T$

1.3 Solve the following system of equations manually. Check the solution.

$$\begin{bmatrix} 20 & 1 & -10 \\ -10 & 3 & 10 \\ 5 & 3 & 5 \end{bmatrix} \begin{bmatrix} a_1 \\ a_2 \\ a_3 \end{bmatrix} = \begin{bmatrix} -2 \\ 4 \\ 9 \end{bmatrix}$$

1.4 Solve the following systems of equations manually and check the solutions.

(a)
$$\begin{bmatrix} 4 & -2 & -2 \\ -2 & 5 & -3 \\ -2 & -3 & 5 \end{bmatrix} \begin{bmatrix} 0 \\ 0 \\ a_3 \end{bmatrix} = \begin{bmatrix} f_1 \\ f_2 \\ 10 \end{bmatrix}$$

(b)
$$\begin{bmatrix} 6 & -4 & -2 \\ -4 & 12 & -8 \\ -2 & -8 & 10 \end{bmatrix} \begin{bmatrix} 1 \\ a_2 \\ a_3 \end{bmatrix} = \begin{bmatrix} f_1 \\ 16 \\ -6 \end{bmatrix}$$

(c) $$\begin{bmatrix} 4 & -4 & 0 & 0 & 0 \\ -4 & 7 & -2 & -1 & 0 \\ 0 & -2 & 5 & -3 & 0 \\ 0 & -1 & -3 & 7 & -3 \\ 0 & 0 & 0 & -3 & 3 \end{bmatrix} \begin{bmatrix} -3 \\ a_2 \\ 0 \\ a_4 \\ 3 \end{bmatrix} = \begin{bmatrix} f_1 \\ 4 \\ f_3 \\ -1 \\ f_5 \end{bmatrix}$$

1.5 Begin with the matrices

$$A = \begin{bmatrix} 2 & 1 & 0 & 3 \\ 6 & 4 & 1 & -2 \\ 0 & 3 & 4 & 1 \\ 1 & 2 & -4 & 6 \end{bmatrix}; \quad B = \begin{bmatrix} 3 & 4 & 1 & -2 \\ 6 & 8 & 1 & 0 \\ 2 & 2 & 3 & -2 \\ 1 & 4 & 0 & 4 \end{bmatrix};$$

$$C = \begin{bmatrix} -4 \\ 2 \\ 3 \\ 1 \end{bmatrix}; \quad D = \begin{bmatrix} 1 & 4 & -3 & 6 \end{bmatrix}$$

and perform the following matrix operations with CALFEM. For the sub-exercises with more than one matrix operation, compare and comment on the results.

(a) $A + B$ and $B + A$
(b) AB and BA
(c) $(AB)^T$, $(BA)^T$ and $B^T A^T$
(d) CD and DC
(e) $C^T AC$
(f) $\det A$, A^{-1} and AA^{-1}

1.6 Compute the determinant of the matrices in the following systems of equations with CALFEM. If possible, solve the systems of equations and check the solutions. If any of the systems is unsolvable, explain why.

(a) $$\begin{bmatrix} -4 & 3 & 0 & 1 \\ 1 & 2 & -1 & 4 \\ 0 & 1 & -1 & 2 \\ 2 & 0 & 2 & 2 \end{bmatrix} \begin{bmatrix} a_1 \\ a_2 \\ a_3 \\ a_4 \end{bmatrix} = \begin{bmatrix} -2 \\ -1 \\ -3 \\ 2 \end{bmatrix}$$

(b) $$\begin{bmatrix} 4 & -4 & 0 \\ -4 & 6 & -2 \\ 0 & -2 & 2 \end{bmatrix} \begin{bmatrix} a_1 \\ a_2 \\ a_3 \end{bmatrix} = \begin{bmatrix} 0 \\ 0 \\ 0 \end{bmatrix}$$

(c) $$\begin{bmatrix} 8 & -3 & -5 \\ -3 & 5 & -2 \\ -5 & -2 & 7 \end{bmatrix} \begin{bmatrix} a_1 \\ a_2 \\ a_3 \end{bmatrix} = \begin{bmatrix} 4 \\ 2 \\ -6 \end{bmatrix}$$

1.7 Consider the eigenvalue problem $(A - \lambda I)a = 0$, where

$$A = \begin{bmatrix} 10 & -3 \\ -3 & 2 \end{bmatrix}$$

(a) Compute the eigenvalues.
(b) Compute the eigenvectors and check that they are orthogonal.

2

Systems of Connected Springs

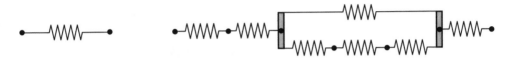

Figure 2.1 Elastic spring and a system of connected springs

A system of connected springs is a set of *discrete material points* connected by *springs* (Figure 2.1). Of the different building blocks of structural mechanics, the spring is the simplest one. The study of systems built up of springs only can therefore be an instructive way to describe and explain the models at the system level.

In structural mechanics, a system is basically composed of two components: *nodes with degrees of freedom* and *elements*. Here, we choose to study a system of connected springs that carries load only in one direction and we let this direction be the *x*-axis (Figure 2.2). A number of reference points or *nodes* are introduced. In each node, there can be an arbitrary number of *global degrees of freedom*. These degrees of freedom represent different possible movements for the ends of the elements connected to a node. Here, we choose to allow only one possible movement for each node, the displacement in a certain direction. The nodes and the degrees of freedom also form locations and directions where external forces (prescribed loads or arising support forces) can be applied and equilibrium equations can be set up.

Between two nodes, we can create a potential force path by inserting an elastic spring. The tendency of a spring to carry load depends on its *spring stiffness*. In a system with several different force paths, the stiffer ones carry the greatest load.

As was the case at the system level, the description of a single spring can be based on discrete nodes; here, they comprise the end points of the spring (Figure 2.3). To these *local nodes*, we can associate *local degrees of freedom*, which describe the possible movements of the nodes and also enable forces to act on the spring. Based on the degrees of freedom, defined for the element, a matrix that represents the stiffness properties of the spring is formed. This matrix can be placed between global degrees of freedom and constitutes then a force path in the global spring system.

Structural Mechanics: Modelling and Analysis of Frames and Trusses, First Edition.
Karl-Gunnar Olsson and Ola Dahlblom.
© 2016 John Wiley & Sons, Ltd. Published 2016 by John Wiley & Sons, Ltd.

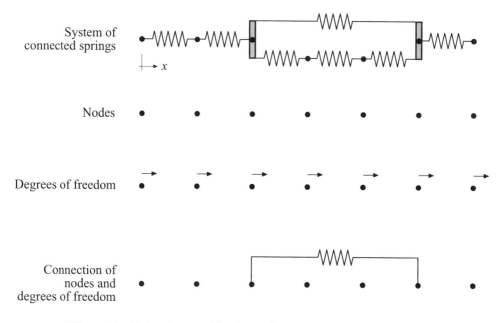

Figure 2.2   Nodes, degrees of freedom and connection of degrees of freedom

Figure 2.3   A spring element with two degrees of freedom

In structural mechanics, every system contains three basic quantities – force, stiffness and deformation – which can be considered at different scale levels. Figure 2.4 shows a map, which summarises the quantities and relations of a system of connected springs. The map has the following structure:

- a scale with three levels: the elastic spring, the systematically described spring element and the system of connected springs;
- three types of quantities: force measure, stiffness measure and displacement measure;
- for force measures: relations between force measures at different scale levels – equilibrium/static equivalence;
- for displacement measures: relations between displacement measures at different scale levels – kinematics/compatibility;
- at each level: a constitutive relation between the force measure and corresponding displacement measure.

At the lowermost level, there is a relation between force and deformation for an *elastic spring*, $N = k\,\delta$. This relation is called the *constitutive relation* and is the basis for the derivation of corresponding relations at higher scale levels. The spring relation is further described in Section 2.1.

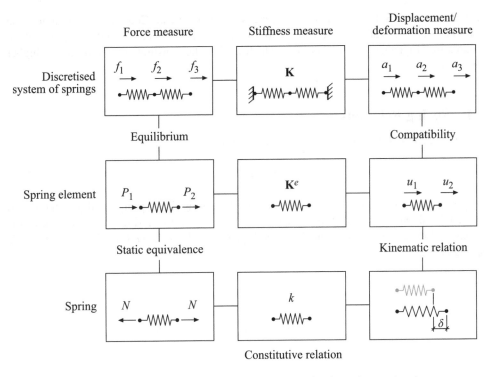

Figure 2.4 The quantities and relations of structural mechanics for springs and spring systems

By systematically introducing local degrees of freedom and expressing the deformation and the forces of the spring in connection to them, we can reformulate the constitutive relation of the spring to a corresponding constitutive relation for a *spring element*. This intermediate level, which is described in Section 2.2, is a preparatory step for the uppermost level of the scale, the model of a spring system.

The uppermost level deals with the systematic construction of computational models for global load-carrying structures. The methodology introduced here for a *system of connected springs* is general and is applied for all the systems considered in this book. The methodology consists of six steps, which are described in Section 2.3.

Each level in the map represents a constitutive relation between forces and deformations. Such a constitutive relation is always derived from a lower level to a higher one. We, in terms of six steps, introduce the general principle for such derivations.

- Start from the *constitutive relation* of the lower level (1).
- Define the deformation measure of the higher level, *kinematic quantities (2)*.
- Formulate a relation between the kinematic quantities of the lower and the higher level – *the kinematic relation* (3).
- Define the loading on the body/structure at the higher level, *force quantities* (4).
- Formulate a relation between the forces of the lower and the higher level – *equilibrium/static equivalence* (5).
- Determine a *constitutive relation* for the higher level using the three relations (6).

In Sections 2.1–2.3, the numbers of these steps recur in the text. Consistently throughout the textbook, each derivation from a lower to a higher level is concluded with a figure, which summarises Equations (1), (3) and (5), which lead to the constitutive relation of the higher level (6).

2.1   Spring Relations

The basic action of a spring is given by the relation

$$N = k \, \delta \tag{2.1}$$

which describes the resistance to deformation of a spring. The spring relation (Figure 2.5) consists of three types of quantities: the *force N* acting on the spring, the *stiffness k* of the spring and the *deformation δ* which arises. Equation (2.1) is the *constitutive relation* of the spring (1).

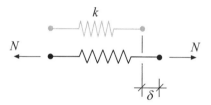

Figure 2.5   A spring with the stiffness k is loaded with the force N and thereby it is elongated by a distance δ

2.2   Spring Element

A discretised spring element (Figure 2.6) has two nodes, each with one displacement degree of freedom, u_1 and u_2. The displacements u_1 and u_2 are referred to as the *nodal displacements* of the element (2) and we choose here to define them as positive when they have the same direction as the *x*-axis. The forces acting at the nodes are denoted P_1 and P_2, and referred to as *element forces* (4). These are also defined to be positive in the direction of the *x*-axis.

We are now able to formulate a kinematic relation (3) by expressing the deformation δ of the spring as a function of the nodal displacements,

$$\delta = u_2 - u_1 \tag{2.2}$$

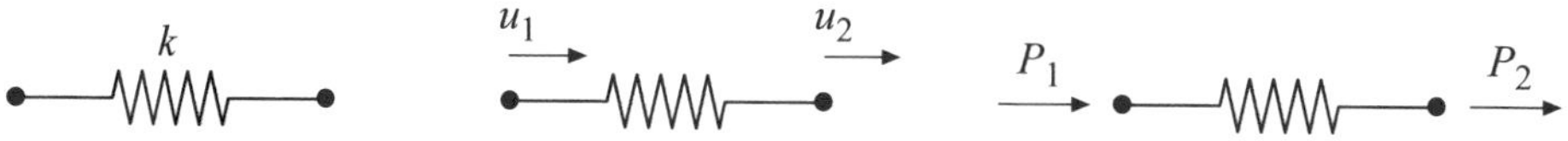

Figure 2.6   A discretised spring element

$$P_1 = -N \qquad (2.4)$$
$$P_2 = N$$
$$N = k\,\delta \qquad (2.1)$$
$$\delta = u_2 - u_1 \qquad (2.2)$$

$$\Rightarrow \mathbf{K}^e \mathbf{a}^e = \mathbf{f}^e \qquad (2.7)$$

where

$$\mathbf{K}^e = k \begin{bmatrix} 1 & -1 \\ -1 & 1 \end{bmatrix} ; \quad \mathbf{a}^e = \begin{bmatrix} u_1 \\ u_2 \end{bmatrix} ; \quad \mathbf{f}^e = \begin{bmatrix} P_1 \\ P_2 \end{bmatrix}$$

Figure 2.7   From spring to spring element

By substituting (2.2) into (2.1), we can express the spring force as

$$N = k(u_2 - u_1) \qquad (2.3)$$

For a spring to be in equilibrium, two forces that are equal in magnitude and opposite in direction must be acting on the spring, one at each end. If we compare the definition of the spring force N (Figure 2.5) with the definition of the element forces P_1 and P_2 (Figure 2.6), we observe (5) that

$$P_1 = -N; \; P_2 = N \qquad (2.4)$$

Substituting (2.4) into (2.3) gives

$$P_1 = -k(u_2 - u_1) \qquad (2.5)$$
$$P_2 = k(u_2 - u_1) \qquad (2.6)$$

or, in matrix form

$$\mathbf{K}^e \mathbf{a}^e = \mathbf{f}^e \qquad (2.7)$$

where

$$\mathbf{f}^e = \begin{bmatrix} P_1 \\ P_2 \end{bmatrix} ; \quad \mathbf{K}^e = k \begin{bmatrix} 1 & -1 \\ -1 & 1 \end{bmatrix} ; \quad \mathbf{a}^e = \begin{bmatrix} u_1 \\ u_2 \end{bmatrix} \qquad (2.8)$$

The relation (2.7) is the constitutive relation (6) of the spring and is referred to as the *element equation* of the spring where $\mathbf{K}^e$ is the *element stiffness matrix*, $\mathbf{a}^e$ the *element displacement vector* and $\mathbf{f}^e$ the *element force vector*. The index e is used to denote that the relation is for a single element.

A summary of the relations, which lead to the element equation of the spring, is shown in Figure 2.7.

2.3   Systems of Springs

With the spring element (1) described in (2.7), we can now construct and analyse complex systems of connected springs (Figure 2.8). The aim is to establish a *constitutive relation* for the entire spring system. We begin by defining the *degrees of freedom* in a global coordinate

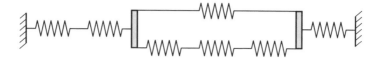

Figure 2.8   A system of connected springs

system, and introducing a global numbering of all the degrees of freedom, from 1 to n. These displacements are gathered in a global *displacement vector* **a** (2).

$$\mathbf{a} = \begin{bmatrix} a_1 \\ \cdot \\ a_i \\ a_j \\ \cdot \\ a_n \end{bmatrix} \tag{2.9}$$

The next step is to put each of the spring elements into the global system. In a given global system, each element has its defined position with defined connections to the degrees of freedom of the global system. For example, the local displacements u_1 and u_2 of the element β correspond to the global degrees of freedom a_i and a_j (Figure 2.9), that is

$$u_1 = a_i \tag{2.10}$$

$$u_2 = a_j \tag{2.11}$$

These relations describe how the spring elements are connected physically in the global system. The relations are a type of kinematic relations referred to as *compatibility requirements* (3). These compatibility requirements can be written in matrix form as

$$\mathbf{a}^e = \mathbf{H}\mathbf{a} \tag{2.12}$$

where $\mathbf{a}^e$ is defined in (2.8), **a** in (2.9) and where

$$\mathbf{H} = \begin{bmatrix} 0 & \cdot & 1 & 0 & \cdot & 0 \\ 0 & \cdot & 0 & 1 & \cdot & 0 \end{bmatrix} \tag{2.13}$$

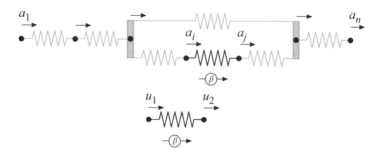

Figure 2.9   Global and local displacements

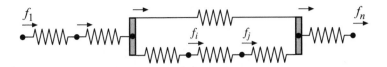

Figure 2.10   External forces at nodes

The elements of the matrix $\mathbf{H}$ that are not printed out are zero. Through the matrix form of the compatibility requirements we have established a matrix $\mathbf{H}$, which describes a transformation between two different sets of degrees of freedom.

In the global system, *external forces* can be introduced at the nodes (Figure 2.10); it may be external loads or support forces that act at the external supports of the system. We choose to denote these forces by f_i, where i is the degree of freedom in which the force acts, and we gather them as components of a global *force vector* $\mathbf{f}$ (4).

$$\mathbf{f} = \begin{bmatrix} f_1 \\ \cdot \\ f_i \\ f_j \\ \cdot \\ f_n \end{bmatrix} \tag{2.14}$$

The element forces, that is the components of $\mathbf{f}^e$ in (2.8), in the global system, act in the global degrees of freedom i and j (Figure 2.11). Therefore, it is a good idea to write the element forces as components of an *expanded element force vector* $\hat{\mathbf{f}}^e$. For element β,

$$\hat{\mathbf{f}}^\beta = \begin{bmatrix} 0 \\ \cdot \\ P_1^{(\beta)} \\ P_2^{(\beta)} \\ \cdot \\ 0 \end{bmatrix} \tag{2.15}$$

Between $\hat{\mathbf{f}}^e$ and $\mathbf{f}^e$ we then have the following relation:

$$\hat{\mathbf{f}}^e = \mathbf{H}^T \mathbf{f}^e \tag{2.16}$$

where the matrix $\mathbf{H}^T$ is the transpose of the transformation matrix $\mathbf{H}$ (2.13).

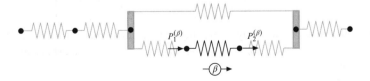

Figure 2.11   The free-body diagram of a spring element

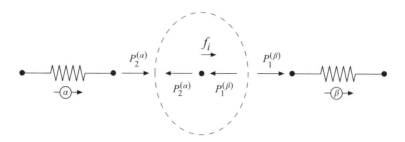

Figure 2.12 Equilibrium for degree of freedom i

The spring system is at rest. This means that each node is also at rest and with this an equilibrium equation can be established for each degree of freedom. The sum of the forces acting on the node should be equal to zero. The forces are both the internal forces with which the spring ends act on the node, $P_2^{(\alpha)}$ and $P_1^{(\beta)}$, and the external forces f_i, which may act at the node (Figure 2.12). A spring system with all the springs in the same direction has only one displacement degree of freedom at each node. This means that one equilibrium equation is sufficient to formulate the equilibrium of the node. The equilibrium for degree of freedom i can be written as[1] (5)

$$- P_2^{(\alpha)} - P_1^{(\beta)} + f_i = 0 \tag{2.17}$$

or

$$\sum_{e=1}^{m} f_i^e = f_i \tag{2.18}$$

where we formally sum over all the included elements, but where f_i^e has values different from zero only for the elements connecting to degree of freedom i.

For each of the introduced degrees of freedom, we can formulate an equilibrium equation, that is for the entire system the number of equilibrium equations is the same as the number of degrees of freedom (n). These equilibrium equations can together be written in matrix form

$$\sum_{e=1}^{m} \hat{\mathbf{f}}^e = \mathbf{f} \tag{2.19}$$

or

$$
\begin{bmatrix} P_1^{(1)} \\ \cdot \\ 0 \\ 0 \\ \cdot \\ 0 \end{bmatrix}
+ \cdot +
\begin{bmatrix} 0 \\ \cdot \\ P_2^{(\alpha)} \\ 0 \\ \cdot \\ 0 \end{bmatrix}
+
\begin{bmatrix} 0 \\ \cdot \\ P_1^{(\beta)} \\ P_2^{(\beta)} \\ \cdot \\ 0 \end{bmatrix}
+ \cdot +
\begin{bmatrix} 0 \\ \cdot \\ 0 \\ 0 \\ \cdot \\ P_2^{(m)} \end{bmatrix}
=
\begin{bmatrix} f_1 \\ \cdot \\ f_i \\ f_j \\ \cdot \\ f_n \end{bmatrix}
\tag{2.20}
$$

It turns out that the systematic way of writing that we have introduced in (2.15) results in that the left-hand side of (2.19), or (2.20), consists of the sum of the expanded element force vectors for all the elements.

---

[1] An alternative way to derive this expression is to note that the external forces acting at a node are statically equivalent to the sum of the internal forces (the element forces) at the same node.

The expanded element force vector for an element is related to the local element force vector through (2.16). Substituting the constitutive relation (2.7) and the compatibility requirements (2.12) into (2.16), we can write the expanded element force vector as

$$\hat{\mathbf{f}}^e = \hat{\mathbf{K}}^e \mathbf{a} \tag{2.21}$$

where

$$\hat{\mathbf{K}}^e = \mathbf{H}^T \mathbf{K}^e \mathbf{H} \tag{2.22}$$

$\hat{\mathbf{K}}^e$ is referred to as the *expanded element stiffness matrix* and shows where in a global system the stiffness of an element should be placed. For a spring element β, the expanded element stiffness matrix is obtained from

$$\hat{\mathbf{K}}^\beta = k \begin{bmatrix} 0 & 0 \\ \cdot & \cdot \\ 1 & 0 \\ 0 & 1 \\ \cdot & \cdot \\ 0 & 0 \end{bmatrix} \begin{bmatrix} 1 & -1 \\ -1 & 1 \end{bmatrix} \begin{bmatrix} 0 & \cdot & 1 & 0 & \cdot & 0 \\ 0 & \cdot & 0 & 1 & \cdot & 0 \end{bmatrix} = k \begin{bmatrix} 0 & \cdot & 0 & 0 & \cdot & 0 \\ \cdot & \cdot & & \cdot & \cdot & \cdot \\ 0 & \cdot & 1 & -1 & \cdot & 0 \\ 0 & \cdot & -1 & 1 & \cdot & 0 \\ \cdot & \cdot & & \cdot & \cdot & \cdot \\ 0 & \cdot & 0 & 0 & \cdot & 0 \end{bmatrix} \tag{2.23}$$

Substituting (2.21) into the equilibrium relation (2.19) gives

$$\sum_{e=1}^{m} \hat{\mathbf{K}}^e \mathbf{a} = \mathbf{f} \tag{2.24}$$

or

$$\boxed{\mathbf{Ka} = \mathbf{f}} \tag{2.25}$$

where

$$\mathbf{K} = \sum_{e=1}^{m} \hat{\mathbf{K}}^e \tag{2.26}$$

The sequence of relations, from the local element relation (2.7) to the global relation for the spring system (2.25), shows a general structure that will appear throughout the textbook. Even if the contents are different for different types of problems, the same matrix notations are used. Thus, in summary,

1. We have started from the element relation of the spring (2.7).
2. We have introduced a global displacement vector (2.9).
3. We have related local displacements to global ones using compatibility (2.12).
4. We have introduced a global vector for external loads that act on the nodes of the system (2.14).
5. With an expanded way of writing (2.16) and using equilibrium conditions, we have related local internal forces to global external forces (2.19).
6. With the compatibility requirements (2.12), the constitutive relation of the element (2.7) and the expanded way of writing element forces (2.16), we have derived an expression that describes a single element in a global system (2.21). By substituting the expanded element relations (2.21) into the global equilibrium relations (2.19) for all the elements, we derived a global constitutive relation for the spring system (2.25).

$$\left.\begin{array}{l} \hat{\mathbf{f}}^e = \mathbf{H}^T \mathbf{f}^e \quad (2.16) \\ \mathbf{f}^e = \mathbf{K}^e \mathbf{a}^e \quad (2.7) \\ \mathbf{a}^e = \mathbf{H}\mathbf{a} \quad (2.12) \end{array}\right\} \Rightarrow \left.\begin{array}{l} \mathbf{f} = \sum_{e=1}^{m} \hat{\mathbf{f}}^e \quad (2.19) \\ \hat{\mathbf{f}}^e = \hat{\mathbf{K}}^e \mathbf{a} \quad (2.21) \\ \text{where} \\ \hat{\mathbf{K}}^e = \mathbf{H}^T \mathbf{K}^e \mathbf{H} \end{array}\right\} \Rightarrow \begin{array}{l} \mathbf{f} = \mathbf{K}\mathbf{a} \quad (2.25) \\ \text{where} \\ \mathbf{K} = \sum_{e=1}^{m} \hat{\mathbf{K}}^e \end{array}$$

Figure 2.13   From element relations to system relations

All the steps are summarised in Figure 2.13. Note that the matrices $\mathbf{K}^e$, $\hat{\mathbf{K}}^e$ and $\mathbf{K}$ are symmetric.

From this structure, we get not only a solvable system of equations (2.25), but also a simple and practical method for establishing (constructing) this system of equations. The method follows from Equation (2.26) and is referred to as *assembling*. A basis is that the expanded element stiffness matrix $\hat{\mathbf{K}}^\beta$ has components different from zero only at positions that correspond to the global degrees of freedom for the element, see (2.23). With this knowledge, the procedure of summing the element stiffness matrices, (2.26), can be simplified to the following steps:

- A *topology matrix* is introduced. It describes in a compact way how a single element is related to the degrees of freedom for the global system.

$$\text{topology} = \begin{bmatrix} 1 & 1 & \cdot \\ \cdot & \cdot & \cdot \\ \alpha & \cdot & i \\ \beta & i & j \\ \cdot & \cdot & \cdot \\ m & \cdot & n \end{bmatrix} \qquad (2.27)$$

- A matrix $\mathbf{K}$ is created and filled with zeros. The matrix $\mathbf{K}$ is given the dimensions $n \times n$, where n is the number of degrees of freedom of the spring system. When the process of summation is completed, this matrix will be the global stiffness matrix.
- An element stiffness matrix $\mathbf{K}^e$ is created for each of the single elements.
- Using the topology matrix, the coefficients from an element stiffness matrix are added to the correct positions in the global stiffness matrix $\mathbf{K}$ (Figure 2.14). This procedure is repeated for each of the elements.

In the topology matrix, each row contains information for one element. The first column contains the element number and the following columns list the global degrees of freedom for that element. Here, the local orientation of the element determines in what order the degrees of freedom are given. Using the information of the topology matrix, the components of the local element matrices are added to the correct positions of the matrix of the global system.

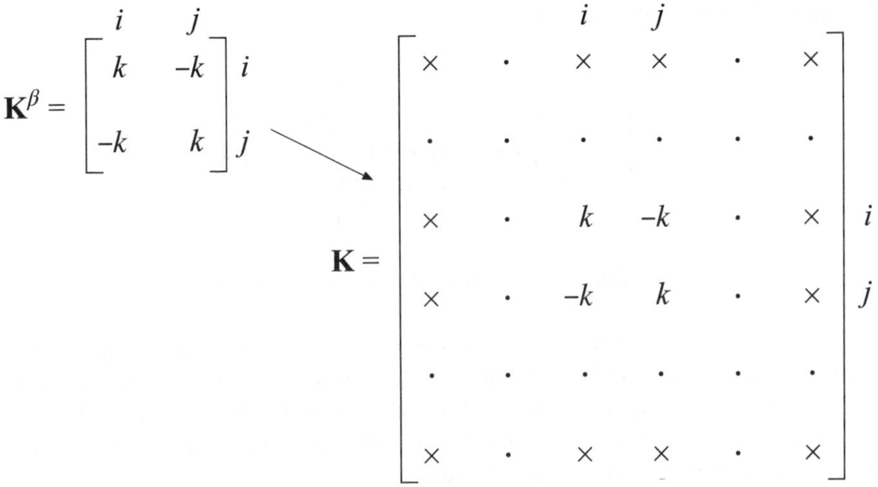

$$
\mathbf{K}^\beta =
\begin{array}{cc}
 & \begin{array}{cc} i & j \end{array} \\
\begin{bmatrix} k & -k \\ -k & k \end{bmatrix} & \begin{array}{c} i \\ j \end{array}
\end{array}
$$

$$
\mathbf{K} =
\begin{array}{c}
\begin{array}{cccccc} i & & & j & & \end{array} \\
\begin{bmatrix}
\times & \cdot & \times & \times & \cdot & \times \\
\cdot & \cdot & \cdot & \cdot & \cdot & \cdot \\
\times & \cdot & k & -k & \cdot & \times \\
\times & \cdot & -k & k & \cdot & \times \\
\cdot & \cdot & \cdot & \cdot & \cdot & \cdot \\
\times & \cdot & \times & \times & \cdot & \times
\end{bmatrix}
\begin{array}{c} \\ \\ i \\ j \\ \\ \end{array}
\end{array}
$$

Figure 2.14 Placement of the element stiffness matrix for element β into the global stiffness matrix according to the topology matrix in Equation (2.27)

When the stiffness matrix has been created, displacements and support forces can be determined from the system of equations (2.25) with consideration of the given loads and prescribed displacements. When the displacements $\mathbf{a}$ have been computed, the displacements $\mathbf{a}^e$ for one element can be determined from (2.12). The spring force can finally be determined using (2.3).

Beginning from the relations in matrix form that have been derived, a systematic method for modelling and analysis of spring systems has been established. The method is general and with some small modifications it will be used later for trusses and frames as well. It consists of two parts with a total of seven separate steps[2].

Formulation of a computational model:

- define the computational model;
- formulate element matrices;
- establish compatibility conditions;
- assemble element matrices by establishing equilibria.

Analysis of response for different influences on the computational model:

- define boundary conditions and nodal forces;
- solve the system of equations;
- determine the internal forces.

---

[2] Boundary conditions as well as loads can alternatively be considered as a part of the computational model, but here we have chosen a computational model based on the unconstrained non-loaded material body. Different possibilities for the support conditions and different load cases are considered as a part of the analysis.

Example 2.1 A system of springs

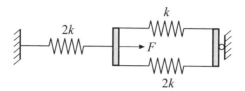

Figure 1   A system of three connected springs

Consider a system of springs connected in series and in parallel, where the spring stiffness is different for different springs (Figure 1). The system is fixed at its external ends and loaded with a force F at the midpoint. For the stiffness $k = 1500$ N/m and the external force $F = 100$ N, determine the displacement of the midpoint, the spring forces in all the springs and the support forces at the two supports.

Define a computational model

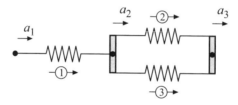

Figure 2   The computational model

A computational model is defined by naming (numbering) the degrees of freedom and the elements and by giving the positive directions. For the system considered, three displacement degrees of freedom are defined, a_1, a_2 and a_3, with positive direction in the direction of the global x-axis. The three spring elements are numbered from 1 to 3 and each element is given a local positive direction marked with an arrow at the element number; see Figure 2.

Formulate element matrices

For each spring element, we have a local element relation (2.7)

$$\mathbf{K}^e \mathbf{a}^e = \mathbf{f}^e \tag{1}$$

With $k = 1500$ N/m, we obtain for the three elements:

Element 1:

$$\begin{bmatrix} P_1^{(1)} \\ P_2^{(1)} \end{bmatrix} = \begin{bmatrix} 3000 & -3000 \\ -3000 & 3000 \end{bmatrix} \begin{bmatrix} u_1^{(1)} \\ u_2^{(1)} \end{bmatrix} \tag{2}$$

Element 2:

$$\begin{bmatrix} P_1^{(2)} \\ P_2^{(2)} \end{bmatrix} = \begin{bmatrix} 1500 & -1500 \\ -1500 & 1500 \end{bmatrix} \begin{bmatrix} u_1^{(2)} \\ u_2^{(2)} \end{bmatrix} \tag{3}$$

Element 3:

$$\begin{bmatrix} P_1^{(3)} \\ P_2^{(3)} \end{bmatrix} = \begin{bmatrix} 3000 & -3000 \\ -3000 & 3000 \end{bmatrix} \begin{bmatrix} u_1^{(3)} \\ u_2^{(3)} \end{bmatrix} \tag{4}$$

Compatibility conditions

The spring system has the global displacement vector

$$\mathbf{a} = \begin{bmatrix} a_1 \\ a_2 \\ a_3 \end{bmatrix} \tag{5}$$

The local degrees of freedom for Elements 1–3 correspond to global degrees of freedom according to the following:

Element 1:

$$u_1^{(1)} = a_1; \ u_2^{(1)} = a_2 \tag{6}$$

Element 2:

$$u_1^{(2)} = a_2; \ u_2^{(2)} = a_3 \tag{7}$$

Element 3:

$$u_1^{(3)} = a_2; \ u_2^{(3)} = a_3 \tag{8}$$

From the compatibility conditions, we have now obtained a description of how the elements of the spring system are connected to the degrees of freedom for the system. This description is summarised in a topology matrix

$$\text{topology} = \begin{bmatrix} 1 & 1 & 2 \\ 2 & 2 & 3 \\ 3 & 2 & 3 \end{bmatrix} \tag{9}$$

Using the compatibility conditions (2.12) and the expanded element matrices, (2.7) can be written in expanded form (2.21). For the three spring elements, the following is obtained:

Element 1:

$$\begin{bmatrix} P_1^{(1)} \\ P_2^{(1)} \\ 0 \end{bmatrix} = \begin{bmatrix} 3000 & -3000 & 0 \\ -3000 & 3000 & 0 \\ 0 & 0 & 0 \end{bmatrix} \begin{bmatrix} a_1 \\ a_2 \\ a_3 \end{bmatrix} \tag{10}$$

Element 2:

$$\begin{bmatrix} 0 \\ P_1^{(2)} \\ P_2^{(2)} \end{bmatrix} = \begin{bmatrix} 0 & 0 & 0 \\ 0 & 1500 & -1500 \\ 0 & -1500 & 1500 \end{bmatrix} \begin{bmatrix} a_1 \\ a_2 \\ a_3 \end{bmatrix} \tag{11}$$

Element 3:

$$\begin{bmatrix} 0 \\ P_1^{(3)} \\ P_2^{(3)} \end{bmatrix} = \begin{bmatrix} 0 & 0 & 0 \\ 0 & 3000 & -3000 \\ 0 & -3000 & 3000 \end{bmatrix} \begin{bmatrix} a_1 \\ a_2 \\ a_3 \end{bmatrix} \tag{12}$$

The local element matrix $\mathbf{K}^e$ can directly be placed in the expanded stiffness matrix $\hat{\mathbf{K}}^e$ using the information given by the topology matrix. The placement of the components of a single spring element into an expanded element matrix is one of the steps of the procedure called assembling.

Assemble element matrices

For each of the three nodes, equilibrium in the direction of the degree of freedom is required:

Degree of freedom 1:

$$P_1^{(1)} = f_1 \tag{13}$$

Degree of freedom 2:

$$P_2^{(1)} + P_1^{(2)} + P_1^{(3)} = f_2 \tag{14}$$

Degree of freedom 3:

$$P_2^{(2)} + P_2^{(3)} = f_3 \tag{15}$$

By using expanded element force vectors, the three equilibria can be written:

$$\hat{\mathbf{f}}^1 + \hat{\mathbf{f}}^2 + \hat{\mathbf{f}}^3 = \mathbf{f} \tag{16}$$

where

$$\mathbf{f} = \begin{bmatrix} f_1 \\ f_2 \\ f_3 \end{bmatrix} \tag{17}$$

Substitution of the relation (2.21) for respective spring element gives

$$\mathbf{Ka} = \mathbf{f} \tag{18}$$

where
$$\mathbf{K} = \hat{\mathbf{K}}^1 + \hat{\mathbf{K}}^2 + \hat{\mathbf{K}}^3 \tag{19}$$

With the matrix components printed out, we obtain

$$\mathbf{K} = \begin{bmatrix} 3000 & -3000 & 0 \\ -3000 & 3000 & 0 \\ 0 & 0 & 0 \end{bmatrix} + \begin{bmatrix} 0 & 0 & 0 \\ 0 & 1500 & -1500 \\ 0 & -1500 & 1500 \end{bmatrix} + \begin{bmatrix} 0 & 0 & 0 \\ 0 & 3000 & -3000 \\ 0 & -3000 & 3000 \end{bmatrix} \tag{20}$$

or

$$\mathbf{K} = \begin{bmatrix} 3000 & -3000 & 0 \\ -3000 & 7500 & -4500 \\ 0 & -4500 & 4500 \end{bmatrix} \tag{21}$$

By establishing equilibria for all nodes, the stiffness matrix $\mathbf{K}$ of the spring system is obtained as a sum of the expanded element stiffness matrices, (2.26). This summation can be developed into a systematic process for adding local element matrices to a matrix for the global system. We can now express the system of equations (18) as

$$\begin{bmatrix} 3000 & -3000 & 0 \\ -3000 & 7500 & -4500 \\ 0 & -4500 & 4500 \end{bmatrix} \begin{bmatrix} a_1 \\ a_2 \\ a_3 \end{bmatrix} = \begin{bmatrix} f_1 \\ f_2 \\ f_3 \end{bmatrix} \tag{22}$$

Define boundary conditions and loads

So far, the computational model has developed into a complete description of the properties of the spring system through the fact that the components of the stiffness matrix $\mathbf{K}$ are now determined. The model has also defined a possibility to prescribe different displacements $\mathbf{a}$ and different external loadings $\mathbf{f}$ for which the response of the spring system can be examined. As long as no displacements have been prescribed, the model describes a system of springs that is not fixed to its surroundings, the system floats freely in a one-dimensional space. The determinant of the system matrix $\mathbf{K}$ is zero. For a computation of displacements and internal forces, boundary conditions and loads have to be defined. Our spring system is fixed at its outer ends, that is $a_1 = 0$ and $a_3 = 0$. This can be described by the boundary condition matrix

$$\text{boundary conditions (bc)} = \begin{bmatrix} 1 & 0 \\ 3 & 0 \end{bmatrix} \tag{23}$$

where the first column gives the degree of freedom at which the displacement should be prescribed and the second column gives the value it should be given. By splitting the force vector $\mathbf{f}$ and expressing it as the sum of two vectors, we can distinguish *loads* $\mathbf{f}_l$ (l is an abbreviation of load) from *support forces* $\mathbf{f}_b$ (b is an abbreviation of boundary).

$$\mathbf{f} = \mathbf{f}_l + \mathbf{f}_b \tag{24}$$

At the degrees of freedom where displacement is prescribed, support forces will arise, which we denote by $f_{b,1}$ and $f_{b,3}$. The spring system is loaded with the force 100 N in degree of freedom 2, that is $f_{l,2} = 100$. The two vectors $\mathbf{f}_l$ and $\mathbf{f}_b$ are given by

$$\mathbf{f}_l = \begin{bmatrix} 0 \\ 100 \\ 0 \end{bmatrix}; \ \mathbf{f}_b = \begin{bmatrix} f_{b,1} \\ 0 \\ f_{b,3} \end{bmatrix} \tag{25}$$

Solving the system of equations

With the prescribed displacements, external loads and unknown support forces introduced, the system of equations can be written as

$$\begin{bmatrix} 3000 & -3000 & 0 \\ -3000 & 7500 & -4500 \\ 0 & -4500 & 4500 \end{bmatrix} \begin{bmatrix} 0 \\ a_2 \\ 0 \end{bmatrix} = \begin{bmatrix} 0 \\ 100 \\ 0 \end{bmatrix} + \begin{bmatrix} f_{b,1} \\ 0 \\ f_{b,3} \end{bmatrix} \tag{26}$$

The system of equations contains three unknowns: the displacement a_2 and the support forces $f_{b,1}$ and $f_{b,3}$. When considering the prescribed displacements, the system of equations can be reduced to

$$7500\, a_2 = 100 \tag{27}$$

and the displacement a_2 can now be determined to be

$$a_2 = \frac{100}{7500} = 0.01333 \tag{28}$$

This means that the connecting point in the middle of the spring system is displaced 13.33 mm to the right; see Figure 3.

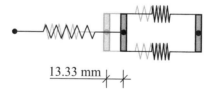

13.33 mm

Figure 3   Computed displacement

With a_2, the support forces $f_{b,1}$ and $f_{b,3}$ can be computed as

$$f_{b,1} = -3000\, a_2 = -40 \tag{29}$$

$$f_{b,2} = -4500\, a_2 = -60 \tag{30}$$

The support force at the left fixing is consequently equal to 40 N and directed leftwards and at the right fixing the support force is 60 N, this one directed leftwards as well; see Figure 4.

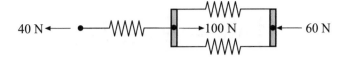

Figure 4   External equilibrium

Internal forces

Using the compatibility conditions, the local displacements for Elements 1–3 can be determined:

Element 1:
$$u_1^{(1)} = 0; \quad u_2^{(1)} = 0.01333 \tag{31}$$

Element 2:
$$u_1^{(2)} = 0.01333; \quad u_2^{(2)} = 0 \tag{32}$$

Element 3:
$$u_1^{(3)} = 0.01333; \quad u_2^{(3)} = 0 \tag{33}$$

after which the spring forces can be determined from (2.3):

Element 1:
$$N^{(1)} = 3000(0.01333 - 0) = 40 \tag{34}$$

Element 2:
$$N^{(2)} = 1500(0 - 0.01333) = -20 \tag{35}$$

Element 3:
$$N^{(3)} = 3000(0 - 0.01333) = -40 \tag{36}$$

Element 1 is consequently exposed to a tensile force of 40 N and Elements 2 and 3 to compressive forces of 20 N and 40 N, respectively; see Figure 5.

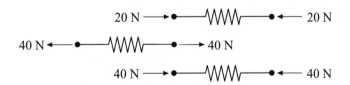

Figure 5   Spring forces

Exercises

2.1

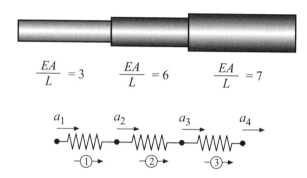

Perform manually an analysis of the spring system in the figure using the same methodology that is shown in Example 2.1.

2.2 Perform an analysis of the spring system according to Exercise 2.1 using CALFEM. Follow the method of computation for linear spring systems in the example section in the CALFEM manual. Let $k = 1$ and $F = 1$.

2.3

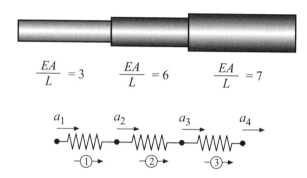

The figure shows a bar structure and its corresponding computational model. Each part of the bar has been modelled as a spring with spring stiffness $k = \frac{EA}{L}$.

(a) Establish the element stiffness matrices and assemble them into the global stiffness matrix $\mathbf{K}$ manually.

(b) Establish the system of equations $\mathbf{Ka} = \mathbf{f}$ and show that there is no unique solution by checking the value of det $\mathbf{K}$ using CALFEM.

(c) How many a-values have to be prescribed at least for the system of equations to be solvable? Compare with the behaviour of the construction.

(d) Determine a_2 and a_3 for the case when $a_1 = a_4 = 0, f_2 = 0.5$ and $f_3 = 1.0$. What are f_1 and f_4 equal to? (To be performed manually.)

(e) Determine a_2 and a_3 for the case when $a_1 = 0.02$, $a_4 = 0.05$ and $f_2 = f_3 = 0$. What are f_1 and f_4 equal to? (To be performed manually.)

2.4

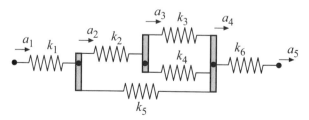

Establish a topology matrix for the system of connected springs in the figure. Substitute element matrices into a global stiffness matrix using the topology matrix.

3

Bars and Trusses

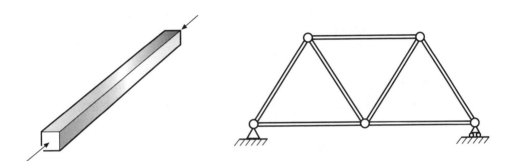

Figure 3.1 An axially loaded bar and a two-dimensional (plane) truss

A *bar* is a long body that carries only axial load (bar action). A *truss* is a load-carrying structure consisting of bars connected at frictionless hinges. By these two definitions, we have introduced the truss as a *computational model* (Figure 3.1). Here, we limit ourselves to two-dimensional (plane) trusses.

A map summarising the quantities and relations of structural mechanics for bars and systems of bars, that is trusses, is shown in Figure 3.2. The map has the following structure:

- a scale with six levels: material level, cross-section level, bar action, bar element – local coordinates, bar element – global coordinates and truss/system of bars;
- three types of quantities: force measure, stiffness measure and displacement/deformation measure;
- relations between force measures at different scale levels – equilibrium/static equivalence;
- relations between displacement/deformation measures at different scale levels – kinematics/ compatibility;
- at each level: a constitutive relation between force measure and the corresponding displacement/deformation measure.

Structural Mechanics: Modelling and Analysis of Frames and Trusses, First Edition.
Karl-Gunnar Olsson and Ola Dahlblom.

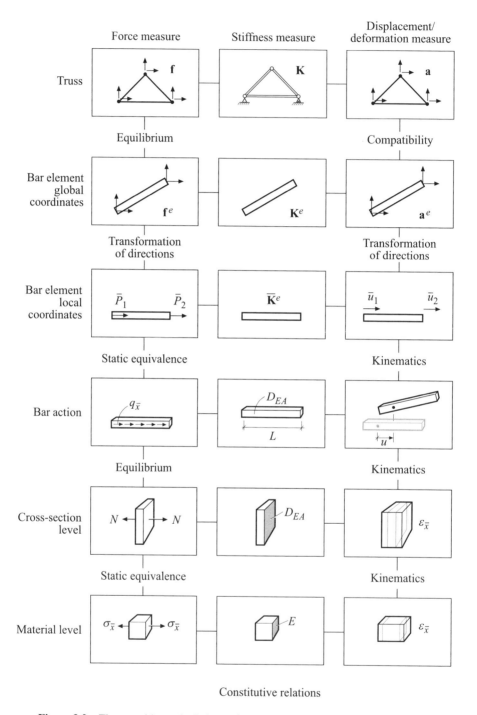

Figure 3.2   The quantities and relations of structural mechanics for bars and trusses

Compare the map for spring systems with three levels (Figure 2.4) with this map. The latter has six levels, which can be divided into three groups: truss (one level), bar element (two levels) and bar action (three levels). At the lowermost level, the material level, we have the material stiffness relating the loading of the material (stress) to its deformation (strain). This relation is usually referred to as *Hooke's law* or *the constitutive relation of the material*. With kinematic assumptions and equilibrium/static equivalence, the constitutive relation of the material is developed to a relation between axial loading and axial deformation – the differential equation for bar action (3.25). We perform this derivation in Section 3.1.

At the intermediate level of the map, the differential equation for bar action is reformulated to an element relation written in matrix form, where force and displacement measures are expressed at the end points of the bar in a systematic manner and where the stiffness of the bar is related to these measures by the element stiffness matrix. This is a preparation for placing the bar between discrete nodes in a global system of bars. The reformulation consists of two steps: first, an element relation in matrix form is established in a local coordinate system oriented in the direction of the bar (3.60). Then, the relation is transformed so that it is expressed in the coordinate system of the global system of bars (3.85). Matrix relations for bar elements at these two levels are discussed in Section 3.2.

The uppermost level concerns the systematics by which we can construct computational models for global load-carrying structures. The systematics for systems of bars includes the same six steps that were introduced for systems of springs in Section 2.3. Section 3.3 repeats this systematics and, with an example, it is shown how general computational models for trusses can be constructed.

3.1   The Differential Equation for Bar Action

We seek an expression to describe the relation between *loading* and *displacement* of a bar (Figure 3.3). The derivation consists of two steps: from the material level to the cross-section level and from the cross-section level to bar action. Each step from a lower to a higher level begins with a definition of the deformation measure of the higher level and is followed by a kinematic assumption relating the higher level to the lower one (Figure 3.4). Thereafter, the kinematic relation is substituted into the constitutive relation of the lower level. Finally, the force measure of the higher level is defined and related to the force measure of the lower level by equilibrium/static equivalence. By successively repeating this procedure, we enable a derivation of the constitutive relations for higher and higher levels.

3.1.1   Definitions

A bar is a body that has its main extension in one dimension. Thus, the formulation of a computational model involves simplifying by introducing force measures, displacement measures and deformation measures that are functions of this single dimension.

To describe the properties of the bar, we introduce a *local coordinate system* $(\bar{x}, \bar{y}, \bar{z})$, where $\bar{x}$ is parallel to, and $\bar{y}$ as well as $\bar{z}$ is perpendicular to, the longitudinal direction of the bar. The quantities of bar theory are illustrated in Figure 3.5. On the material level, all the variables are free to vary in three-dimensional space. We have stress $\sigma_{\bar{x}}(\bar{x}, \bar{y}, \bar{z})$, strain $\varepsilon_{\bar{x}}(\bar{x}, \bar{y}, \bar{z})$ and material stiffness $E(\bar{x}, \bar{y}, \bar{z})$. The main purpose of the step from the material level to cross-section level is to introduce restrictions (simplifications), which result in the variables of the theory

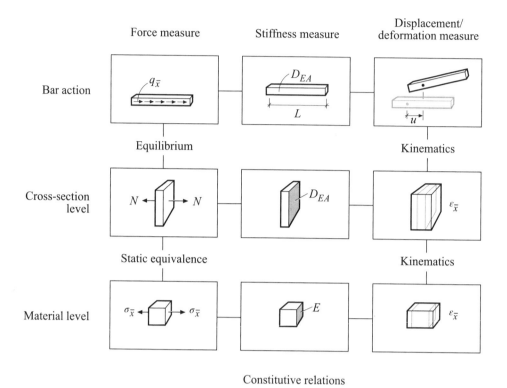

Figure 3.3 From the material level to bar action

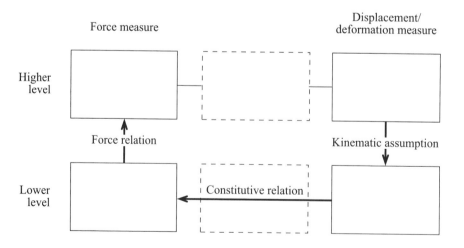

Figure 3.4 The principle for derivation of the constitutive relation of a higher level

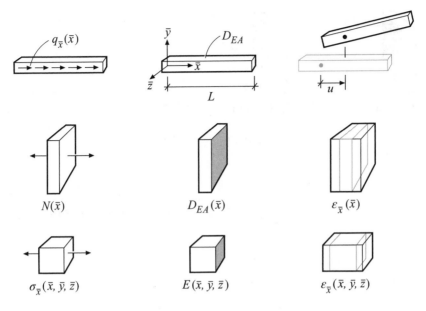

Figure 3.5 The quantities of bar theory

only vary in one dimension, the $\bar{x}$-direction of the bar. This is accomplished by introducing the so-called *generalised force and deformation measures*. The generalised force measure of the bar cross-section is the normal force $N(\bar{x})$, which is the resultant of the stresses of the cross-section. The generalised deformation measure is the normal strain $\varepsilon_{\bar{x}}(\bar{x})$, which describes how a thin lamella of the bar is stretched and contracted. Beginning from these measures, a generalised stiffness measure $D_{EA}(\bar{x})$ can be derived (3.17). At the level describing *bar action*, we introduce an axial loading $q_{\bar{x}}(\bar{x})$ and an axial displacement $u(\bar{x})$. At this level, the bar also has a length L.

The location of the local $\bar{x}$-axis on the surface of the cross-section can in principle be arbitrary. The formulation of the bar theory can, however, be simplified considerably if the position satisfies the conditions $\int_A E \, \bar{y} \, dA = 0$ and $\int_A E \, \bar{z} \, dA = 0$. For an over the cross-section constant elastic modulus, $E(\bar{y}, \bar{z}) = $ constant, this means that $\int_A \bar{y} \, dA = 0$ and $\int_A \bar{z} \, dA = 0$. The latter conditions are satisfied at the centroid of the bar cross-section. When describing systems of bars, each bar is usually represented by a *system line*, which is the local $\bar{x}$-axis of the bar; see Figure 3.6. For constant $E(\bar{y}, \bar{z})$, these are located at the centroid of the cross-section.

3.1.2   The Material Level

Strain

The deformation of a material can be described in different ways depending on the character of the application. For one-dimensional theories, such as bar theory and beam theory, it is

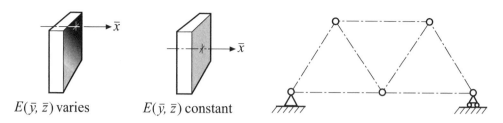

$E(\bar{y}, \bar{z})$ varies $E(\bar{y}, \bar{z})$ constant

Figure 3.6   System lines

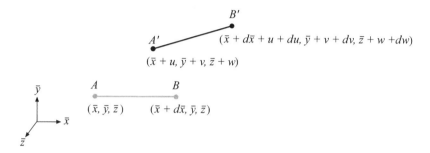

Figure 3.7   Displacement and change in length (deformation) of a material fibre

common to interpret a material as three fibres that are perpendicular to each other in space.[1] The deformation of the material then consists of two parts: the relative change in length of the fibres and the relative change in angle of the fibres. The two parts describe how the volume and the shape of the material are changed, respectively. With the kinematic presumptions (idealizations), which are introduced later for the bar theory, and in Chapter 4 for the beam theory, the only interesting deformation measure will be the relative change in length of an axial fibre.

The deformation of a material fibre can be described using the line AB between two adjacent points, which in the undeformed state have the coordinates $(\bar{x}, \bar{y}, \bar{z})$ and $(\bar{x} + d\bar{x}, \bar{y}, \bar{z})$, respectively; see Figure 3.7. It means that the length $|AB|$ of the line AB is

$$|AB| = d\bar{x} \tag{3.1}$$

In a deformed state, the line AB has been moved and its length has been changed, the resulting new line is denoted $A'B'$. The material points considered have been moved to the coordinates $(\bar{x} + u, \bar{y} + v, \bar{z} + w)$ and $(\bar{x} + d\bar{x} + u + du, \bar{y} + v + dv, \bar{z} + w + dw)$, respectively. The length $|A'B'|$ of the line $A'B'$ is then

$$|A'B'| = \sqrt{(d\bar{x} + du)^2 + (dv)^2 + (dw)^2} \tag{3.2}$$

---

[1] For three-dimensional theories, for example geotechnical applications, other deformation measures may be more relevant. If the deformation measures are based on the same fundamental conditions, there is always a possibility for translation.

With assumption of *small displacements*, that is du, dv and dw are assumed to be small compared with $d\bar{x}$, (3.2) can be written as

$$|A'B'| = d\bar{x} + du \tag{3.3}$$

A material point has no extension. Therefore, we need to introduce a deformation measure that is independent of length. Such a measure is the relative change in length of a fibre. This deformation measure is referred to as *normal strain*, or just *strain*, and is denoted ε. For small relative changes in length (*small strains*), the change in length $|A'B'| - |AB|$ is usually normalised with respect to the original length $|AB|$, that is the strain $\varepsilon_{\bar{x}}$ is given as

$$\varepsilon_{\bar{x}} = \frac{|A'B'| - |AB|}{|AB|} \tag{3.4}$$

Substitution of (3.1) and (3.3) into (3.4) gives

$$\varepsilon_{\bar{x}} = \frac{d\bar{x} + du - d\bar{x}}{d\bar{x}} \tag{3.5}$$

that is

$$\boxed{\varepsilon_{\bar{x}} = \frac{du}{d\bar{x}}} \tag{3.6}$$

This strain measure is often referred to as *engineering strain* and is positive when the fibre is lengthened.

Stress

The loading of the material depends on the size of the area over which a loading force can be distributed. To get a size-independent force measure for a material point, we introduce a relative force measure – force per unit of area. Here, we can formulate the relative force measure *stress* by describing the material as a rectangular cuboid and letting an arbitrarily directed tensile force act on a section surface dA, with the normal vector $n_{\bar{x}}$ parallel to the $\bar{x}$-axis (Figure 3.8). The force $d\mathbf{P}$ is divided into three components $dP_{\bar{x}}$, $dP_{\bar{y}}$ and $dP_{\bar{z}}$. With stress defined as force per unit of area, we obtain three stress components acting on the section surface:

$$\sigma_{\bar{x}\bar{x}} = \frac{dP_{\bar{x}}}{dA}; \quad \sigma_{\bar{x}\bar{y}} = \frac{dP_{\bar{y}}}{dA}; \quad \sigma_{\bar{x}\bar{z}} = \frac{dP_{\bar{z}}}{dA} \tag{3.7}$$

where the first index of the stress component gives the direction of the normal vector of the section surface dA and the second index gives the direction of the force component and with

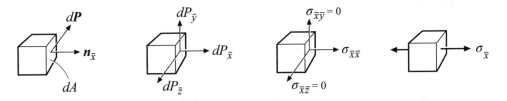

Figure 3.8 The concept of stress

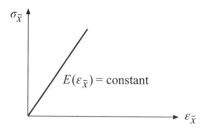

Figure 3.9   Linear elastic material relation

that also the direction of the stress. From the concept of bar action, we have an assumption of axial loading. With this and an assumption of isotropic material or orthotropic material with one principal direction along the local $\bar{x}$-axis, it follows from equilibrium that $dP_{\bar{y}}$ and $dP_{\bar{z}}$ are equal to zero. With that $\sigma_{\bar{x}\bar{y}}$ and $\sigma_{\bar{x}\bar{z}}$ are equal to zero as well; the remaining stress $\sigma_{\bar{x}\bar{x}}$, in the longitudinal direction of the bar, is

$$\sigma_{\bar{x}} = \frac{dP_{\bar{x}}}{dA} \tag{3.8}$$

where we have chosen to print only one of the indices $\bar{x}\bar{x}$. This stress is parallel to the normal vector of the section surface considered and is referred to as *normal stress*. For the rectangular cuboid representing the material to be in equilibrium, two stresses that are equal in magnitude and opposite in direction have to be acting on the opposite sides of the cuboid. These two stress components together define the normal stress $\sigma_{\bar{x}}$. Normal stress is defined to be positive in tension and negative in compression.

The Constitutive Relation of the Material

The material is assumed to be *linear elastic*. This means that the stress $\sigma_{\bar{x}}$ is proportional to the strain $\varepsilon_{\bar{x}}$ (Figure 3.9), that is

$$\sigma_{\bar{x}}(\bar{x}, \bar{y}, \bar{z}) = E(\bar{x}, \bar{y}, \bar{z}) \ \varepsilon_{\bar{x}}(\bar{x}, \bar{y}, \bar{z}) \tag{3.9}$$

where E is the elastic modulus, or Young's modulus, of the material. The material can be *isotropic* or *orthotropic*. For an orthotropic material, E refers to the elastic modulus in the longitudinal direction of the bar. Equation (3.9) is the constitutive relation of the material and is referred to as *Hooke's law* after the English researcher Robert Hooke, who in the year 1660 formulated a relation for elastic deformation of a spring.

3.1.3   The Cross-Section Level

Kinematics

The description of the kinematics of bar action begins from the reference axis of the bar, the local $\bar{x}$-axis. Each point on the axis has an original position $\bar{x}$ and, when loaded, it gets a

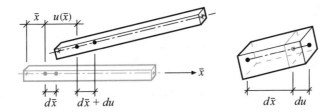

Figure 3.10   The displacement $u(\bar{x})$ and the deformation du of the reference axis

displacement $u(\bar{x})$. The deformation of the bar can be related to the change in displacement, du, which arises between two adjacent points, originally with distance dx between them (Figure 3.10). Provided that the rotation of the bar is small ($\cos\theta \approx 1$), the strain of the reference axis can be written as

$$\varepsilon_{\bar{x}}(\bar{x}) = \frac{du}{d\bar{x}} \tag{3.10}$$

where the same strain definition as in Equation (3.6) is used.

A cross-section lamella is associated with each $d\bar{x}$ along the reference axis. The lamella has in its undeformed state the volume $A(\bar{x})\,d\bar{x}$. When such a lamella is deformed, it is normally assumed that deformation occurs only in the direction of the $\bar{x}$-axis, that is in the $\bar{y}, \bar{z}$-plane the shape and size of the cross-section remain unchanged. For deformation in the $\bar{x}$-direction, each material point $(\bar{y}, \bar{z})$ on a cross-section lamella positioned at $\bar{x}$ has a strain $\varepsilon_{\bar{x}}(\bar{x}, \bar{y}, \bar{z})$ which, with use of kinematic assumptions, can be written in the form

$$\varepsilon_{\bar{x}}(\bar{x}, \bar{y}, \bar{z}) = f(\bar{x})g(\bar{y}, \bar{z}) \tag{3.11}$$

The strains of the cross-section lamella are divided into two parts: a summarising measure of the strains of the cross-section lamella $f(\bar{x})$, referred to as *generalised strain*, and a description of the shape of the deformation of the cross-section $g(\bar{y}, \bar{z})$, which is referred to as *strain mode* or *deformation mode*. Specific shapes of different strain modes are justified by a series of kinematic assumptions.[2] In ordinary bar theory and beam theory, one assumes that each plane cross-section remains plane during the deformation. In bar action, we also have the assumption that the cross-section planes remain perpendicular to the system line; see Figure 3.11. For bar action, this implies that each point on the cross-sectional plane has the same displacement as the reference axis,

$$u(\bar{x}, \bar{y}, \bar{z}) = u(\bar{x}) \tag{3.12}$$

and with that we have

$$\boxed{\varepsilon_{\bar{x}}(\bar{x}, \bar{y}, \bar{z}) = \varepsilon_{\bar{x}}(\bar{x})} \tag{3.13}$$

If we compare Equation (3.11) with (3.13), we find that the kinematic assumptions we have made imply that $f(\bar{x}) = \varepsilon_{\bar{x}}(\bar{x})$ and $g(\bar{y}, \bar{z}) = 1$. Thereby, we get the *normal strain* $\varepsilon_{\bar{x}}(\bar{x})$ of the reference axis as the generalised strain measure for bar action.

---

[2] If one wants to understand the relevance of a theory, the understanding of the reasonableness of the present choice of strain mode is essential. A prescribed strain mode implies a kinematic restriction on the studied body; the body becomes stiffer and the computed deformations slightly smaller than what would have been the case in a model where $\varepsilon_{\bar{x}}(\bar{x}, \bar{y}, \bar{z})$ can vary freely.

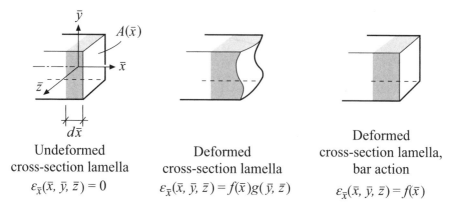

Figure 3.11   Undeformed and deformed cross-section lamella

Force Relations

The force acting on a small part dA of a cross-sectional surface is $\sigma_{\bar{x}}dA$. The resulting normal force $N(\bar{x})$ on the whole cross-section is then the integral (the sum) of the stresses $\sigma_{\bar{x}}$ over the area A of the cross-section (Figure 3.12)

$$N(\bar{x}) = \int_A \sigma_{\bar{x}}(\bar{x}, \bar{y}, \bar{z})\, dA \tag{3.14}$$

From the definition of stress, it follows that the normal force is positive in tension and with the conditions $\int_A E\bar{y}dA = 0$ and $\int_A E\bar{z}dA = 0$, the local $\bar{x}$-axis (the system line) will coincide with the line of action of the normal force.

The Constitutive Relation at the Cross-Section Level

The expression for the resultant (3.14), the kinematic relation (3.13) and the material relation (3.9) can be combined to a constitutive relation at the cross-section level

$$N(\bar{x}) = \int_A E(\bar{x}, \bar{y}, \bar{z})\, \varepsilon_{\bar{x}}(\bar{x})\, dA \tag{3.15}$$

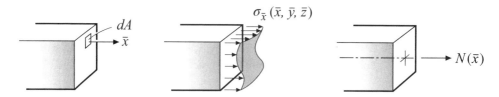

Figure 3.12   Normal stress and normal force

The generalised strain $\varepsilon_{\bar{x}}(\bar{x})$ describes the strain of the cross-section lamella. It is independent of $\bar{y}$ and $\bar{z}$ and can therefore be moved outside the integral, which gives

$$\boxed{N(\bar{x}) = D_{EA}(\bar{x})\varepsilon_{\bar{x}}(\bar{x})} \qquad (3.16)$$

where

$$D_{EA}(\bar{x}) = \int_A E(\bar{x}, \bar{y}, \bar{z})dA \qquad (3.17)$$

is the *axial stiffness* of the cross-section lamella of the bar, which depends on the material stiffness E and the shape of the cross-section. If the elastic modulus is constant over the cross-section, that is independent of $\bar{y}$ and $\bar{z}$, then

$$D_{EA}(\bar{x}) = E(\bar{x})A(\bar{x}) \qquad (3.18)$$

Figure 3.13 gives a summary of the relations for the cross-section level.

$$
\left.
\begin{aligned}
N(\bar{x}) &= \int_A \sigma_{\bar{x}}(\bar{x}, \bar{y}, \bar{z})\, dA & (3.14)\\
\sigma_{\bar{x}}(\bar{x}, \bar{y}, \bar{z}) &= E(\bar{x}, \bar{y}, \bar{z})\, \varepsilon_{\bar{x}}(\bar{x}, \bar{y}, \bar{z}) & (3.9)\\
\varepsilon_{\bar{x}}(\bar{x}, \bar{y}, \bar{z}) &= \varepsilon_{\bar{x}}(\bar{x}) & (3.13)
\end{aligned}
\right\}
\Rightarrow
\begin{aligned}
& N(\bar{x}) = D_{EA}(\bar{x})\,\varepsilon_{\bar{x}}(\bar{x}) \quad (3.16)\\
& \text{where}\\
& D_{EA}(\bar{x}) = \int_A E(\bar{x}, \bar{y}, \bar{z})\, dA
\end{aligned}
$$

Figure 3.13   From the material level to the cross-section level

3.1.4   Bar Action

Kinematics

The deformation of a bar is described by the axial displacement $u(\bar{x})$, which arises along the $\bar{x}$-axis (system line) of the bar. In (3.10), we have a relation between displacement and strain of the system. Furthermore, in (3.13), we have that this strain measure is the generalised strain for bar action. This means that (3.10)

$$\boxed{\varepsilon_{\bar{x}}(\bar{x}) = \frac{du}{d\bar{x}}} \qquad (3.19)$$

is also the kinematic relation between the deformation measure of bar action $u(\bar{x})$ and the deformation measure at the cross-section level $\varepsilon_{\bar{x}}(\bar{x})$.

The kinematic assumptions introduced here and in the previous sections can be summarised to

- small displacements;
- small strains;
- plane cross-sections remain plane and perpendicular to the system line.

Equilibrium

Consider a small part $d\bar{x}$ of a bar loaded with an external axial load $q_{\bar{x}}(\bar{x})$ according to Figure 3.14. For the part considered here, the equilibrium relation is

$$-N(\bar{x}) + (N(\bar{x}) + dN) + q_{\bar{x}}(\bar{x})d\bar{x} = 0 \tag{3.20}$$

where $N(\bar{x})$ is the normal force at $\bar{x}$ and $N(\bar{x}) + dN$ is the normal force at $\bar{x} + d\bar{x}$. Here, the equilibrium relation is established for the undeformed position of the bar part. In Chapter 9, the corresponding equilibrium relation is established for the deformed position of the bar part. The relation can be simplified to

$$dN + q_{\bar{x}}(\bar{x})d\bar{x} = 0 \tag{3.21}$$

or

$$\boxed{\frac{dN}{d\bar{x}} + q_{\bar{x}}(\bar{x}) = 0} \tag{3.22}$$

which is the equilibrium relation relating the loading $N(\bar{x})$ of the cross-section lamella to the loading $q_{\bar{x}}(\bar{x})$ of the bar.

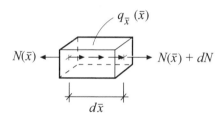

Figure 3.14 Equilibrium for a slice $d\bar{x}$ of a bar

The Differential Equation for Bar Action

The kinematic relation (3.19) substituted into Equation (3.16) gives

$$\boxed{N(\bar{x}) = D_{EA}(\bar{x}) \frac{du}{d\bar{x}}} \tag{3.23}$$

Substitution into the equilibrium relation (3.22) leads to

$$\frac{d}{d\bar{x}}\left(D_{EA}(\bar{x}) \frac{du}{d\bar{x}}\right) + q_{\bar{x}}(\bar{x}) = 0 \tag{3.24}$$

This differential equation describes the relation between axial loading $q_{\bar{x}}(\bar{x})$ and axial displacement $u(\bar{x})$ for bar action. If the axial stiffness D_{EA} is constant along the bar, the expression can be rewritten as

$$\boxed{D_{EA} \frac{d^2u}{d\bar{x}^2} + q_{\bar{x}}(\bar{x}) = 0} \tag{3.25}$$

If the elastic modulus is constant over the cross-section, that is independent of $\bar{y}$ and $\bar{z}$, the following expression is obtained by using (3.18):

$$EA\frac{d^2u}{d\bar{x}^2} + q_{\bar{x}}(\bar{x}) = 0 \qquad (3.26)$$

where the stiffness of the bar is the product of the elastic modulus E and the area A of the cross-section. In Figure 3.15, it is shown how to combine the kinematic relation, the constitutive relation and the equilibrium relation to a relation for bar action.

$$
\left.\begin{aligned}
\frac{dN}{d\bar{x}} + q_{\bar{x}}(\bar{x}) &= 0 \qquad (3.22)\\[4pt]
N(\bar{x}) &= D_{EA}(\bar{x})\,\varepsilon_{\bar{x}}(\bar{x}) \qquad (3.16)\\[4pt]
\varepsilon_{\bar{x}}(\bar{x}) &= \frac{du}{d\bar{x}} \qquad (3.19)
\end{aligned}\right\}
\Rightarrow
D_{EA}\frac{d^2u}{d\bar{x}^2} + q_{\bar{x}}(\bar{x}) = 0 \quad (3.25)
$$
for constant D_{EA}

Figure 3.15 From cross-section level to bar action

For a bar without distributed load ($q_{\bar{x}} = 0$), (3.25) becomes the homogeneous equation

$$D_{EA}\frac{d^2u}{d\bar{x}^2} = 0 \qquad (3.27)$$

The boundary conditions necessary to solve the differential equation can be prescribed displacement u or prescribed normal force N at the end points of the bar.

3.2 Bar Element

Beginning from the differential equation for bar action (3.25), the relation between forces and displacements of a bar element can be derived (Figure 3.16). This derivation consists of two steps. First, a relation in the local coordinate $\bar{x}$ is established. Thereafter, a transformation of coordinates is performed, which allows a bar element to be placed with arbitrary orientation in a two-dimensional truss.

3.2.1 Definitions

The bar element in Figure 3.17 has two displacement degrees of freedom: $\bar{u}_1$ and $\bar{u}_2$. These describe the axial displacement of the *nodes* of the bar, that is $u(\bar{x})$ at $\bar{x} = 0$ and at $\bar{x} = L$. The forces $\bar{P}_1$ and $\bar{P}_2$ acting in these points are referred to as *nodal forces* and are defined to be positive when they have the same direction as the $\bar{x}$-axis.

3.2.2 Solving the Differential Equation

The general solution $u(\bar{x})$ to the differential equation (3.25) can be written as the sum of the solution $u_h(\bar{x})$ to the homogeneous differential equation and an arbitrary particular solution $u_p(\bar{x})$

$$u(\bar{x}) = u_h(\bar{x}) + u_p(\bar{x}) \qquad (3.28)$$

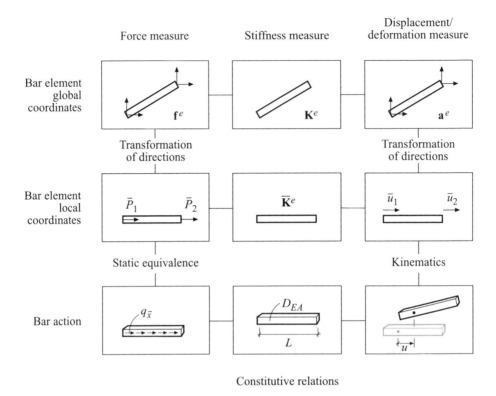

Figure 3.16   From bar action to bar element

Figure 3.17   A bar element

The constants of integration are usually determined from the general solution $u(\bar{x})$. In order to establish a systematic procedure to reach a system of equations, we instead determine the constants of integration from the solution to the homogeneous differential equation $u_h(\bar{x})$. This enables us to express the solution as a function of the displacements $\bar{u}_1$ and $\bar{u}_2$.

If the homogeneous differential equation (3.27) is divided by the stiffness D_{EA}, we obtain

$$\frac{d^2u}{d\bar{x}^2} = 0 \tag{3.29}$$

Integrating twice gives

$$u_h(\bar{x}) = \alpha_1 + \alpha_2\bar{x} \tag{3.30}$$

or in matrix form

$$u_h(\bar{x}) = \bar{\mathbf{N}}\boldsymbol{\alpha} \tag{3.31}$$

where $\bar{\mathbf{N}} = \bar{\mathbf{N}}(\bar{x})$ describes how the solution varies along the $\bar{x}$-axis and $\boldsymbol{\alpha}$ contains the constants of integration,

$$\bar{\mathbf{N}} = \begin{bmatrix} 1 & \bar{x} \end{bmatrix}; \quad \boldsymbol{\alpha} = \begin{bmatrix} \alpha_1 \\ \alpha_2 \end{bmatrix} \tag{3.32}$$

At the nodes of the bar, at $\bar{x} = 0$ and $\bar{x} = L$, the boundary conditions

$$u_h(0) = \bar{u}_1 \tag{3.33}$$

$$u_h(L) = \bar{u}_2 \tag{3.34}$$

apply. These conditions when substituted into (3.31) give

$$\bar{u}_1 = \alpha_1 \tag{3.35}$$

$$\bar{u}_2 = \alpha_1 + \alpha_2 L \tag{3.36}$$

or in matrix form

$$\bar{\mathbf{a}}^e = \mathbf{C}\boldsymbol{\alpha} \tag{3.37}$$

where

$$\bar{\mathbf{a}}^e = \begin{bmatrix} \bar{u}_1 \\ \bar{u}_2 \end{bmatrix}; \quad \mathbf{C} = \begin{bmatrix} 1 & 0 \\ 1 & L \end{bmatrix} \tag{3.38}$$

By inverting $\mathbf{C}$, we can express the constants of integration $\boldsymbol{\alpha}$ as functions of the displacement degrees of freedom of the element $\bar{\mathbf{a}}^e$, that is as

$$\boldsymbol{\alpha} = \mathbf{C}^{-1}\bar{\mathbf{a}}^e \tag{3.39}$$

where

$$\mathbf{C}^{-1} = \begin{bmatrix} 1 & 0 \\ -\frac{1}{L} & \frac{1}{L} \end{bmatrix} \tag{3.40}$$

Substituting (3.39) into (3.31), we get the solution $u_h(\bar{x})$

$$u_h(\bar{x}) = \mathbf{N}\bar{\mathbf{a}}^e \tag{3.41}$$

where

$$\mathbf{N} = \bar{\mathbf{N}}\mathbf{C}^{-1} = \begin{bmatrix} 1 & \bar{x} \end{bmatrix} \begin{bmatrix} 1 & 0 \\ -\frac{1}{L} & \frac{1}{L} \end{bmatrix} = \begin{bmatrix} 1 - \frac{\bar{x}}{L} & \frac{\bar{x}}{L} \end{bmatrix} \tag{3.42}$$

With that, we have reformulated $u_h(\bar{x})$ written as a general polynomial (3.30) to a solution in the form

$$u_h(\bar{x}) = \mathbf{N}\bar{\mathbf{a}}^e = N_1(\bar{x})\bar{u}_1 + N_2(\bar{x})\bar{u}_2 \tag{3.43}$$

where

$$N_1(\bar{x}) = 1 - \frac{\bar{x}}{L} \tag{3.44}$$

$$N_2(\bar{x}) = \frac{\bar{x}}{L} \tag{3.45}$$

The functions $N_1(\bar{x})$ and $N_2(\bar{x})$ describe how the solution varies with $\bar{x}$ and are referred to as *base functions* or *shape functions*.[3] We have in (3.43) an expression where the product $N_i(\bar{x})\,\bar{u}_i$ contributes to $u_h(\bar{x})$ from the displacement $\bar{u}_i$ and where $N_i(\bar{x})$ states its shape and $\bar{u}_i$ its size. Substitution of (3.43) into the general solution (3.28) gives

$$\boxed{u(\bar{x}) = \mathbf{N}\bar{\mathbf{a}}^e + u_p(\bar{x})} \tag{3.46}$$

where the particular solution $u_p(\bar{x})$ is different for different axial loadings of the bar.

Since we have chosen the constants of integration of the general solution to be equal to the constants of integration of the solution to the homogeneous equation, there is only one possible particular solution, namely the one where the solution is unaffected by the displacements of the nodes, that is these displacements are equal to zero

$$u_p(0) = 0 \tag{3.47}$$

$$u_p(L) = 0 \tag{3.48}$$

With this systematics, the general solution $u(\bar{x})$ can be understood as the sum of the displacement $u_h(\bar{x})$ of a bar displaced at its ends, but otherwise non-loaded and the displacement $u_p(\bar{x})$ of an axially loaded bar fixed at both ends (Figure 3.18). In Example 3.1, it is shown how to find the particular solution for a bar with a uniformly distributed load.

Differentiation of (3.46) gives

$$\frac{du}{d\bar{x}} = \mathbf{B}\bar{\mathbf{a}}^e + \frac{du_p}{d\bar{x}} \tag{3.49}$$

where

$$\mathbf{B} = \frac{d\mathbf{N}}{d\bar{x}} = \frac{d\bar{\mathbf{N}}}{d\bar{x}}\mathbf{C}^{-1} = \begin{bmatrix} 0 & 1 \end{bmatrix} \begin{bmatrix} 1 & 0 \\ -\frac{1}{L} & \frac{1}{L} \end{bmatrix} = \frac{1}{L}\begin{bmatrix} -1 & 1 \end{bmatrix} \tag{3.50}$$

Substitution of (3.49) into the expression for the normal force of the bar element (3.23) gives

$$N(\bar{x}) = D_{EA}\left(\mathbf{B}\bar{\mathbf{a}}^e + \frac{du_p}{d\bar{x}}\right) \tag{3.51}$$

or

$$N(\bar{x}) = D_{EA}\mathbf{B}\bar{\mathbf{a}}^e + N_p(\bar{x}) \tag{3.52}$$

---

[3] The base functions (shape functions) that are components of the matrix $\mathbf{N}$ are denoted $N_1(\bar{x})$ and $N_2(\bar{x})$ and should not be mistaken for the normal force, denoted $N(\bar{x})$.

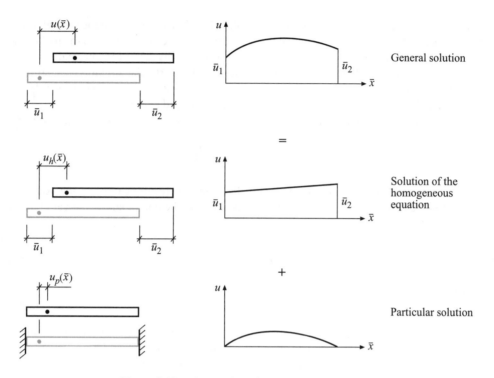

Figure 3.18 The solution of the differential equation

where

$$N_p(\bar{x}) = D_{EA}\frac{du_p}{d\bar{x}} \tag{3.53}$$

The definitions we have introduced for forces acting at the nodes of the element give

$$\bar{P}_1 = -N(0); \quad \bar{P}_2 = N(L) \tag{3.54}$$

Substitution of (3.52) gives the nodal forces

$$\bar{P}_1 = -D_{EA}\mathbf{B}\bar{\mathbf{a}}^e - N_p(0) \tag{3.55}$$

$$\bar{P}_2 = D_{EA}\mathbf{B}\bar{\mathbf{a}}^e + N_p(L) \tag{3.56}$$

With

$$\bar{\mathbf{f}}_b^e = \begin{bmatrix} \bar{P}_1 \\ \bar{P}_2 \end{bmatrix}; \quad \bar{\mathbf{K}}^e = \frac{D_{EA}}{L}\begin{bmatrix} 1 & -1 \\ -1 & 1 \end{bmatrix}; \quad \bar{\mathbf{f}}_p^e = \begin{bmatrix} -N_p(0) \\ N_p(L) \end{bmatrix} \tag{3.57}$$

the Equations (3.55) and (3.56) can be written in matrix form

$$\bar{\mathbf{f}}_b^e = \bar{\mathbf{K}}^e\,\bar{\mathbf{a}}^e + \bar{\mathbf{f}}_p^e \tag{3.58}$$

The left-hand side of the system of equations contains the nodal forces of the element $\bar{\mathbf{f}}_b^e$, that is the normal forces that act on both the ends of the element. On the right-hand side, these normal forces are divided into two parts. The product $\bar{\mathbf{K}}^e\,\bar{\mathbf{a}}^e$ gives the part of the normal forces that is generated by the displacements of the end points and the vector $\bar{\mathbf{f}}^e_p$ gives the part of the normal forces that is generated by the axial load $q_{\bar{x}}(\bar{x})$. The division of the normal forces of the bar into two parts is illustrated in Figure 3.19. Since the particular solution is determined using the conditions that $u_p(0) = 0$ and $u_p(L) = 0$, (3.47) and (3.48), the components of $\bar{\mathbf{f}}^e_p$ can be interpreted as the support forces that arise for a bar clamped at the ends.

To prepare for a systematic handling of *loads*, we now introduce an *element load vector* $\bar{\mathbf{f}}^e_l$,

$$\bar{\mathbf{f}}^e_l = -\bar{\mathbf{f}}^e_p = \begin{bmatrix} N_p(0) \\ -N_p(L) \end{bmatrix} \tag{3.59}$$

where the terms of $\bar{\mathbf{f}}^e_l$ can be interpreted as statically equivalent resulting forces to the axial load $q_{\bar{x}}(\bar{x})$. These resulting forces act on the free-body nodes at the end points of the bar element (Figure 3.20).

Thereby, we can write (3.58) as

$$\boxed{\bar{\mathbf{K}}^e\,\bar{\mathbf{a}}^e = \bar{\mathbf{f}}^e} \tag{3.60}$$

where

$$\bar{\mathbf{f}}^e = \bar{\mathbf{f}}^e_b + \bar{\mathbf{f}}^e_l \tag{3.61}$$

Equation (3.60) is the constitutive relation between forces and displacements of a bar element. The relation is referred to as the element equation for the bar element and $\bar{\mathbf{K}}^e$ is the stiffness matrix of the bar element, $\bar{\mathbf{a}}^e$ its displacement vector and $\bar{\mathbf{f}}^e$ its force vector. A summary of the relations – kinematics, constitutive relation and equilibrium – which lead to the element equation of the bar is shown in Figure 3.21.

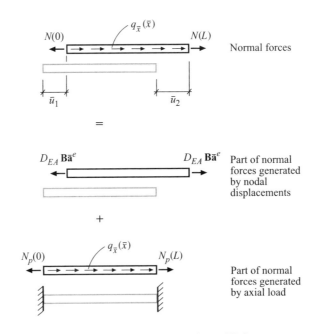

Figure 3.19   A bar element in equilibrium

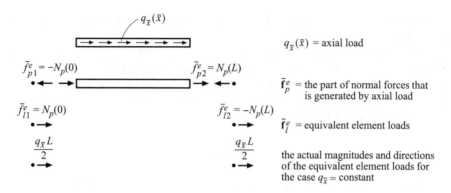

$q_{\bar{x}}(\bar{x})$ = axial load

$\bar{f}_p^e$ = the part of normal forces that is generated by axial load

$\bar{f}_l^e$ = equivalent element loads

the actual magnitudes and directions of the equivalent element loads for the case $q_{\bar{x}}$ = constant

Figure 3.20   Axial load and equivalent element loads

$$\bar{P}_1 = -N(0) \qquad (3.54)$$
$$\bar{P}_2 = N(L)$$
$$N(\bar{x}) = D_{EA}(\bar{x})\,\frac{du}{d\bar{x}} \qquad (3.23)$$
$$u(\bar{x}) = \mathbf{N}\bar{\mathbf{a}}^e + u_p(\bar{x}) \qquad (3.46)$$

$$\Rightarrow \bar{\mathbf{K}}^e\,\bar{\mathbf{a}}^e = \bar{\mathbf{f}}^e \quad (3.60)$$

where

$$\bar{\mathbf{f}}^e = \bar{\mathbf{f}}_b^e + \bar{\mathbf{f}}_l^e$$

$$\bar{\mathbf{K}}^e = \frac{D_{EA}}{L}\begin{bmatrix} 1 & -1 \\ -1 & 1 \end{bmatrix}; \quad \bar{\mathbf{a}}^e = \begin{bmatrix} \bar{u}_1 \\ \bar{u}_2 \end{bmatrix}$$

$$\bar{\mathbf{f}}_b^e = \begin{bmatrix} \bar{P}_1 \\ \bar{P}_2 \end{bmatrix}; \quad \bar{\mathbf{f}}_l^e = \begin{bmatrix} N_p(0) \\ -N_p(L) \end{bmatrix}$$

Figure 3.21   From bar action to bar element

For a non-loaded bar element, that is $\bar{\mathbf{f}}_l^e = \mathbf{0}$, the displacements are directly given by the solution to the homogeneous differential equation. A bar element with a uniformly distributed load is treated in Example 3.1.

Example 3.1 A bar element with uniformly distributed load

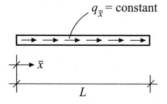

$q_{\bar{x}}$ = constant

Figure 1   A bar with uniformly distributed load

Determine the element load vector $\bar{\mathbf{f}}_l^e$ for a bar element of length L loaded with a uniformly distributed load $q_{\bar{x}}$ (Figure 1).

The element load vector $\bar{\mathbf{f}}_l^e$ is given by (3.59). To be able to determine $N_p(\bar{x})$, which is given by (3.53), we first seek a particular solution $u_p(x)$ to the differential equation for bar action (3.25). The particular solution is required to satisfy (3.25) and the two boundary conditions (3.47) and (3.48); see Figure 3.18. With constant $q_{\bar{x}}$, Equation (3.25) can, for the particular solution, be written as

$$D_{EA}\frac{d^2 u_p}{d\bar{x}^2} + q_{\bar{x}} = 0 \tag{1}$$

Integrating twice, where we choose to put a minus sign before the integration constants, gives

$$D_{EA}\frac{d u_p}{d\bar{x}} + q_{\bar{x}}\bar{x} - C_1 = 0 \tag{2}$$

$$D_{EA}u_p(\bar{x}) + q_{\bar{x}}\frac{\bar{x}^2}{2} - C_1\bar{x} - C_2 = 0 \tag{3}$$

or

$$u_p(\bar{x}) = \frac{1}{D_{EA}}\left(-q_{\bar{x}}\frac{\bar{x}^2}{2} + C_1\bar{x} + C_2\right) \tag{4}$$

Using the boundary conditions (3.47) and (3.48), we obtain

$$u_p(0) = \frac{1}{D_{EA}}\,C_2 = 0; \quad C_2 = 0 \tag{5}$$

$$u_p(L) = \frac{1}{D_{EA}}\left(-q_{\bar{x}}\frac{L^2}{2} + C_1 L + C_2\right) = 0; \quad C_1 = q_{\bar{x}}\frac{L}{2} \tag{6}$$

Substituting the constants C_1 and C_2, the particular solution becomes

$$u_p(\bar{x}) = -\frac{q_{\bar{x}}}{D_{EA}}\left(\frac{\bar{x}^2}{2} - \frac{L\bar{x}}{2}\right) \tag{7}$$

Differentiation gives

$$\frac{d u_p}{d\bar{x}} = -\frac{q_{\bar{x}}}{D_{EA}}\left(\bar{x} - \frac{L}{2}\right) \tag{8}$$

which substituted into (3.53) gives

$$N_p(\bar{x}) = -q_{\bar{x}}\left(\bar{x} - \frac{L}{2}\right) \tag{9}$$

At the end points of the element, we have

$$N_p(0) = q_{\bar{x}}\frac{L}{2}; \quad N_p(L) = -q_{\bar{x}}\frac{L}{2} \tag{10}$$

Substituting $N_p(0)$ and $N_p(L)$ into (3.59), we obtain the element load vector

$$\bar{\mathbf{f}}_l^e = \frac{q_{\bar{x}}L}{2}\begin{bmatrix} 1 \\ 1 \end{bmatrix} \tag{11}$$

Compare the result in (11) with Figure 3.20.

3.2.3 From Local to Global Coordinates

In element relation (3.60), nodal forces $\bar{\mathbf{f}}_b^e$, element displacements $\bar{\mathbf{a}}^e$ and element loads $\bar{\mathbf{f}}_l^e$ are expressed in the local coordinate system $\bar{x}$ of the bar. To be able to place the bar in a truss, we have to derive a corresponding element relation where forces and displacements are expressed in the global coordinate system (x, y) of the plane truss. In the global coordinate system, the bar is described by the displacement degrees of freedom u_1, u_2, u_3 and u_4 and by the nodal forces P_1, P_2, P_3 and P_4 (Figure 3.22).

The change of coordinate system requires the displacements and element forces expressed in the local system to be reformulated to the global system. The concept *direction cosine* is essential for this reformulation. For a moment, we use vector algebra and not matrix algebra to define a direction cosine.

In vector algebra, capitals as well as lower-case letters are set in bold to denote vectors, and by setting them in italics, we distinguish them from the vectors of matrix algebra. A vector A in the $\bar{x}$-direction can be expressed as

$$A = An_{\bar{x}} \tag{3.62}$$

where A denotes the magnitude of the vector and $n_{\bar{x}}$ is a unit vector giving the direction of the vector (Figure 3.23). The vector A can also be expressed as a sum of the components A_x and A_y, directed along the x- and y-axis, respectively,

$$A = A_x + A_y \tag{3.63}$$

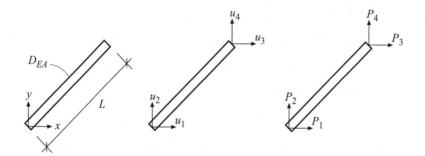

Figure 3.22 A bar element in a global coordinate system

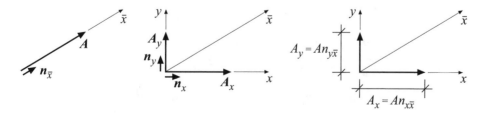

Figure 3.23 Vector components

With unit vectors in the x- and y-direction, these components can be written as

$$A_x = A_x n_x; \quad A_y = A_y n_y \tag{3.64}$$

where A_x and A_y are the magnitudes of the vectors and $\mathbf{n}_x$ and $\mathbf{n}_y$ are referred to as *direction vectors*. Moreover, the magnitudes of A_x and A_y can be expressed in terms of the dot products,

$$A_x = \mathbf{A} \cdot \mathbf{n}_x; \quad A_y = \mathbf{A} \cdot \mathbf{n}_y \tag{3.65}$$

Substitution of (3.62) gives

$$A_x = A\mathbf{n}_{\bar{x}} \cdot \mathbf{n}_x; \quad A_y = A\mathbf{n}_{\bar{x}} \cdot \mathbf{n}_y \tag{3.66}$$

The dot products between the direction vectors in (3.66) can be written as

$$\mathbf{n}_{\bar{x}} \cdot \mathbf{n}_x = |\mathbf{n}_{\bar{x}}||\mathbf{n}_x| \cos \theta_{\bar{x},x} = 1 \cdot 1 \cdot n_{\bar{x}x} = n_{\bar{x}x} \tag{3.67}$$

$$\mathbf{n}_{\bar{x}} \cdot \mathbf{n}_y = |\mathbf{n}_{\bar{x}}||\mathbf{n}_y| \cos \theta_{\bar{x},y} = 1 \cdot 1 \cdot n_{\bar{x}y} = n_{\bar{x}y} \tag{3.68}$$

where $n_{\bar{x},x} = \cos \theta_{\bar{x},x}$ and $n_{\bar{x}y} = \cos \theta_{\bar{x},y}$ are the *direction cosines* and defined as the dot product of two direction vectors (Figure 3.24). For the cosine function, the angle may be given either clockwise or counter-clockwise with equal results, which implies that $n_{\bar{x}x} = n_{x\bar{x}}$ and $n_{\bar{x}y} = n_{y\bar{x}}$. Substituting the direction cosines into (3.66), we obtain

$$A_x = A n_{x\bar{x}}; \quad A_y = A n_{y\bar{x}} \tag{3.69}$$

where, with the relations (3.69), we have derived scalar expressions for the components of a vector in a new coordinate system; see Figure 3.23.

If we begin with the vector components A_x and A_y in the coordinate system xy instead, we can determine the action of these vectors in an arbitrary direction $\bar{x}$ (see Figure 3.25) from

$$A_{x\bar{x}} = \mathbf{A}_x \cdot \mathbf{n}_{\bar{x}} = A_x \mathbf{n}_x \cdot \mathbf{n}_{\bar{x}}; \quad A_{y\bar{x}} = \mathbf{A}_y \cdot \mathbf{n}_{\bar{x}} = A_y \mathbf{n}_y \cdot \mathbf{n}_{\bar{x}} \tag{3.70}$$

The total action in $\bar{x}$-direction is then

$$A_{\bar{x}} = A_{x\bar{x}} + A_{y\bar{x}} = A_x \mathbf{n}_x \cdot \mathbf{n}_{\bar{x}} + A_y \mathbf{n}_y \cdot \mathbf{n}_{\bar{x}} \tag{3.71}$$

Using the direction cosines defined in (3.67) and (3.68), Equation (3.71) can be written as

$$A_{\bar{x}} = A_x n_{x\bar{x}} + A_y n_{y\bar{x}} \tag{3.72}$$

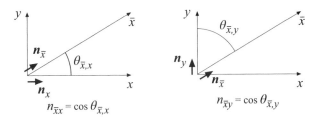

Figure 3.24   Direction cosines

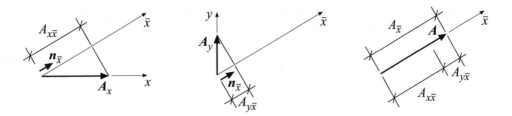

Figure 3.25   Total vector action in the $\bar{x}$-direction

The scalar relations (3.69) and (3.72) will be used for the transformations between different coordinate systems.

With relation (3.72), the displacements $\bar{u}_1$ and $\bar{u}_2$ in the longitudinal direction of the bar can be expressed in the global displacement degrees of freedom u_1, u_2, u_3 and u_4

$$\bar{u}_1 = n_{x\bar{x}}u_1 + n_{y\bar{x}}u_2 \tag{3.73}$$

$$\bar{u}_2 = n_{x\bar{x}}u_3 + n_{y\bar{x}}u_4 \tag{3.74}$$

or in matrix form

$$\boxed{\bar{\mathbf{a}}^e = \mathbf{G}\mathbf{a}^e} \tag{3.75}$$

where

$$\bar{\mathbf{a}}^e = \begin{bmatrix} \bar{u}_1 \\ \bar{u}_2 \end{bmatrix}; \quad \mathbf{G} = \begin{bmatrix} n_{x\bar{x}} & n_{y\bar{x}} & 0 & 0 \\ 0 & 0 & n_{x\bar{x}} & n_{y\bar{x}} \end{bmatrix}; \quad \mathbf{a}^e = \begin{bmatrix} u_1 \\ u_2 \\ u_3 \\ u_4 \end{bmatrix} \tag{3.76}$$

Using the relations (3.69), the components P_1, P_2, P_3 and P_4 of the nodal forces $\bar{P}_1$ and $\bar{P}_2$ can be expressed as

$$P_1 = n_{x\bar{x}}\bar{P}_1 \tag{3.77}$$

$$P_2 = n_{y\bar{x}}\bar{P}_1 \tag{3.78}$$

$$P_3 = n_{x\bar{x}}\bar{P}_2 \tag{3.79}$$

$$P_4 = n_{y\bar{x}}\bar{P}_2 \tag{3.80}$$

or in matrix form

$$\boxed{\mathbf{f}_b^e = \mathbf{G}^T\bar{\mathbf{f}}_b^e} \tag{3.81}$$

where

$$\mathbf{f}_b^e = \begin{bmatrix} P_1 \\ P_2 \\ P_3 \\ P_4 \end{bmatrix}; \quad \mathbf{G}^T = \begin{bmatrix} n_{x\bar{x}} & 0 \\ n_{y\bar{x}} & 0 \\ 0 & n_{x\bar{x}} \\ 0 & n_{y\bar{x}} \end{bmatrix}; \quad \bar{\mathbf{f}}_b^e = \begin{bmatrix} \bar{P}_1 \\ \bar{P}_2 \end{bmatrix} \tag{3.82}$$

In the same manner, a relation between the equivalent element loads $\mathbf{f}_l^e$ in a global system and the equivalent element loads $\bar{\mathbf{f}}_l^e$ in a local system can be written as

$$\mathbf{f}_l^e = \mathbf{G}^T \bar{\mathbf{f}}_l^e \tag{3.83}$$

where

$$\mathbf{f}_l^e = \begin{bmatrix} f_{l1}^e \\ f_{l2}^e \\ f_{l3}^e \\ f_{l4}^e \end{bmatrix} \tag{3.84}$$

The matrices $\mathbf{G}$ and $\mathbf{G}^T$ are referred to as transformation matrices and their purpose is to transform quantities so that they can be expressed in different coordinate systems. The contents of a transformation matrix depend on the type of quantity/quantities to be transformed and between which coordinate systems the transformation is performed.

If we substitute the transformations (3.81), (3.75) and (3.83) into the element relation (3.58), we get a new element relation, one with its quantities expressed in the directions of the global coordinate system,

$$\mathbf{K}^e \mathbf{a}^e = \mathbf{f}^e \tag{3.85}$$

where

$$\mathbf{K}^e = \mathbf{G}^T \bar{\mathbf{K}}^e \mathbf{G}; \quad \mathbf{f}^e = \mathbf{f}_b^e + \mathbf{f}_l^e \tag{3.86}$$

Figure 3.26 shows how transformations of displacements and forces between different coordinate systems lead to a relation for the bar element in global coordinates.

If the matrix multiplication in Equation (3.86) is performed, we obtain the element stiffness matrix $\mathbf{K}^e$ for a bar element in the global system as

$$\mathbf{K}^e = \frac{D_{EA}}{L} \begin{bmatrix} \mathbf{C} & -\mathbf{C} \\ -\mathbf{C} & \mathbf{C} \end{bmatrix}; \quad \mathbf{C} = \begin{bmatrix} n_{x\bar{x}}n_{x\bar{x}} & n_{x\bar{x}}n_{y\bar{x}} \\ n_{y\bar{x}}n_{x\bar{x}} & n_{y\bar{x}}n_{y\bar{x}} \end{bmatrix} \tag{3.87}$$

$$\left. \begin{aligned} \mathbf{f}_b^e &= \mathbf{G}^T \bar{\mathbf{f}}_b^e \quad (3.81) \\ \mathbf{f}_l^e &= \mathbf{G}^T \bar{\mathbf{f}}_l^e \quad (3.83) \\ \bar{\mathbf{K}}^e \bar{\mathbf{a}}^e &= \bar{\mathbf{f}}^e \quad (3.60) \\ \bar{\mathbf{f}}^e &= \bar{\mathbf{f}}_b^e + \bar{\mathbf{f}}_l^e \quad (3.61) \\ \bar{\mathbf{a}}^e &= \mathbf{G}\mathbf{a}^e \quad (3.75) \end{aligned} \right\} \Rightarrow \mathbf{K}^e \mathbf{a}^e = \mathbf{f}^e \quad (3.85)$$

where

$$\mathbf{K}^e = \mathbf{G}^T \bar{\mathbf{K}}^e \mathbf{G}; \quad \mathbf{f}^e = \mathbf{f}_b^e + \mathbf{f}_l^e$$

Figure 3.26   From local coordinates to global coordinates

3.3 Trusses

A truss consists of bars connected with frictionless hinges. The mathematical representation of a truss is called computational model. In this model, the bars are represented by their system lines and the frictionless hinges by nodes (Figure 3.27). The bar element we have formulated can be used to create a computational model for a truss (Figure 3.28).

For a truss, we introduce, in the same manner as with a spring system, a global numbering of all the displacement degrees of freedom and we gather these in a global displacement vector **a**,

$$\mathbf{a} = \begin{bmatrix} a_1 \\ \cdot \\ a_i \\ a_j \\ a_k \\ a_l \\ \cdot \\ a_n \end{bmatrix} \tag{3.88}$$

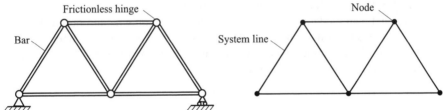

Figure 3.27 A truss and the associated computational model

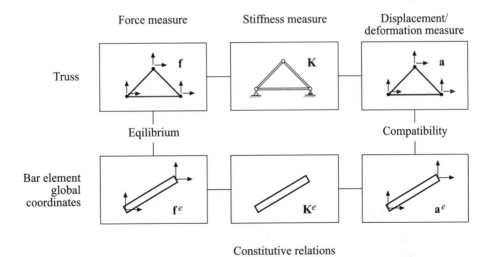

Figure 3.28 From a bar element to a truss

From the element relation of each single bar, we have a local numbering of the displacements, u_1, u_2, u_3 and u_4. By compatibility conditions, each one of these displacements at the element level is associated with a displacement in the global system. For an element associated with the global displacements a_i, a_j, a_k and a_l (Figure 3.29), we obtain the following compatibility conditions:

$$u_1 = a_i \qquad (3.89)$$

$$u_2 = a_j \qquad (3.90)$$

$$u_3 = a_k \qquad (3.91)$$

$$u_4 = a_l \qquad (3.92)$$

The compatibility conditions can be written in matrix form

$$\mathbf{a}^e = \mathbf{Ha} \qquad (3.93)$$

where $\mathbf{a}^e$ is the element nodal displacements in global directions, (3.76), $\mathbf{a}$ the displacement degrees of freedom of the truss (3.88) and $\mathbf{H}$ a transformation matrix with $H_{1,i} = 1, H_{2,j} = 1$, $H_{3,k} = 1, H_{4,l} = 1$ and all other elements equal to 0; cf. (2.13).

In the computational model for a truss, external forces can only be introduced at the nodes. These may be element loads from distributed loads on the bar elements, point loads at the nodes and support forces at the supports of the truss (Figure 3.30). These forces are denoted as f_i and are gathered in a global force vector $\mathbf{f}$,

$$\mathbf{f} = \begin{bmatrix} f_1 \\ \cdot \\ f_i \\ f_j \\ f_k \\ f_l \\ \cdot \\ f_n \end{bmatrix} \qquad (3.94)$$

By the equilibrium conditions, we now relate the nodal forces of the single bar elements to the truss (Figure 3.31). This is done by expressing the nodal forces in a form that enables

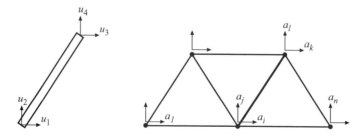

Figure 3.29   The displacements of the bar element and the displacements of the truss

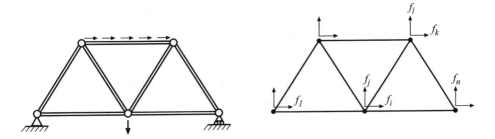

Figure 3.30 External forces that are introduced at the nodes in the computational model

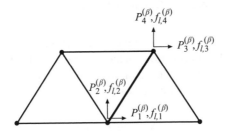

Figure 3.31 A free-body diagram of a bar element

the equilibrium equations in the nodes of the truss to be established easily. From the element relation (3.85), we have the normal forces expressed as nodal forces in global coordinates, $\mathbf{f}_b^e$. As for the spring system, we introduce an expanded nodal force vector $\hat{\mathbf{f}}_b^e$, and moreover, we introduce an expanded vector for element loads $\hat{\mathbf{f}}_l^e$; each of the expanded vectors with a number of rows equal to the number of degrees of freedom in the truss. It turns out that the expanded force vectors can be expressed in matrix form using the same matrix $\mathbf{H}$ as was defined by the compatibility conditions,[4]

$$\hat{\mathbf{f}}_b^e = \mathbf{H}^T \mathbf{f}_b^e \tag{3.95}$$

$$\hat{\mathbf{f}}_l^e = \mathbf{H}^T \mathbf{f}_l^e \tag{3.96}$$

Substituting Equations (3.93), (3.95) and (3.96) into (3.85) gives an element relation in expanded form

$$\hat{\mathbf{f}}_b^e = \hat{\mathbf{K}}^e \mathbf{a} - \hat{\mathbf{f}}_l^e \tag{3.97}$$

where

$$\hat{\mathbf{K}}^e = \mathbf{H}^T \mathbf{K}^e \mathbf{H} \tag{3.98}$$

The matrix $\hat{\mathbf{K}}^e$ will contain the elements of $\mathbf{K}^e$, but placed in rows and columns corresponding to the global degrees of freedom to which that element is associated. By this expanded way of writing, we have a formulation where force components associated with the same global

---

[4] This is an expression for one of the basic assumptions of solid mechanics/structural mechanics, formulated for example by Maxwell's reciprocity theorem, and which means that strain energy only can be transformed, not created.

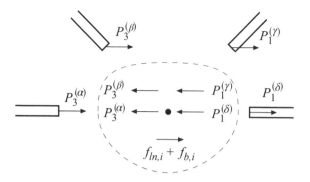

Figure 3.32   Equilibrium for degree of freedom i

degrees of freedom are on the same row in the force vector and in this way the formulation is prepared for global equilibrium equations in the directions of the degrees of freedom.

For a single degree of freedom i, the equilibrium of the node in the direction of the degree of freedom can be written (see Figure 3.32) as

$$\sum_{e=1}^{m} \hat{f}_{b,i}^{e} = f_{ln,i} + f_{b,i} \tag{3.99}$$

where e denotes the element number, $f_{ln,i}$ is a possible *nodal load*, that is a point load acting on the node, and $f_{b,i}$ is a possible support force. By establishing an equilibrium equation for each degree of freedom, we obtain for the entire truss

$$\sum_{e=1}^{m} \hat{\mathbf{f}}_{b}^{e} = \mathbf{f}_{ln} + \mathbf{f}_{b} \tag{3.100}$$

If the expanded element equations (3.97) are substituted into the equilibrium relations, we obtain

$$\sum_{e=1}^{m} \left(\hat{\mathbf{K}}^{e} \mathbf{a} - \hat{\mathbf{f}}_{l}^{e} \right) = \mathbf{f}_{ln} + \mathbf{f}_{b} \tag{3.101}$$

or

$$\boxed{\mathbf{Ka} = \mathbf{f}} \tag{3.102}$$

where

$$\mathbf{K} = \sum_{e=1}^{m} \hat{\mathbf{K}}^{e}; \quad \mathbf{f} = \mathbf{f}_{l} + \mathbf{f}_{b}; \quad \mathbf{f}_{l} = \mathbf{f}_{ln} + \mathbf{f}_{lq}; \quad \mathbf{f}_{lq} = \sum_{e=1}^{m} \hat{\mathbf{f}}_{l}^{e} \tag{3.103}$$

How compatibility conditions, element relations and equilibrium lead to a system of equations for a truss is shown in Figure 3.33.

When considering the present boundary conditions, the displacements and the support forces can be determined from (3.102). Once the displacements $\mathbf{a}$ have been determined, the displacements $\mathbf{a}^{e}$ for one element can be determined from (3.93). After that, the displacements $\bar{\mathbf{a}}^{e}$ in the longitudinal direction of the bar can be determined from (3.75). The

$$\left.\begin{array}{l} \hat{\mathbf{f}}_b^e = \mathbf{H}^T \mathbf{f}_b^e \quad (3.95) \\[4pt] \hat{\mathbf{f}}_l^e = \mathbf{H}^T \mathbf{f}_l^e \quad (3.96) \\[4pt] \mathbf{K}^e \mathbf{a}^e = \mathbf{f}^e \quad (3.85) \\[4pt] \mathbf{f}^e = \mathbf{f}_b^e + \mathbf{f}_l^e \quad (3.86) \\[4pt] \mathbf{a}^e = \mathbf{H}\mathbf{a} \quad (3.93) \end{array}\right\} \Rightarrow$$

$$\left.\begin{array}{l} \displaystyle\sum_{e=1}^{m} \hat{\mathbf{f}}_b^e = \mathbf{f}_{ln} + \mathbf{f}_b \quad (3.100) \\[12pt] \hat{\mathbf{f}}_b^e = \hat{\mathbf{K}}^e \mathbf{a} - \hat{\mathbf{f}}_l^e \quad (3.97) \\[6pt] \text{where} \\[4pt] \hat{\mathbf{K}}^e = \mathbf{H}^T \mathbf{K}^e \mathbf{H} \quad (3.98) \end{array}\right\} \Rightarrow \mathbf{K}\mathbf{a} = \mathbf{f} \quad (3.102)$$

where

$$\mathbf{K} = \sum_{e=1}^{m} \hat{\mathbf{K}}^e$$

$$\mathbf{f} = \mathbf{f}_l + \mathbf{f}_b$$

$$\mathbf{f}_l = \mathbf{f}_{ln} + \sum_{e=1}^{m} \hat{\mathbf{f}}_l^e$$

Figure 3.33   From bar element to truss

displacement distribution along the bar can then be determined using (3.46), and the normal force distribution can be determined using (3.52).

Here, the stiffness matrix $\mathbf{K}$ and the load vector $\mathbf{f}_l$ have been described as sums of expanded matrices $\hat{\mathbf{K}}^e$ and vectors $\hat{\mathbf{f}}_l^e$. Usually, these expanded matrices are not actually created. Instead, the stiffness matrix $\mathbf{K}$ is established directly by defining a matrix with dimensions $n \times n$ filled with zeros after which for each element the coefficients in the element matrix $\mathbf{K}^e$ are added to the positions corresponding to the global degrees of freedom for the element in question. In the same manner, the load vector $\mathbf{f}_l$ is created from a vector where at first the loads acting at the nodes are placed and then the element loads $\mathbf{f}_l^e$ are added to the rows corresponding to the global degrees of freedom of the element (cf. Figure 2.14 and Section 2.3).

Example 3.2 Truss

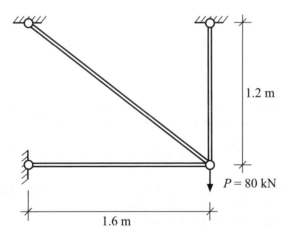

Figure 1   A plane truss consisting of three bars

The truss in Figure 1 consists of three bars with lengths $L_1 = 1.6\,\text{m}$, $L_2 = 1.2\,\text{m}$ and $L_3 = \sqrt{1.6^2 + 1.2^2} = 2.0\,\text{m}$. The cross-sectional areas of the bars are $A_1 = 6.0 \times 10^{-4}\,\text{m}^2$, $A_2 = 3.0 \times 10^{-4}\,\text{m}^2$ and $A_3 = 10.0 \times 10^{-4}\,\text{m}^2$, respectively. The elastic modulus is $E = 200.0\,\text{GPa}$ for all the bars. The load and boundary conditions for the truss are shown in Figure 1. The displacements at the free node of the truss, the support forces and the normal forces in the bars of the truss shall be determined.

Computational model

The truss is built up of three bar elements, denoted as 1, 2 and 3 (Figure 2). The model has the displacement degrees of freedom a_1 to a_8. The downwards directed force acting in degree of freedom 6 implies that $f_6 = -80\,\text{kN}$. In the degrees of freedom a_1, a_2, a_3, a_4, a_7 and a_8, the displacement is prescribed to be zero.

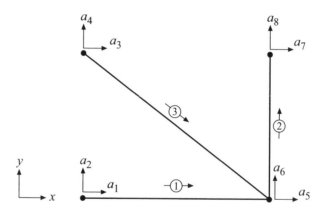

Figure 2   The computational model

Element matrices

For each bar element, an element relation $\mathbf{K}^e \mathbf{a}^e = \mathbf{f}^e_b$ can be established. The element stiffness matrices $\mathbf{K}^e$ for the three elements are given from (3.87).
Element 1:

$$\frac{EA_1}{L_1} = \frac{200.0 \times 10^9 \cdot 6.0 \times 10^{-4}}{1.6} = 75.0 \times 10^6 \tag{1}$$

The local $\bar{x}$-axis coincides with the global x-axis and is perpendicular to the global y-axis. The direction cosines for the angle between these are, therefore, $n_{x\bar{x}} = \cos(x, \bar{x}) = 1$ and $n_{y\bar{x}} = \cos(y, \bar{x}) = 0$, which gives the element stiffness matrix

$$\mathbf{K}^1 = 75.0 \times 10^6 \begin{bmatrix} 1 & 0 & -1 & 0 \\ 0 & 0 & 0 & 0 \\ -1 & 0 & 1 & 0 \\ 0 & 0 & 0 & 0 \end{bmatrix} = \begin{bmatrix} 75 & 0 & -75 & 0 \\ 0 & 0 & 0 & 0 \\ -75 & 0 & 75 & 0 \\ 0 & 0 & 0 & 0 \end{bmatrix} 10^6 \tag{2}$$

Element 2:

$$\frac{EA_2}{L_2} = \frac{200.0 \times 10^9 \cdot 3.0 \times 10^{-4}}{1.2} = 50.0 \times 10^6 \tag{3}$$

The local $\bar{x}$-axis is perpendicular to the global x-axis and coincides with the global y-axis. The direction cosines are, therefore, $n_{x\bar{x}} = \cos(x, \bar{x}) = 0$ and $n_{y\bar{x}} = \cos(y, \bar{x}) = 1$, which gives the element stiffness matrix

$$\mathbf{K}^2 = 50.0 \times 10^6 \begin{bmatrix} 0 & 0 & 0 & 0 \\ 0 & 1 & 0 & -1 \\ 0 & 0 & 0 & 0 \\ 0 & -1 & 0 & 1 \end{bmatrix} = \begin{bmatrix} 0 & 0 & 0 & 0 \\ 0 & 50 & 0 & -50 \\ 0 & 0 & 0 & 0 \\ 0 & -50 & 0 & 50 \end{bmatrix} 10^6 \tag{4}$$

Element 3:

$$\frac{EA_3}{L_3} = \frac{200.0 \times 10^9 \cdot 10.0 \times 10^{-4}}{2.0} = 100.0 \times 10^6 \tag{5}$$

The direction cosines are $n_{x\bar{x}} = \cos(x, \bar{x}) = \frac{1.6}{2.0} = 0.8$ and $n_{y\bar{x}} = \cos(y, \bar{x}) = -\frac{1.2}{2.0} = -0.6$. This gives the element stiffness matrix

$$\mathbf{K}^3 = 100.0 \times 10^6 \begin{bmatrix} 0.64 & -0.48 & -0.64 & 0.48 \\ -0.48 & 0.36 & 0.48 & -0.36 \\ -0.64 & 0.48 & 0.64 & -0.48 \\ 0.48 & -0.36 & -0.48 & 0.36 \end{bmatrix}$$

$$= \begin{bmatrix} 64 & -48 & -64 & 48 \\ -48 & 36 & 48 & -36 \\ -64 & 48 & 64 & -48 \\ 48 & -36 & -48 & 36 \end{bmatrix} 10^6 \tag{6}$$

Compatibility conditions

The local displacement degrees of freedom for Elements 1–3 correspond to global degrees of freedom according to what follows:

Element 1:

$$u_1^{(1)} = a_1 \tag{7}$$

$$u_2^{(1)} = a_2 \tag{8}$$

$$u_3^{(1)} = a_5 \tag{9}$$

$$u_4^{(1)} = a_6 \tag{10}$$

Element 2:

$$u_1^{(2)} = a_5 \tag{11}$$

$$u_2^{(2)} = a_6 \tag{12}$$

$$u_3^{(2)} = a_7 \tag{13}$$

$$u_4^{(2)} = a_8 \tag{14}$$

Element 3:

$$u_1^{(3)} = a_3 \tag{15}$$

$$u_2^{(3)} = a_4 \tag{16}$$

$$u_3^{(3)} = a_5 \tag{17}$$

$$u_4^{(3)} = a_6 \tag{18}$$

which can be summarised in the topology matrix:

$$\text{topology} = \begin{bmatrix} 1 & 1 & 2 & 5 & 6 \\ 2 & 5 & 6 & 7 & 8 \\ 3 & 3 & 4 & 5 & 6 \end{bmatrix} \tag{19}$$

Assembling

The stiffness matrix is established by creating a matrix filled with zeros, after which the coefficients of the element stiffness matrices are added to the positions in the stiffness matrix given by the topology matrix. This yields

$$\mathbf{K} = \begin{bmatrix} 75 & 0 & 0 & 0 & -75 & 0 & 0 & 0 \\ 0 & 0 & 0 & 0 & 0 & 0 & 0 & 0 \\ 0 & 0 & 64 & -48 & -64 & 48 & 0 & 0 \\ 0 & 0 & -48 & 36 & 48 & -36 & 0 & 0 \\ -75 & 0 & -64 & 48 & 139 & -48 & 0 & 0 \\ 0 & 0 & 48 & -36 & -48 & 86 & 0 & -50 \\ 0 & 0 & 0 & 0 & 0 & 0 & 0 & 0 \\ 0 & 0 & 0 & 0 & 0 & -50 & 0 & 50 \end{bmatrix} 10^6 \tag{20}$$

Boundary conditions and nodal loads

The only load on the truss is a downwards directed force $P = 80$ kN acting in degree of freedom 6. Because the force is directed downwards, it is directed opposite to the positive y-direction and, therefore, negative. With that, the load vector $\mathbf{f}_l$ becomes

$$\mathbf{f}_l = \begin{bmatrix} 0 \\ 0 \\ 0 \\ 0 \\ 0 \\ -80 \\ 0 \\ 0 \end{bmatrix} 10^3 \tag{21}$$

The displacements are prescribed to be zero in the degrees of freedom where the truss is fixed, that is $a_1 = 0$, $a_2 = 0$, $a_3 = 0$, $a_4 = 0$, $a_7 = 0$ and $a_8 = 0$. This can be described by the boundary condition matrix

$$\text{boundary conditions} = \begin{bmatrix} 1 & 0 \\ 2 & 0 \\ 3 & 0 \\ 4 & 0 \\ 7 & 0 \\ 8 & 0 \end{bmatrix} \tag{22}$$

The only degrees of freedom where the displacement is not prescribed are a_5 and a_6. In the degrees of freedom where the displacement is prescribed, support forces arise. These are at present unknown and denoted as $f_{b,1}, f_{b,2}, f_{b,3}, f_{b,4}, f_{b,7}$ and $f_{b,8}$. The displacement vector $\mathbf{a}$ and the boundary force vector $\mathbf{f}_b$ can with that be written as

$$\mathbf{a} = \begin{bmatrix} 0 \\ 0 \\ 0 \\ 0 \\ a_5 \\ a_6 \\ 0 \\ 0 \end{bmatrix} ; \quad \mathbf{f}_b = \begin{bmatrix} f_{b,1} \\ f_{b,2} \\ f_{b,3} \\ f_{b,4} \\ 0 \\ 0 \\ f_{b,7} \\ f_{b,8} \end{bmatrix} \tag{23}$$

Solving the system of equations

We can now establish a system of equations $\mathbf{Ka} = \mathbf{f}_l + \mathbf{f}_b$ for the truss,

$$10^6 \begin{bmatrix} 75 & 0 & 0 & 0 & -75 & 0 & 0 & 0 \\ 0 & 0 & 0 & 0 & 0 & 0 & 0 & 0 \\ 0 & 0 & 64 & -48 & -64 & 48 & 0 & 0 \\ 0 & 0 & -48 & 36 & 48 & -36 & 0 & 0 \\ -75 & 0 & -64 & 48 & 139 & -48 & 0 & 0 \\ 0 & 0 & 48 & -36 & -48 & 86 & 0 & -50 \\ 0 & 0 & 0 & 0 & 0 & 0 & 0 & 0 \\ 0 & 0 & 0 & 0 & 0 & -50 & 0 & 50 \end{bmatrix} \begin{bmatrix} 0 \\ 0 \\ 0 \\ 0 \\ a_5 \\ a_6 \\ 0 \\ 0 \end{bmatrix} = \begin{bmatrix} 0 \\ 0 \\ 0 \\ 0 \\ 0 \\ -80 \\ 0 \\ 0 \end{bmatrix} 10^3 + \begin{bmatrix} f_{b,1} \\ f_{b,2} \\ f_{b,3} \\ f_{b,4} \\ 0 \\ 0 \\ f_{b,7} \\ f_{b,8} \end{bmatrix} \tag{24}$$

The system of equations contains eight equations and eight unknowns; the displacements a_5 and a_6 and the support forces $f_{b,1}, f_{b,2}, f_{b,3}, f_{b,4}, f_{b,7}$ and $f_{b,8}$. Considering the prescribed displacements, the system of equations can be reduced to

$$10^6 \begin{bmatrix} 139 & -48 \\ -48 & 86 \end{bmatrix} \begin{bmatrix} a_5 \\ a_6 \end{bmatrix} = \begin{bmatrix} 0 \\ -80 \end{bmatrix} 10^3 \tag{25}$$

and the displacements a_5 and a_6 be determined

$$\begin{bmatrix} a_5 \\ a_6 \end{bmatrix} = \begin{bmatrix} -0.3979 \\ -1.1523 \end{bmatrix} 10^{-3} \tag{26}$$

This means that the free node is displaced 0.40 mm leftwards and 1.15 mm downwards. The displacements are shown in Figure 3.

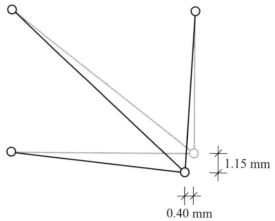

Figure 3   The computed displacements drawn in a magnified scale

When a_5 and a_6 have been determined, all nodal displacements are known and the support forces $f_{b,1}$, $f_{b,2}$, $f_{b,3}$, $f_{b,4}$, $f_{b,7}$ and $f_{b,8}$ can be determined from the global system of equations,

$$\begin{bmatrix} f_{b,1} \\ f_{b,2} \\ f_{b,3} \\ f_{b,4} \\ f_{b,7} \\ f_{b,8} \end{bmatrix} = 10^6 \begin{bmatrix} -75 & 0 \\ 0 & 0 \\ -64 & 48 \\ 48 & -36 \\ 0 & 0 \\ 0 & -50 \end{bmatrix} \begin{bmatrix} a_5 \\ a_6 \end{bmatrix} = \begin{bmatrix} 29.84 \\ 0 \\ -29.84 \\ 22.38 \\ 0 \\ 57.62 \end{bmatrix} 10^3 \tag{27}$$

The external load and the support forces computed are shown in Figure 4. We can conclude that the sum of the horizontal forces as well as the sum of vertical forces is zero. Consequently, equilibria of external forces are fulfilled. By establishing an equation of moments, we can also show that the equilibrium of moments is fulfilled.

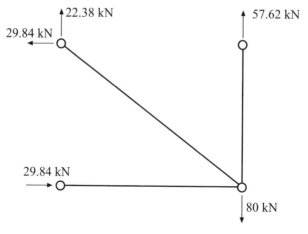

Figure 4   The external load and the computed support forces

Compute internal forces

Beginning from the compatibility relations, the displacements for Element 1 can be determined

$$\mathbf{a}^1 = \begin{bmatrix} 0 \\ 0 \\ -0.3979 \\ -1.1523 \end{bmatrix} 10^{-3} \tag{28}$$

Furthermore, using (3.75), the displacements can be expressed in the local coordinate system of the element

$$\bar{\mathbf{a}}^1 = \mathbf{Ga}^1 = \begin{bmatrix} 1 & 0 & 0 & 0 \\ 0 & 0 & 1 & 0 \end{bmatrix} \begin{bmatrix} 0 \\ 0 \\ -0.3979 \\ -1.1523 \end{bmatrix} 10^{-3} = \begin{bmatrix} 0 \\ -0.3979 \end{bmatrix} 10^{-3} \tag{29}$$

and the normal force in the element can be computed from (3.52)

$$N^{(1)} = EA_1 \mathbf{B} \bar{\mathbf{a}}^1$$

$$= 200.0 \times 10^9 \cdot 6.0 \times 10^{-4} \frac{1}{1.6} \begin{bmatrix} -1 & 1 \end{bmatrix} \begin{bmatrix} 0 \\ -0.3979 \end{bmatrix} 10^{-3}$$

$$= -29.84 \times 10^3 \tag{30}$$

For Element 2, we have in the same manner:

$$\mathbf{a}^2 = \begin{bmatrix} -0.3979 \\ -1.1523 \\ 0 \\ 0 \end{bmatrix} 10^{-3} \tag{31}$$

$$\bar{\mathbf{a}}^2 = \mathbf{Ga}^2 = \begin{bmatrix} 0 & 1 & 0 & 0 \\ 0 & 0 & 0 & 1 \end{bmatrix} \begin{bmatrix} -0.3979 \\ -1.1523 \\ 0 \\ 0 \end{bmatrix} 10^{-3} = \begin{bmatrix} -1.1523 \\ 0 \end{bmatrix} 10^{-3} \tag{32}$$

$$N^{(2)} = EA_2 \mathbf{B} \bar{\mathbf{a}}^2$$

$$= 200.0 \times 10^9 \cdot 3.0 \times 10^{-4} \frac{1}{1.2} \begin{bmatrix} -1 & 1 \end{bmatrix} \begin{bmatrix} -1.1523 \\ 0 \end{bmatrix} 10^{-3}$$

$$= 57.62 \times 10^3 \tag{33}$$

and for Element 3

$$\mathbf{a}^3 = \begin{bmatrix} 0 \\ 0 \\ -0.3979 \\ -1.1523 \end{bmatrix} 10^{-3} \tag{34}$$

$$\mathbf{\bar{a}}^3 = \mathbf{Ga}^3 = \begin{bmatrix} 0.8 & -0.6 & 0 & 0 \\ 0 & 0 & 0.8 & -0.6 \end{bmatrix} \begin{bmatrix} 0 \\ 0 \\ -0.3979 \\ -1.1523 \end{bmatrix} 10^{-3} = \begin{bmatrix} 0 \\ 0.3730 \end{bmatrix} 10^{-3} \quad (35)$$

$$N^{(3)} = EA_3 \mathbf{B\bar{a}}^3$$

$$= 200.0 \times 10^9 \cdot 10.0 \times 10^{-4} \frac{1}{2.0} \begin{bmatrix} -1 & 1 \end{bmatrix} \begin{bmatrix} 0 \\ 0.3730 \end{bmatrix} 10^{-3}$$

$$= 37.30 \times 10^3 \quad (36)$$

This result means that the normal forces in the three elements are -29.84 kN, 57.62 kN and 37.30 kN, respectively; see Figure 5.

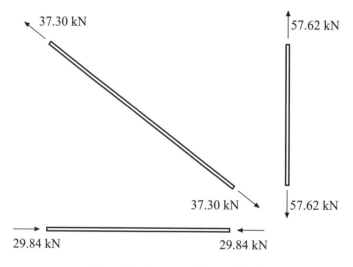

Figure 5   The normal forces in the bars

Exercises

3.1

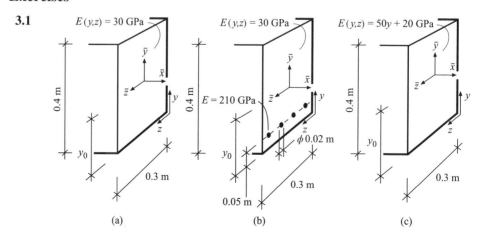

(a)

(b)

(c)

Among the three cross-sections mentioned here, one is homogeneous and two are non-homogeneous. The position of the local $\bar{x}$-axis in the respective cross-section is determined by the condition $\int_A E \, \bar{y} \, dA = 0$. With the coordinate systems given in the figures, this yields the following three expressions, where $\bar{y} = y - y_0$

(a) $\displaystyle\int_0^{0.3} \int_0^{0.4} 30 \, \bar{y} \, dy \, dz = 0$

(b) $\displaystyle\int_0^{0.3} \int_0^{0.4} 30 \, \bar{y} \, dy \, dz + 4 \, (210 - 30)\frac{\pi \, 0.02^2}{4} \, (0.05 - y_0) = 0$

(c) $\displaystyle\int_0^{0.3} \int_0^{0.4} (50 \, y + 20) \, \bar{y} \, dy \, dz = 0$

Determine y_0 for the three cross-sections.

3.2 Consider the cross-sections mentioned in Exercise 3.1. For the locations of the local $\bar{x}$-axes found in the exercise,

(a) determine the stiffness D_{EA} of the cross-section.

(b) determine the normal force N for the generalised strain $\varepsilon_{\bar{x}} = 0.001$.

3.3 Consider a non-loaded bar of length $L = 1$. Begin from (3.43) and let the displacements of the ends of the bar be $\bar{u}_1 = 0.001$ and $\bar{u}_2 = 0.002$.

(a) Draw the shape functions N_1 and N_2 as functions of $\bar{x}$.

(b) Draw $N_1 \bar{u}_1$ and $N_2 \bar{u}_2$.

(c) Draw $u_h(\bar{x}) = \mathbf{N}\bar{\mathbf{a}}^e = N_1 \bar{u}_1 + N_2 \bar{u}_2$. Compare with Figure 3.18.

(d) For $D_{EA} = 1.0 \times 10^9$, determine the normal force $N(\bar{x})$.

3.4 Consider the differential equation (3.25) for linearly varying load $q_{\bar{x}}(\bar{x}) = q_0 \frac{\bar{x}}{L}$. With the method mentioned in Example 3.1,

(a) determine the particular solution $u_p(\bar{x})$.

(b) draw $u_p(\bar{x})$ and compare with Figure 3.18.

(c) determine and draw $N_p(\bar{x})$

(d) determine element loads $\bar{\mathbf{f}}_l^e$.

3.5

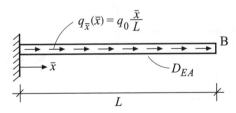

Consider a bar fixed at its left end and unconstrained at its right. The bar is loaded with a linearly varying load $q_{\bar{x}}(\bar{x}) = q_0 \frac{\bar{x}}{L}$. With $\bar{\mathbf{f}}_l^e$ obtained in Exercise 3.4 and using the element equation (3.60), determine the displacement of point B.

3.6 For the bar in Exercise 3.5, with $D_{EA} = 400$ MN, $L = 2$ m and $q_0 = 300$ kN/m, determine the displacement distribution $u(\bar{x})$ and the distribution of the normal force $N(\bar{x})$.

3.7

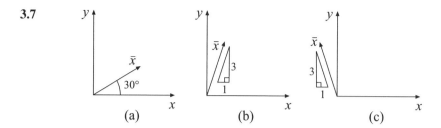

(a)   (b)   (c)

Determine the direction cosines $n_{x\bar{x}}$ and $n_{y\bar{x}}$ for the three cases shown in the figure.

3.8

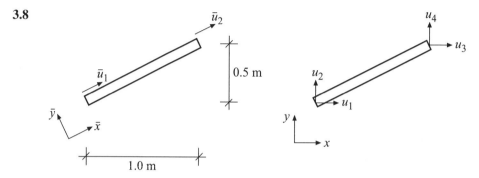

The bar element in the figure has the degrees of freedom $\bar{u}_1$ and $\bar{u}_2$ in a local $\bar{x}\bar{y}$-system. Express $\bar{u}_1$ and $\bar{u}_2$ as functions of u_1, u_2, u_3 and u_4 in a global xy-system, that is determine the coefficients in the matrix $\mathbf{G}$ in expression (3.75).

3.9

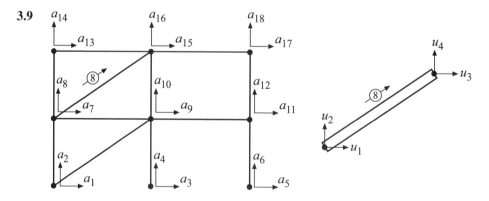

The element equations for Element 8 in the truss in the figure has the form

$$\begin{bmatrix} K_{11}^e & K_{12}^e & K_{13}^e & K_{14}^e \\ K_{21}^e & K_{22}^e & K_{23}^e & K_{24}^e \\ K_{31}^e & K_{32}^e & K_{33}^e & K_{34}^e \\ K_{41}^e & K_{42}^e & K_{43}^e & K_{44}^e \end{bmatrix} \begin{bmatrix} u_1 \\ u_2 \\ u_3 \\ u_4 \end{bmatrix} = \begin{bmatrix} P_1 \\ P_2 \\ P_3 \\ P_4 \end{bmatrix}$$

Give the position in the stiffness matrix $\mathbf{K}$, where elements K_{11}^e, K_{24}^e and K_{32}^e are added at the assembling.

3.10

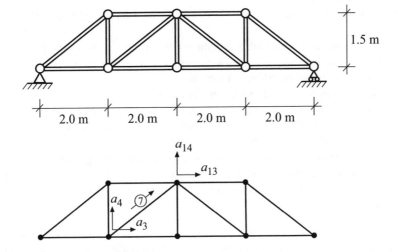

For the truss in the figure, $A = 1.0 \times 10^{-3}$ m^2 and $E = 200$ GPa. The displacements a_3, a_4, a_{13} and a_{14} have been computed to $a_3 = 0.960$ mm, $a_4 = -8.160$ mm, $a_{13} = 2.880$ mm and $a_{14} = -13.220$ mm.
(a) Determine the displacements $\bar{u}_1^{(7)}$ and $\bar{u}_2^{(7)}$ for Element 7.
(b) Determine the axial deformation, the normal force and the stress of the bar.
(c) Draw the bar in its original (undeformed) state and in its deformed state. Draw the displacements in global and local directions.

3.11

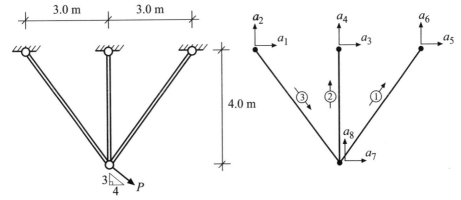

For the truss in the figure to the left, $A = 0.25 \times 10^{-3}$ m^2, $E = 200$ GPa and $P = 50$ kN. The figure to the right shows the element subdivision and the definition of the degrees of freedom. Perform the following tasks manually.

(a) Compute the element stiffness matrix $\mathbf{K}^e$ for Element 1 in the global coordinate system by establishing the matrices $\bar{\mathbf{K}}^e$ and $\mathbf{G}$ and performing the matrix multiplications in (3.86).

(b) Establish the element stiffness matrices $\mathbf{K}^e$ for Elements 2 and 3. Use expression (3.87).

(c) Summarise the compatibility conditions by setting up a topology matrix.

(d) Use the information in the topology matrix to assemble the three element stiffness matrices to a global stiffness matrix $\mathbf{K}$.

(e) Define boundary conditions and nodal loads.

(f) Compute unknown displacements and support forces by solving the system of equations. Check the external equilibrium of forces horizontally and vertically, and check also the equilibrium of moments.

(g) Determine the nodal displacements of the elements in the global coordinate system.

(h) Determine the nodal displacements of each of the three elements in the local coordinate system of the element. Compute also the normal force in each of the three elements.

3.12 Follow the method of computation for trusses in the example section in the CALFEM manual and analyse the truss in Exercise 3.11. Print out the matrices and compare with the corresponding matrices in the computations done manually.

4

Beams and Frames

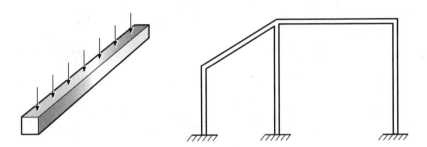

Figure 4.1 Transversely loaded beam and a two-dimensional (plane) frame

A *beam* is defined as a long body that can carry axial load by bar action and transverse load by beam action. By a *frame* we mean a structure that is built up from several beam members connected to each other, which can carry load by bar action and beam action; see Figure 4.1.

In Chapter 3, the bar and its mode of action, bar action, were presented. This chapter deals with the straight two-dimensional beam. It has two modes of action – bar action and beam action. Figure 4.2 shows the quantities and relations of structural mechanics for beam action and for two-dimensional beams and frames. In the same manner as for bars and trusses (Figure 3.2), this map has a scale with six levels divided into three groups. The group ranging from the material level to beam action leads to the differential equation for beam action, one of the two differential equations of the beam. The derivation of this differential equation is performed in Section 4.1. The other differential equation, the one for bar action, has already been derived in Chapter 3. From the two differential equations of the beam, we can derive a beam element in global coordinates. This is discussed in Section 4.1. Finally, the beam element can, in a systematic manner, be placed in a frame. The derivation of this systematics, which leads to systems of equations for frames, is discussed in Section 4.3.

Structural Mechanics: Modelling and Analysis of Frames and Trusses, First Edition.
Karl-Gunnar Olsson and Ola Dahlblom.
© 2016 John Wiley & Sons, Ltd. Published 2016 by John Wiley & Sons, Ltd.

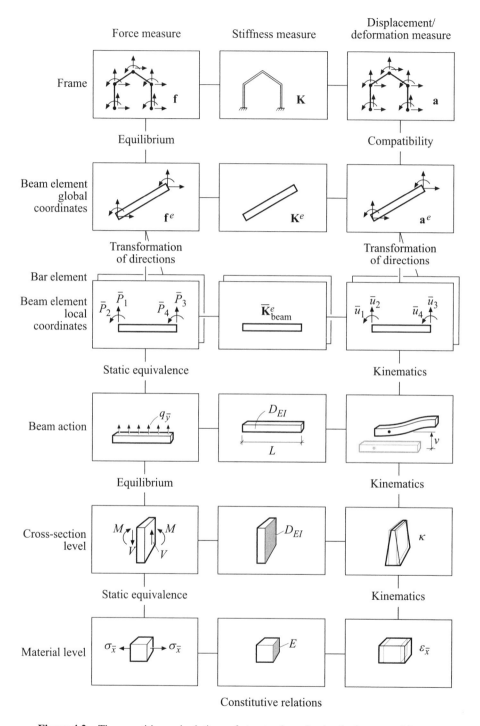

Figure 4.2   The quantities and relations of structural mechanics for beams and frames

4.1   The Differential Equation for Beam Action

We seek an expression that describes the relation between load and displacement (deflection) for beam action (Figure 4.3). In the same manner as for bar action, the basis is a constitutive relation, or a material relation, which relates strain to stress. Via kinematic relations, we can establish relations between the strain of the material and the deflection of the beam. Via force equivalence and equilibrium, the stress acting on the material is related to the external load that acts on the beam.

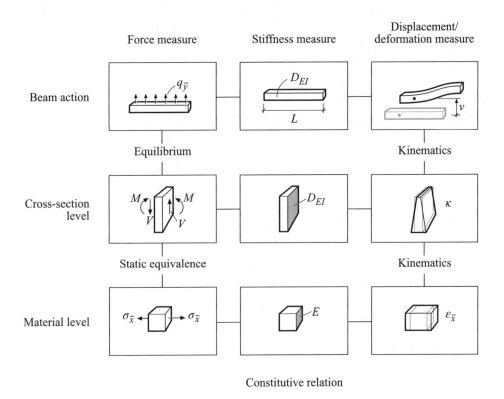

Figure 4.3   From the material level to beam action

4.1.1   Definitions

The beam has, similar to the bar, its main extension in one dimension, and we let this dimension be the $\bar{x}$-axis in a local coordinate system $(\bar{x}, \bar{y}, \bar{z})$. The quantities of beam theory are illustrated in Figure 4.4. We have, at the material level, normal stresses $\sigma_{\bar{x}}(\bar{x}, \bar{y}, \bar{z})$ as well as shear stresses $\sigma_{\bar{x}\bar{y}}(\bar{x}, \bar{y}, \bar{z})$, normal strains $\varepsilon_{\bar{x}}(\bar{x}, \bar{y}, \bar{z})$ and shear strains $\gamma_{\bar{x}\bar{y}}(\bar{x}, \bar{y}, \bar{z})$ and, finally, we also have a material stiffness $E(\bar{x}, \bar{y}, \bar{z})$. At the cross-section level, we summarise the stresses and strains of the material to the generalised force and deformation measures. The generalised forces that act on a cross-section lamella are the bending moment $M(\bar{x})$ and the shear force $V(\bar{x})$. The corresponding generalised deformation measures are the curvature $\kappa(\bar{x})$ and shear

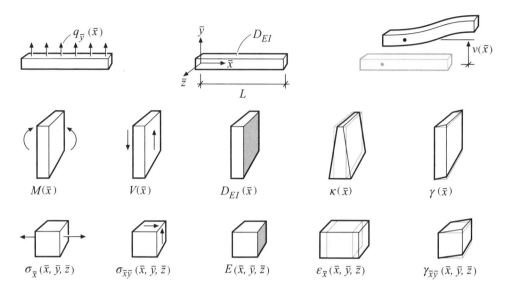

Figure 4.4   The quantities of beam theory

$\gamma(\bar{x})$ of the cross-section. In the beam theory discussed here, shear deformations are neglected and with that, the generalised shear $\gamma(\bar{x})$ is neglected as well. On the system line of the beam, the local $\bar{x}$-axis, a transverse loading $q_{\bar{y}}(\bar{x})$ acts in the $\bar{y}$-direction, which leads to transverse displacements $v(\bar{x})$. For a material stiffness $E(\bar{y}, \bar{z})$ constant across the beam cross-section, the system line of the beam coincides with the centroid of the cross-section; cf. Section 3.1.1. The two-dimensional (plane) beam action we study here presumes a cross-section with one symmetry axis (or two symmetry axes) and that one symmetry axis coincides with the local $\bar{y}$-axis.[1]

4.1.2   The Material Level

Strain

In (3.5) in Section 3.1.2, the normal strain for a fibre has been defined as

$$\varepsilon_{\bar{x}} = \frac{d\bar{x} + du - d\bar{x}}{d\bar{x}} = \frac{du}{d\bar{x}} \tag{4.1}$$

The definition assumes small strains and small displacements.

Another type of strain, *shear strain*, arises in beam action, but it has usually a negligible influence on the transverse displacements $v(\bar{x})$ of the beam. We assume that shear strains are neglected, which, for rectangular beam cross-sections, is reasonable if the order of magnitude of the height/length relation h/L is less than 1/5 and for I-sections if the order of magnitude of h/L is less than 1/10. This beam theory, which assumes that shear strains can be neglected, is usually referred to as the Bernoulli–Euler beam theory.

---

[1] For other cross-section shapes, a more complex definition of the reference axis of the beam is required. See also the comment in Section 7.3.1.

Stress

In Section 3.1.2, the three stress components

$$\sigma_{\bar{x}} = \frac{dP_{\bar{x}}}{dA}; \quad \sigma_{\bar{x}\bar{y}} = \frac{dP_{\bar{y}}}{dA}; \quad \sigma_{\bar{x}\bar{z}} = \frac{dP_{\bar{z}}}{dA} \tag{4.2}$$

have been defined in (3.7). The stress component $\sigma_{\bar{x}}$, directed perpendicular to the considered section surface is referred to as normal stress. In addition to this, there are two more stress components in beam action; $\sigma_{\bar{x}\bar{y}}$ and $\sigma_{\bar{x}\bar{z}}$. These are directed parallel to the section surface and are referred to as shear stresses.[2]

The Constitutive Relation of the Material

The material is assumed to be linear elastic. This means that there is a linear relation between normal stress and normal strain; cf. Equation (3.9) and Figure 3.9

$$\sigma_{\bar{x}}(\bar{x}, \bar{y}, \bar{z}) = E(\bar{x}, \bar{y}, \bar{z})\, \varepsilon_{\bar{x}}(\bar{x}, \bar{y}, \bar{z}) \tag{4.3}$$

where E is the elastic modulus. The material can be isotropic or orthotropic. For an orthotropic material, E denotes the elastic modulus in the longitudinal direction of the beam. Under the assumption that shear strains are neglected, a material relation for shear stresses is unnecessary.

4.1.3   The Cross-Section Level

Kinematics

The description of the kinematics of beam action is based on the reference axis of the beam, the local $\bar{x}$-axis. Each point on the axis has an original position $\bar{x}$. A loading of the beam leads to a displacement $v(\bar{x})$ perpendicular to the axis and a rotation $\theta(\bar{x})$. The deformation is, in beam action, related to this rotation, or more specifically to the change in rotation, $d\theta$, which arises between two adjacent points with a distance of $d\bar{x}$ between them (Figure 4.5). Under the assumptions that plane cross-sections remain plane and perpendicular to the displaced reference axis and that the rotation of the beam is small ($\cos\theta \approx 1$), the displacement du in the $\bar{x}$-direction of a fibre end at an arbitrary position of a cross-section lamella will be determined from the magnitude of $d\theta$ and will be proportional to the distance $\bar{y}$ from the reference axis,

$$du = -d\theta\,\bar{y} \tag{4.4}$$

The strain $\varepsilon_{\bar{x}}(\bar{x}, \bar{y}, \bar{z})$ of a fibre can then be written as

$$\varepsilon_{\bar{x}}(\bar{x}, \bar{y}, \bar{z}) = \frac{du}{d\bar{x}} = -\frac{d\theta}{d\bar{x}}\bar{y} \tag{4.5}$$

We introduce the notation $\kappa(\bar{x})$ for the derivative of the rotation

$$\kappa(\bar{x}) = \frac{d\theta}{d\bar{x}} \tag{4.6}$$

---

[2] Other common notations for the shear stresses are $\tau_{\bar{x}\bar{y}}$ and $\tau_{\bar{x}\bar{z}}$.

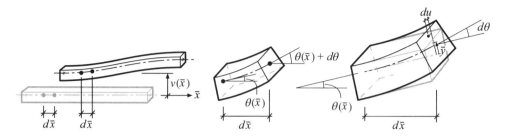

Figure 4.5   The deflection $v(\bar{x})$ of the reference axis

which on substitution into (4.5) gives

$$\varepsilon_{\bar{x}}(\bar{x}, \bar{y}, \bar{z}) = -\kappa(\bar{x})\,\bar{y} \tag{4.7}$$

With that, we have that for beam action $\varepsilon_{\bar{x}}(\bar{x}, \bar{y}, \bar{z}) = f(\bar{x})\,g(\bar{y}, \bar{z})$, where $f(\bar{x}) = \kappa(\bar{x})$ is the generalised strain measure, and $g(\bar{y}, \bar{z}) = -\bar{y}$ describes the strain mode of beam action. The generalised strain measure $\kappa(\bar{x})$ is referred to as *curvature*.

Force Relations

On a small part dA of the cross-sectional area, the forces $\sigma_{\bar{x}}\, dA$ and $\sigma_{\bar{x}\bar{y}}\, dA$ act, perpendicular and parallel, respectively, to the cross-sectional surface. The resulting bending moment $M(\bar{x})$ (Figure 4.6) is given by the integral

$$M(\bar{x}) = -\int_A \sigma_{\bar{x}}(\bar{x}, \bar{y}, \bar{z})\,\bar{y}\, dA \tag{4.8}$$

where $\sigma_{\bar{x}}(\bar{x}, \bar{y}, \bar{z})\bar{y}dA$ is the force perpendicular to the cross-sectional surface multiplied by the distance $\bar{y}$ (the moment arm) to the reference axis. The resulting shear force $V(\bar{x})$ (Figure 4.7) is given by the integral

$$V(\bar{x}) = \int_A \sigma_{\bar{x}\bar{y}}(\bar{x}, \bar{y}, \bar{z})\, dA \tag{4.9}$$

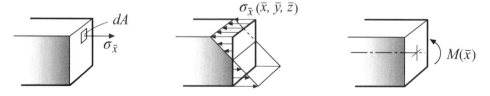

Figure 4.6   Normal stress and bending moment

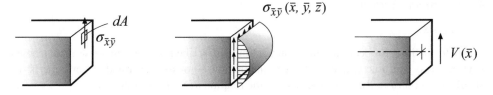

Figure 4.7 Shear stress and shear force

The Constitutive Relation at the Cross-Section Level

The material relation (4.3) and the kinematic relation (4.7) substituted into the expression for the bending moment (4.8) gives

$$M(\bar{x}) = -\int_A E(\bar{x}, \bar{y}, \bar{z})\, \varepsilon_{\bar{x}}(\bar{x}, \bar{y}, \bar{z})\, \bar{y}\, dA$$

$$= -\int_A E(\bar{x}, \bar{y}, \bar{z})\, (-\kappa(\bar{x})\, \bar{y})\, \bar{y}\, dA \tag{4.10}$$

Since $\kappa(\bar{x})$ represents the whole cross-section lamella and does not vary with $\bar{y}$ or $\bar{z}$, Equation (4.10) can be written as

$$M(\bar{x}) = \kappa(\bar{x}) \int_A E(\bar{x}, \bar{y}, \bar{z})\, \bar{y}^2\, dA \tag{4.11}$$

or

$$\boxed{M(\bar{x}) = D_{EI}(\bar{x})\kappa(\bar{x})} \tag{4.12}$$

where

$$D_{EI}(\bar{x}) = \int_A E(\bar{x}, \bar{y}, \bar{z})\bar{y}^2\, dA \tag{4.13}$$

is the *bending stiffness* of the cross-section lamella. If the elastic modulus E is assumed to be constant across the cross-section, then

$$D_{EI}(\bar{x}) = E(\bar{x})I(\bar{x}) \tag{4.14}$$

where $I = \int_A \bar{y}^2 dA$ is the moment of inertia.

In Figure 4.8, the relations for the cross-section are summarised.

$$
\left.
\begin{aligned}
M(\bar{x}) &= -\int_A \sigma_{\bar{x}}(\bar{x}, \bar{y}, \bar{z})\bar{y}dA && (4.8)\\
\sigma_{\bar{x}}(\bar{x}, \bar{y}, \bar{z}) &= E(\bar{x}, \bar{y}, \bar{z})\,\varepsilon_{\bar{x}}(\bar{x}, \bar{y}, \bar{z}) && (4.3)\\
\varepsilon_{\bar{x}}(\bar{x}, \bar{y}, \bar{z}) &= -\kappa(\bar{x})\bar{y} && (4.7)
\end{aligned}
\right\}
\Rightarrow M(\bar{x}) = D_{EI}(\bar{x})\kappa(\bar{x}) \quad (4.12)
$$

where

$$D_{EI}(\bar{x}) = \int_A E(\bar{x}, \bar{y}, \bar{z})\bar{y}^2 dA$$

Figure 4.8 From the material level to the cross-section level

4.1.4 Beam Action

Kinematics

The deformation of a beam in pure beam action is described by the transverse displacement $v(\bar{x})$ of the system line (Figure 4.5). The inclination $\frac{dv}{d\bar{x}}$ of the deformed system line is, for small angles ($\tan\theta \approx \theta$), also a measure of the rotation $\theta(\bar{x})$ of the system line

$$\frac{dv}{d\bar{x}} = \tan(\theta(\bar{x})) = \theta(\bar{x}) \tag{4.15}$$

Differentiation of (4.15) with respect to $\bar{x}$ and using (4.6) gives

$$\boxed{\frac{d^2v}{d\bar{x}^2} = \frac{d\theta}{d\bar{x}} = \kappa(\bar{x})} \tag{4.16}$$

that is a relation between the deformation measure $\kappa(\bar{x})$ of the cross-section level and the deflection $v(\bar{x})$ of the beam.

The kinematic assumptions introduced here and in the previous section can be summarised according to the following:

- small displacements;
- small strains;
- plane cross-sections remain plane and perpendicular to the system line.

Equilibrium

Consider a small part of the undeformed beam of length $d\bar{x}$. In this part, forces act according to the depiction in Figure 4.9. Equilibrium perpendicular to the system line of the beam gives

$$-V(\bar{x}) + (V(\bar{x}) + dV) + q_{\bar{y}}(\bar{x})d\bar{x} = 0 \tag{4.17}$$

where $V(\bar{x})$ is the shear force at $\bar{x}$, $V(\bar{x}) + dV$ is the shear force at $\bar{x} + d\bar{x}$ and $q_{\bar{y}}(\bar{x})d\bar{x}$ is the distributed load perpendicular to the beam. The expression can be simplified to

$$dV + q_{\bar{y}}(\bar{x})d\bar{x} = 0 \tag{4.18}$$

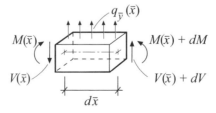

Figure 4.9 Equilibrium for a small part $d\bar{x}$ of a beam

or

$$\frac{dV}{d\bar{x}} + q_{\bar{y}}(\bar{x}) = 0 \tag{4.19}$$

Moment equilibrium about an axis parallel with the $\bar{z}$-axis at the right end of the beam part in Figure 4.9 gives

$$-M(\bar{x}) + (M(\bar{x}) + dM) + V(\bar{x})d\bar{x} - q_{\bar{y}}(\bar{x})d\bar{x}\frac{d\bar{x}}{2} = 0 \tag{4.20}$$

Simplifying the expression and considering that the last term is negligible compared with the others give

$$dM + V(\bar{x})d\bar{x} = 0 \tag{4.21}$$

or

$$\frac{dM}{d\bar{x}} + V(\bar{x}) = 0 \tag{4.22}$$

The two relations (4.19) and (4.22) can be combined to

$$\boxed{\frac{d^2M}{d\bar{x}^2} - q_{\bar{y}}(\bar{x}) = 0} \tag{4.23}$$

With the equilibrium relations (4.19) and (4.23), the force action, $V(\bar{x})$ and $M(\bar{x})$, on a cross-section lamella is related to the loading $q_{\bar{y}}(\bar{x})$ of the beam.

The differential equation for beam action

Substituting the kinematic relation (4.16) into (4.12) gives

$$\boxed{M(\bar{x}) = D_{EI}(\bar{x})\frac{d^2v}{d\bar{x}^2}} \tag{4.24}$$

Substitution of (4.24) into (4.22) then gives

$$\boxed{V(\bar{x}) = -D_{EI}(\bar{x})\frac{d^3v}{d\bar{x}^3}} \tag{4.25}$$

Substitution of (4.24) into the equilibrium relation (4.23) then gives

$$\frac{d^2}{d\bar{x}^2}\left(D_{EI}(\bar{x})\frac{d^2v}{d\bar{x}^2}\right) - q_{\bar{y}}(\bar{x}) = 0 \tag{4.26}$$

If the bending stiffness D_{EI} is constant along the beam, (4.26) can be written as

$$\boxed{D_{EI}\frac{d^4v}{d\bar{x}^4} - q_{\bar{y}}(\bar{x}) = 0} \tag{4.27}$$

and if the elastic modulus is constant across the cross-section

$$EI\frac{d^4v}{d\bar{x}^4} - q_{\bar{y}}(\bar{x}) = 0 \tag{4.28}$$

where the bending stiffness of the beam is the product of the elastic modulus E and the moment of inertia I. Figure 4.10 shows how the equilibrium relation, the constitutive relation and the kinematic relation are combined to form a relation for beam action.

$$\left.\begin{array}{ll} \dfrac{d^2M}{d\bar{x}^2} - q_{\bar{y}}(\bar{x}) = 0 & (4.23) \\[2mm] M(\bar{x}) = D_{EI}(\bar{x})\,\kappa(\bar{x}) & (4.12) \\[2mm] \kappa(\bar{x}) = \dfrac{d^2v}{d\bar{x}^2} & (4.16) \end{array}\right\} \Rightarrow \quad D_{EI}\dfrac{d^4v}{d\bar{x}^4} - q_{\bar{y}}(\bar{x}) = 0 \;\; (4.27)$$

$$\text{for constant } D_{EI}$$

Figure 4.10 From the cross-section level to beam action

For a beam without distributed load ($q_{\bar{y}} = 0$), (4.27) becomes the homogeneous equation

$$D_{EI}\frac{d^4v}{d\bar{x}^4} = 0 \tag{4.29}$$

For the differential equation for beam action to be solvable, a total of four boundary conditions are required, two at each end point of the beam. These are translation v or shear force V in combination with rotation θ or bending moment M.

The two modes of action of the beam, beam action and bar action, can be described as uncoupled relations only if the location of the system line is chosen such that the condition $\int_A E\,\bar{y}\,dA = 0$ is fulfilled.[3]

4.2 Beam Element

Starting from the two differential equations, (4.27) for beam action and (3.25) for bar action, the relations between forces and deformations for a beam element will be derived. First, a relation for a beam element with four degrees of freedom is established in the local coordinates of the beam $\bar{x}$ and $\bar{y}$ (Figure 4.11). This is combined with the element for bar action, from Chapter 3, which yields a beam element with six degrees of freedom. With these six degrees of freedom, beam action as well as bar action is considered. After that, a transformation of coordinates is performed, which enables the beam element to be positioned with an arbitrary orientation in a two-dimensional frame.

---

[3] Normal force caused by beam action is given by $N = \int_A \sigma_{\bar{x}}\,dA = -\kappa(\bar{x})\int_A E\,\bar{y}\,dA$. With $\int_A E\,\bar{y}\,dA = 0$, bending of the beam does not give rise to any normal force. Bending moment caused by bar action is given by $M = -\int_A \sigma_{\bar{x}}\,\bar{y}\,dA = -\varepsilon_{\bar{x}}(\bar{x})\int_A E\,\bar{y}\,dA$. With $\int_A E\,\bar{y}\,dA = 0$, axial strain in the bar does not give rise to any moment. The two differential equations of the beam can thus be treated independent of each other.

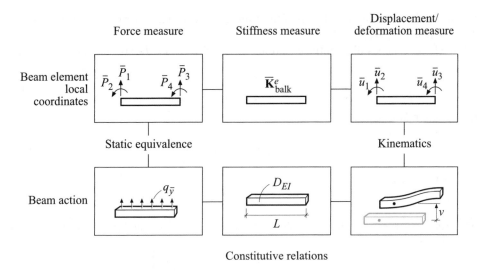

Figure 4.11 From beam action to a beam element in local coordinates

4.2.1 Definitions

We start by formulating a beam element for pure beam action. Such an element has four displacement degrees of freedom, $\bar{u}_1$, $\bar{u}_2$, $\bar{u}_3$ and $\bar{u}_4$, as shown in Figure 4.12. Degrees of freedom $\bar{u}_1$ and $\bar{u}_3$ describe the translation of the nodes in the $\bar{y}$-direction, that is $v(\bar{x})$ at $\bar{x} = 0$ and $\bar{x} = L$, respectively, while degrees of freedom $\bar{u}_2$ and $\bar{u}_4$ describe the rotations of the nodes about the $\bar{z}$-axis, that is $\frac{dv}{d\bar{x}}$ at $\bar{x} = 0$ and $\bar{x} = L$, respectively. The forces acting in the $\bar{y}$-direction at $\bar{x} = 0$ and $\bar{x} = L$ are denoted as $\bar{P}_1$ and $\bar{P}_3$, respectively, and are defined to be positive in the direction of the $\bar{y}$-axis. The moments acting at $\bar{x} = 0$ and $\bar{x} = L$ are denoted as $\bar{P}_2$ and $\bar{P}_4$, respectively, and are defined to be positive when directed as the rotations $\bar{u}_2$ and $\bar{u}_4$, that is counterclockwise.

4.2.2 Solving the Differential Equation for Beam Action

The general solution $v(\bar{x})$ to the differential equation (4.27) can be written as the sum of the solution $v_h(\bar{x})$ to the homogeneous equation and an arbitrary particular solution $v_p(\bar{x})$

$$v(\bar{x}) = v_h(\bar{x}) + v_p(\bar{x}) \tag{4.30}$$

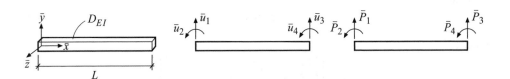

Figure 4.12 A beam element with four degrees of freedom

We first seek a form of the solution of the homogeneous equation written as a function of the displacements $\bar{u}_1, \bar{u}_2, \bar{u}_3$ and $\bar{u}_4$.

If the homogeneous differential equation (4.29) is divided by the stiffness D_{EI}, we obtain

$$\frac{d^4 v}{d\bar{x}^4} = 0 \tag{4.31}$$

Integrating four times gives

$$v_h(\bar{x}) = \alpha_1 + \alpha_2 \bar{x} + \alpha_3 \bar{x}^2 + \alpha_4 \bar{x}^3 \tag{4.32}$$

or in matrix form

$$v_h(\bar{x}) = \bar{\mathbf{N}} \boldsymbol{\alpha} \tag{4.33}$$

where $\bar{\mathbf{N}} = \bar{\mathbf{N}}(\bar{x})$ describes how the solution varies along the $\bar{x}$-axis and $\boldsymbol{\alpha}$ contains the constants of integration,

$$\bar{\mathbf{N}} = \begin{bmatrix} 1 & \bar{x} & \bar{x}^2 & \bar{x}^3 \end{bmatrix} ; \quad \boldsymbol{\alpha} = \begin{bmatrix} \alpha_1 \\ \alpha_2 \\ \alpha_3 \\ \alpha_4 \end{bmatrix} \tag{4.34}$$

At the nodes of the beam, at $\bar{x} = 0$ and $\bar{x} = L$, we have the boundary conditions

$$v_h(0) = \bar{u}_1 \tag{4.35}$$

$$\left(\frac{dv_h}{d\bar{x}} \right)_{\bar{x}=0} = \bar{u}_2 \tag{4.36}$$

$$v_h(L) = \bar{u}_3 \tag{4.37}$$

$$\left(\frac{dv_h}{d\bar{x}} \right)_{\bar{x}=L} = \bar{u}_4 \tag{4.38}$$

Substitution of these conditions into (4.33) gives

$$\bar{u}_1 = \alpha_1 \tag{4.39}$$

$$\bar{u}_2 = \alpha_2 \tag{4.40}$$

$$\bar{u}_3 = \alpha_1 + \alpha_2 L + \alpha_3 L^2 + \alpha_4 L^3 \tag{4.41}$$

$$\bar{u}_4 = \alpha_2 + 2\alpha_3 L + 3\alpha_4 L^2 \tag{4.42}$$

or in matrix form

$$\bar{\mathbf{a}}^e = \mathbf{C} \boldsymbol{\alpha} \tag{4.43}$$

where

$$\bar{\mathbf{a}}^e = \begin{bmatrix} \bar{u}_1 \\ \bar{u}_2 \\ \bar{u}_3 \\ \bar{u}_4 \end{bmatrix} ; \quad \mathbf{C} = \begin{bmatrix} 1 & 0 & 0 & 0 \\ 0 & 1 & 0 & 0 \\ 1 & L & L^2 & L^3 \\ 0 & 1 & 2L & 3L^2 \end{bmatrix} \tag{4.44}$$

By inverting $\mathbf{C}$, we can express the constants of integration $\boldsymbol{\alpha}$ as a functions of the displacement degrees of freedom of the element $\bar{\mathbf{a}}^e$,

$$\boldsymbol{\alpha} = \mathbf{C}^{-1}\bar{\mathbf{a}}^e \tag{4.45}$$

where

$$\mathbf{C}^{-1} = \begin{bmatrix} 1 & 0 & 0 & 0 \\ 0 & 1 & 0 & 0 \\ -\frac{3}{L^2} & -\frac{2}{L} & \frac{3}{L^2} & -\frac{1}{L} \\ \frac{2}{L^3} & \frac{1}{L^2} & -\frac{2}{L^3} & \frac{1}{L^2} \end{bmatrix} \tag{4.46}$$

Substituting (4.45) into (4.33), we obtain the solution $v_h(\bar{x})$ as

$$v_h(\bar{x}) = \mathbf{N}\bar{\mathbf{a}}^e \tag{4.47}$$

where

$$\mathbf{N} = \bar{\mathbf{N}}\mathbf{C}^{-1} = \begin{bmatrix} 1 & \bar{x} & \bar{x}^2 & \bar{x}^3 \end{bmatrix} \begin{bmatrix} 1 & 0 & 0 & 0 \\ 0 & 1 & 0 & 0 \\ -\frac{3}{L^2} & -\frac{2}{L} & \frac{3}{L^2} & -\frac{1}{L} \\ \frac{2}{L^3} & \frac{1}{L^2} & -\frac{2}{L^3} & \frac{1}{L^2} \end{bmatrix} \tag{4.48}$$

which gives

$$\mathbf{N} = \begin{bmatrix} 1 - 3\frac{\bar{x}^2}{L^2} + 2\frac{\bar{x}^3}{L^3} & \bar{x} - 2\frac{\bar{x}^2}{L} + \frac{\bar{x}^3}{L^2} & 3\frac{\bar{x}^2}{L^2} - 2\frac{\bar{x}^3}{L^3} & -\frac{\bar{x}^2}{L} + \frac{\bar{x}^3}{L^2} \end{bmatrix} \tag{4.49}$$

With that, we have reformulated $v_h(\bar{x})$ written as a general polynomial (4.32) to a solution in the form

$$v_h(\bar{x}) = \mathbf{N}\bar{\mathbf{a}}^e = N_1(\bar{x})\,\bar{u}_1 + N_2(\bar{x})\,\bar{u}_2 + N_3(\bar{x})\,\bar{u}_3 + N_4(\bar{x})\,\bar{u}_4 \tag{4.50}$$

where

$$N_1(\bar{x}) = 1 - 3\frac{\bar{x}^2}{L^2} + 2\frac{\bar{x}^3}{L^3} \tag{4.51}$$

$$N_2(\bar{x}) = \bar{x} - 2\frac{\bar{x}^2}{L} + \frac{\bar{x}^3}{L^2} \tag{4.52}$$

$$N_3(\bar{x}) = 3\frac{\bar{x}^2}{L^2} - 2\frac{\bar{x}^3}{L^3} \tag{4.53}$$

$$N_4(\bar{x}) = -\frac{\bar{x}^2}{L} + \frac{\bar{x}^3}{L^2} \tag{4.54}$$

The functions $N_1(\bar{x})$–$N_4(\bar{x})$ describe how the solution varies with $\bar{x}$ and are referred to as *base functions* or *shape functions*; cf. Chapter 3. We have in (4.50) an expression where the product $N_i(\bar{x})\,\bar{u}_i$ gives the contribution to $v_h(\bar{x})$ from the displacement $\bar{u}_i$ and where $N_i(\bar{x})$ states its shape and $\bar{u}_i$ its size. Substitution of (4.47) into the general solution of the differential equation (4.30) gives

$$\boxed{v(\bar{x}) = \mathbf{N}\bar{\mathbf{a}}^e + v_p(\bar{x})} \tag{4.55}$$

where the particular solution $v_p(\bar{x})$ is different for different shapes of transverse load on the beam.

Since we have chosen to determine the constants of integration in the general solution from the homogeneous differential equation, only one possible particular solution $v_p(\bar{x})$ remains; the one where the displacements of the nodes are equal to zero,

$$v_p(0) = 0 \tag{4.56}$$

$$\left(\frac{dv_p}{d\bar{x}}\right)_{\bar{x}=0} = 0 \tag{4.57}$$

$$v_p(L) = 0 \tag{4.58}$$

$$\left(\frac{dv_p}{d\bar{x}}\right)_{\bar{x}=L} = 0 \tag{4.59}$$

All other choices for $v_p(\bar{x})$ imply that the constants of integration, which are functions of $\bar{\mathbf{a}}^e$, Change; cf. (4.45). With the choices we have made, the general solution $v(\bar{x})$ can be understood as the sum of a beam displaced at its end points, but otherwise non-loaded, $v_h(\bar{x})$, and a transversely loaded beam fixed at both ends, $v_p(\bar{x})$ (Figure 4.13).

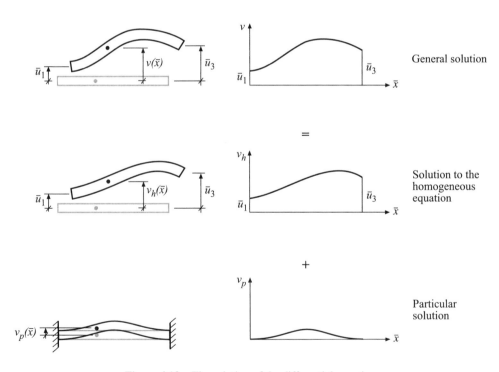

Figure 4.13 The solution of the differential equation

Differentiating (4.55) three times gives

$$\frac{dv}{d\bar{x}} = \frac{d\mathbf{N}}{d\bar{x}}\bar{\mathbf{a}}^e + \frac{dv_p}{d\bar{x}} \tag{4.60}$$

$$\frac{d^2v}{d\bar{x}^2} = \mathbf{B}\bar{\mathbf{a}}^e + \frac{d^2v_p}{d\bar{x}^2} \tag{4.61}$$

$$\frac{d^3v}{d\bar{x}^3} = \frac{d\mathbf{B}}{d\bar{x}}\bar{\mathbf{a}}^e + \frac{d^3v_p}{d\bar{x}^3} \tag{4.62}$$

where

$$\frac{d\mathbf{N}}{d\bar{x}} = \frac{d\bar{\mathbf{N}}}{d\bar{x}}\mathbf{C}^{-1} = \begin{bmatrix} 0 & 1 & 2\bar{x} & 3\bar{x}^2 \end{bmatrix} \begin{bmatrix} 1 & 0 & 0 & 0 \\ 0 & 1 & 0 & 0 \\ -\frac{3}{L^2} & -\frac{2}{L} & \frac{3}{L^2} & -\frac{1}{L} \\ \frac{2}{L^3} & \frac{1}{L^2} & -\frac{2}{L^3} & \frac{1}{L^2} \end{bmatrix} \tag{4.63}$$

$$\mathbf{B} = \frac{d^2\mathbf{N}}{d\bar{x}^2} = \frac{d^2\bar{\mathbf{N}}}{d\bar{x}^2}\mathbf{C}^{-1} = \begin{bmatrix} 0 & 0 & 2 & 6\bar{x} \end{bmatrix} \begin{bmatrix} 1 & 0 & 0 & 0 \\ 0 & 1 & 0 & 0 \\ -\frac{3}{L^2} & -\frac{2}{L} & \frac{3}{L^2} & -\frac{1}{L} \\ \frac{2}{L^3} & \frac{1}{L^2} & -\frac{2}{L^3} & \frac{1}{L^2} \end{bmatrix} \tag{4.64}$$

$$\frac{d\mathbf{B}}{d\bar{x}} = \frac{d^3\mathbf{N}}{d\bar{x}^3} = \frac{d^3\bar{\mathbf{N}}}{d\bar{x}^3}\mathbf{C}^{-1} = \begin{bmatrix} 0 & 0 & 0 & 6 \end{bmatrix} \begin{bmatrix} 1 & 0 & 0 & 0 \\ 0 & 1 & 0 & 0 \\ -\frac{3}{L^2} & -\frac{2}{L} & \frac{3}{L^2} & -\frac{1}{L} \\ \frac{2}{L^3} & \frac{1}{L^2} & -\frac{2}{L^3} & \frac{1}{L^2} \end{bmatrix} \tag{4.65}$$

which gives

$$\frac{d\mathbf{N}}{d\bar{x}} = \begin{bmatrix} -6\frac{\bar{x}}{L^2} + 6\frac{\bar{x}^2}{L^3} & 1 - 4\frac{\bar{x}}{L} + 3\frac{\bar{x}^2}{L^2} & 6\frac{\bar{x}}{L^2} - 6\frac{\bar{x}^2}{L^3} & -2\frac{\bar{x}}{L} + 3\frac{\bar{x}^2}{L^2} \end{bmatrix} \tag{4.66}$$

$$\mathbf{B} = \begin{bmatrix} -\frac{6}{L^2} + 12\frac{\bar{x}}{L^3} & -\frac{4}{L} + 6\frac{\bar{x}}{L^2} & \frac{6}{L^2} - 12\frac{\bar{x}}{L^3} & -\frac{2}{L} + 6\frac{\bar{x}}{L^2} \end{bmatrix} \tag{4.67}$$

$$\frac{d\mathbf{B}}{d\bar{x}} = \begin{bmatrix} \frac{12}{L^3} & \frac{6}{L^2} & -\frac{12}{L^3} & \frac{6}{L^2} \end{bmatrix} \tag{4.68}$$

Substituting (4.55) into (4.24) and (4.22), we obtain expressions for moments and shear forces as functions of the displacements of the nodes,

$$M(\bar{x}) = D_{EI}\left(\mathbf{B}\bar{\mathbf{a}}^e + \frac{d^2v_p}{d\bar{x}^2} \right) \tag{4.69}$$

$$V(\bar{x}) = -\frac{dM}{d\bar{x}} = -D_{EI}\left(\frac{d\mathbf{B}}{d\bar{x}}\bar{\mathbf{a}}^e + \frac{d^3v_p}{d\bar{x}^3} \right) \tag{4.70}$$

or

$$M(\bar{x}) = D_{EI}\mathbf{B}\bar{\mathbf{a}}^e + M_p(\bar{x}) \tag{4.71}$$

$$V(\bar{x}) = -\frac{dM}{d\bar{x}} = -D_{EI}\frac{d\mathbf{B}}{d\bar{x}}\bar{\mathbf{a}}^e + V_p(\bar{x}) \tag{4.72}$$

where

$$M_p(\bar{x}) = D_{EI}\frac{d^2 v_p}{d\bar{x}^2} \tag{4.73}$$

$$V_p(\bar{x}) = -D_{EI}\frac{d^3 v_p}{d\bar{x}^3} \tag{4.74}$$

The definitions we have introduced for forces and moments acting at the nodes of the element give

$$\boxed{\bar{P}_1 = -V(0); \quad \bar{P}_2 = -M(0); \quad \bar{P}_3 = V(L); \quad \bar{P}_4 = M(L)} \tag{4.75}$$

Substitution of (4.71) and (4.72) gives the nodal forces

$$\bar{P}_1 = D_{EI}\left(\frac{d\mathbf{B}}{d\bar{x}}\right)_{\bar{x}=0}\bar{\mathbf{a}}^e - V_p(0) \tag{4.76}$$

$$\bar{P}_2 = -D_{EI}\mathbf{B}_{\bar{x}=0}\bar{\mathbf{a}}^e - M_p(0) \tag{4.77}$$

$$\bar{P}_3 = -D_{EI}\left(\frac{d\mathbf{B}}{d\bar{x}}\right)_{\bar{x}=L}\bar{\mathbf{a}}^e + V_p(L) \tag{4.78}$$

$$\bar{P}_4 = D_{EI}\mathbf{B}_{\bar{x}=L}\bar{\mathbf{a}}^e + M_p(L) \tag{4.79}$$

The two parts of the nodal forces, corresponding to the solution of the homogeneous differential equation and the particular solution, are shown in Figure 4.14.

With

$$\bar{\mathbf{f}}_b^e = \begin{bmatrix} \bar{P}_1 \\ \bar{P}_2 \\ \bar{P}_3 \\ \bar{P}_4 \end{bmatrix}; \quad \bar{\mathbf{K}}^e = \frac{D_{EI}}{L^3}\begin{bmatrix} 12 & 6L & -12 & 6L \\ 6L & 4L^2 & -6L & 2L^2 \\ -12 & -6L & 12 & -6L \\ 6L & 2L^2 & -6L & 4L^2 \end{bmatrix}; \quad \bar{\mathbf{f}}_l^e = \begin{bmatrix} V_p(0) \\ M_p(0) \\ -V_p(L) \\ -M_p(L) \end{bmatrix} \tag{4.80}$$

Equations (4.76)–(4.79) can be written in matrix form

$$\bar{\mathbf{f}}_b^e = \bar{\mathbf{K}}^e\bar{\mathbf{a}}^e - \bar{\mathbf{f}}_l^e \tag{4.81}$$

or

$$\boxed{\bar{\mathbf{K}}^e\bar{\mathbf{a}}^e = \bar{\mathbf{f}}^e} \tag{4.82}$$

where

$$\bar{\mathbf{f}}^e = \bar{\mathbf{f}}_b^e + \bar{\mathbf{f}}_l^e \tag{4.83}$$

Equation (4.82) is the element equation for a beam element in pure beam action. The left-hand side contains the element stiffness matrix $\bar{\mathbf{K}}^e$ and the element displacement vector $\bar{\mathbf{a}}^e$, while

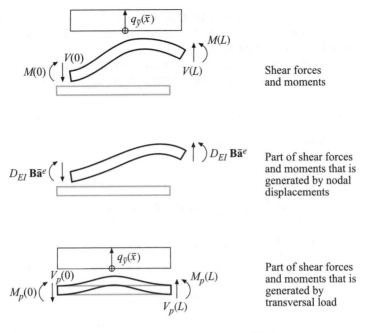

Figure 4.14   A beam element in equilibrium

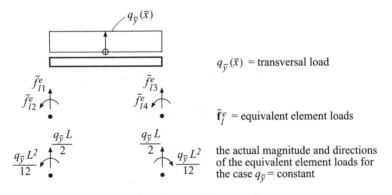

Figure 4.15   Transverse load and equivalent element loads

the right-hand side consists of the element force vector $\bar{\mathbf{f}}^e$. The element force vector $\bar{\mathbf{f}}^e$ is the sum of nodal forces $\bar{\mathbf{f}}^e_b$ and element loads $\bar{\mathbf{f}}^e_l$ of which the latter considers the effect of load distributed along the beam.

For the particular solution $v_p(\bar{x})$, we have assumed that the displacements at the nodes are zero. Therefore, the element loads with reversed sign can be interpreted as the support forces for a beam fixed at both ends; cf. the corresponding discussion in Chapter 3. The equivalent element loads are illustrated in Figure 4.15.

For a non-loaded beam element, that is for $\bar{\mathbf{f}}_l^e = \mathbf{0}$, the deformation of the beam element is described by just the solution to the homogeneous differential equation. The case with uniformly distributed load is discussed in Example 4.1.

A summary of the relations – kinematics, constitutive relation and equilibrium – which lead to the element equation for a beam element with four degrees of freedom is shown in Figure 4.16.

$$\left.\begin{array}{l}
\bar{P}_1 = -V(0) \qquad (4.75) \\[4pt]
\bar{P}_3 = V(L) \\[4pt]
\bar{P}_2 = -M(0) \\[4pt]
\bar{P}_4 = M(L) \\[4pt]
M(\bar{x}) = D_{EI}(\bar{x})\,\dfrac{d^2 v}{d\bar{x}^2} \qquad (4.24) \\[8pt]
V(\bar{x}) = -D_{EI}(\bar{x})\,\dfrac{d^3 v}{d\bar{x}^3} \qquad (4.25) \\[8pt]
v(\bar{x}) = \mathbf{N}\bar{\mathbf{a}}^e + v_p(\bar{x}) \qquad (4.55)
\end{array}\right\} \Rightarrow \bar{\mathbf{K}}^e \bar{\mathbf{a}}^e = \bar{\mathbf{f}}^e \qquad (4.82)$$

where

$$\bar{\mathbf{f}}^e = \bar{\mathbf{f}}_b^e + \bar{\mathbf{f}}_l^e$$

$$\bar{\mathbf{K}}^e = \frac{D_{EI}}{L^3}\begin{bmatrix} 12 & 6L & -12 & 6L \\ 6L & 4L^2 & -6L & 2L^2 \\ -12 & -6L & 12 & -6L \\ 6L & 2L^2 & -6L & 4L^2 \end{bmatrix}$$

$$\bar{\mathbf{a}}^e = \begin{bmatrix} \bar{u}_1 \\ \bar{u}_2 \\ \bar{u}_3 \\ \bar{u}_4 \end{bmatrix}; \quad \bar{\mathbf{f}}_b^e = \begin{bmatrix} \bar{P}_1 \\ \bar{P}_2 \\ \bar{P}_3 \\ \bar{P}_4 \end{bmatrix}; \quad \bar{\mathbf{f}}_l^e = \begin{bmatrix} V_p(0) \\ M_p(0) \\ -V_p(L) \\ -M_p(L) \end{bmatrix}$$

Figure 4.16   From beam action to beam element

Example 4.1   A beam element with a uniformly distributed load

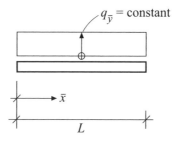

$q_{\bar{y}} = \text{constant}$

$\bar{x}$

L

Figure 1   A beam element with a uniformly distributed load

Consider a beam element of length L loaded with a uniformly distributed load $q_{\bar{y}}$ (Figure 4.1) and determine the element load vector $\bar{f}_f^e$ for the beam. To be able to determine $V_p(0)$, $M_p(0)$, $V_p(L)$ and $M_p(L)$ in (4.80), we seek first a particular solution $v_p(\bar{x})$ to (4.27). The particular solution should satisfy the differential equation (4.27) and the boundary conditions (4.56)–(4.59); see Figure 4.13. With $q_{\bar{y}}$ constant, (4.27) can be written as

$$D_{EI}\frac{d^4 v_p}{d\bar{x}^4} - q_{\bar{y}} = 0 \tag{1}$$

Integrating four times gives

$$D_{EI}\frac{d^3 v_p}{d\bar{x}^3} - q_{\bar{y}}\bar{x} - C_1 = 0 \tag{2}$$

$$D_{EI}\frac{d^2 v_p}{d\bar{x}^2} - q_{\bar{y}}\frac{\bar{x}^2}{2} - C_1\bar{x} - C_2 = 0 \tag{3}$$

$$D_{EI}\frac{d v_p}{d\bar{x}} - q_{\bar{y}}\frac{\bar{x}^3}{6} - C_1\frac{\bar{x}^2}{2} - C_2\bar{x} - C_3 = 0 \tag{4}$$

$$D_{EI}v_p(\bar{x}) - q_{\bar{y}}\frac{\bar{x}^4}{24} - C_1\frac{\bar{x}^3}{6} - C_2\frac{\bar{x}^2}{2} - C_3\bar{x} - C_4 = 0 \tag{5}$$

or

$$\frac{d v_p}{d\bar{x}} = \frac{1}{D_{EI}}\left(q_{\bar{y}}\frac{\bar{x}^3}{6} + C_1\frac{\bar{x}^2}{2} + C_2\bar{x} + C_3\right) \tag{6}$$

$$v_p(\bar{x}) = \frac{1}{D_{EI}}\left(q_{\bar{y}}\frac{\bar{x}^4}{24} + C_1\frac{\bar{x}^3}{6} + C_2\frac{\bar{x}^2}{2} + C_3\bar{x} + C_4\right) \tag{7}$$

The boundary conditions (4.56)–(4.59) give

$$\left(\frac{d v_p}{d\bar{x}}\right)_{\bar{x}=0} = \frac{1}{D_{EI}}C_3 = 0 \tag{8}$$

$$\left(\frac{d v_p}{d\bar{x}}\right)_{\bar{x}=L} = \frac{1}{D_{EI}}\left(q_{\bar{y}}\frac{L^3}{6} + C_1\frac{L^2}{2} + C_2 L + C_3\right) = 0 \tag{9}$$

$$v_p(0) = \frac{1}{D_{EI}}C_4 = 0 \tag{10}$$

$$v_p(L) = \frac{1}{D_{EI}}\left(q_{\bar{y}}\frac{L^4}{24} + C_1\frac{L^3}{6} + C_2\frac{L^2}{2} + C_3 L + C_4\right) = 0 \tag{11}$$

that is

$$C_1 = -q_{\bar{y}}\frac{L}{2}; \quad C_2 = q_{\bar{y}}\frac{L^2}{12}; \quad C_3 = 0; \quad C_4 = 0 \tag{12}$$

Substitution of the constants C_1, C_2, C_3 and C_4 gives the particular solution

$$v_p(\bar{x}) = \frac{q_{\bar{y}}}{D_{EI}}\left(\frac{\bar{x}^4}{24} - \frac{L\bar{x}^3}{12} + \frac{L^2\bar{x}^2}{24}\right) \tag{13}$$

Differentiating three times gives

$$\frac{dv_p}{d\bar{x}} = \frac{q_{\bar{y}}}{D_{EI}}\left(\frac{\bar{x}^3}{6} - \frac{L\bar{x}^2}{4} + \frac{L^2\bar{x}}{12}\right) \tag{14}$$

$$\frac{d^2v_p}{d\bar{x}^2} = \frac{q_{\bar{y}}}{D_{EI}}\left(\frac{\bar{x}^2}{2} - \frac{L\bar{x}}{2} + \frac{L^2}{12}\right) \tag{15}$$

$$\frac{d^3v_p}{d\bar{x}^3} = \frac{q_{\bar{y}}}{D_{EI}}\left(\bar{x} - \frac{L}{2}\right) \tag{16}$$

Substitution into (4.73) and (4.74) gives

$$M_p(\bar{x}) = q_{\bar{y}}\left(\frac{\bar{x}^2}{2} - \frac{L\bar{x}}{2} + \frac{L^2}{12}\right) \tag{17}$$

$$V_p(\bar{x}) = -q_{\bar{y}}\left(\bar{x} - \frac{L}{2}\right) \tag{18}$$

after which (4.80) gives

$$\bar{\mathbf{f}}_l^e = q_{\bar{y}}\begin{bmatrix} \dfrac{L}{2} \\[1ex] \dfrac{L^2}{12} \\[1ex] \dfrac{L}{2} \\[1ex] -\dfrac{L^2}{12} \end{bmatrix} \tag{19}$$

Compare the result in (19) with Figure 4.15.

4.2.3   Beam Element with Six Degrees of Freedom

In Section 3.2.2, we have derived the element equations for a bar element with two degrees of freedom, $\bar{u}_{1,\text{bar}}$ and $\bar{u}_{2,\text{bar}}$ (bar action). In the corresponding manner, we have in Section 4.2.2 derived the element equations for a beam element with four degrees of freedom, $\bar{u}_{1,\text{beam}} - \bar{u}_{4,\text{beam}}$ (beam action). We have also noted that if the location of the local $\bar{x}$-axis (the system line) is chosen so that the condition $\int_A E\,\bar{y}\,dA = 0$ is satisfied, these two modes of action are independent of each other. This enables us to introduce a new beam element with six degrees of freedom $(\bar{u}_1 - \bar{u}_6)$, which includes both bar and beam action (Figure 4.17). This will be the element that we will use to model frames. Figure 4.18 shows how the elements for bar action and beam action are merged to a beam element with six degrees of freedom.

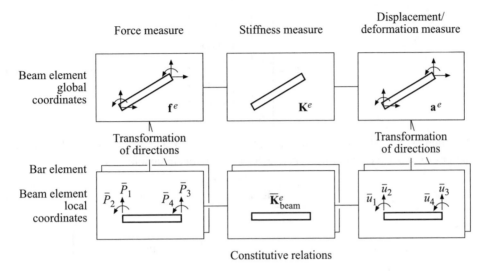

Figure 4.17 From bar and beam elements in local coordinates to a beam element with six degrees of freedom in global coordinates

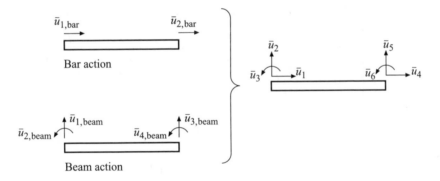

Figure 4.18 A beam element with six degrees of freedom

The merge can be expressed as a kinematic condition (compatibility) and a force relation (static equivalence),

$$
\bar{\mathbf{a}}^e =
\begin{bmatrix}
\bar{u}_1 \\
\bar{u}_2 \\
\bar{u}_3 \\
\bar{u}_4 \\
\bar{u}_5 \\
\bar{u}_6
\end{bmatrix}
=
\begin{bmatrix}
\bar{u}_{1,\text{bar}} \\
\bar{u}_{1,\text{beam}} \\
\bar{u}_{2,\text{beam}} \\
\bar{u}_{2,\text{bar}} \\
\bar{u}_{3,\text{beam}} \\
\bar{u}_{4,\text{beam}}
\end{bmatrix}
\tag{4.84}
$$

$$\bar{\mathbf{f}}^e = \begin{bmatrix} \bar{f}_1 \\ \bar{f}_2 \\ \bar{f}_3 \\ \bar{f}_4 \\ \bar{f}_5 \\ \bar{f}_6 \end{bmatrix} = \begin{bmatrix} \bar{f}_{1,\text{bar}} \\ \bar{f}_{1,\text{beam}} \\ \bar{f}_{2,\text{beam}} \\ \bar{f}_{2,\text{bar}} \\ \bar{f}_{3,\text{beam}} \\ \bar{f}_{4,\text{beam}} \end{bmatrix} \tag{4.85}$$

Substituting the relations (3.60) and (4.82) into (4.85) and using (4.84), the element equations for a beam element with six degrees of freedom are obtained as

$$\boxed{\bar{\mathbf{K}}^e \bar{\mathbf{a}}^e = \bar{\mathbf{f}}^e} \tag{4.86}$$

where

$$\bar{\mathbf{K}}^e = \begin{bmatrix} \dfrac{D_{EA}}{L} & 0 & 0 & -\dfrac{D_{EA}}{L} & 0 & 0 \\ 0 & \dfrac{12D_{EI}}{L^3} & \dfrac{6D_{EI}}{L^2} & 0 & -\dfrac{12D_{EI}}{L^3} & \dfrac{6D_{EI}}{L^2} \\ 0 & \dfrac{6D_{EI}}{L^2} & \dfrac{4D_{EI}}{L} & 0 & -\dfrac{6D_{EI}}{L^2} & \dfrac{2D_{EI}}{L} \\ -\dfrac{D_{EA}}{L} & 0 & 0 & \dfrac{D_{EA}}{L} & 0 & 0 \\ 0 & -\dfrac{12D_{EI}}{L^3} & -\dfrac{6D_{EI}}{L^2} & 0 & \dfrac{12D_{EI}}{L^3} & -\dfrac{6D_{EI}}{L^2} \\ 0 & \dfrac{6D_{EI}}{L^2} & \dfrac{2D_{EI}}{L} & 0 & -\dfrac{6D_{EI}}{L^2} & \dfrac{4D_{EI}}{L} \end{bmatrix} \tag{4.87}$$

and where

$$\bar{\mathbf{f}}^e = \bar{\mathbf{f}}_b^e + \bar{\mathbf{f}}_l^e \tag{4.88}$$

with

$$\bar{\mathbf{f}}_b^e = \begin{bmatrix} \bar{P}_1 \\ \bar{P}_2 \\ \bar{P}_3 \\ \bar{P}_4 \\ \bar{P}_5 \\ \bar{P}_6 \end{bmatrix}; \quad \bar{\mathbf{f}}_l^e = \begin{bmatrix} N_p(0) \\ V_p(0) \\ M_p(0) \\ -N_p(L) \\ -V_p(L) \\ -M_p(L) \end{bmatrix} \tag{4.89}$$

4.2.4   From Local to Global Directions

In the element relations (4.86) for the beam, the nodal force vector $\bar{\mathbf{f}}_b^e$, the element displacement vector $\bar{\mathbf{a}}^e$ and the element load vector $\bar{\mathbf{f}}_l^e$ are expressed in the local coordinate system $(\bar{x}, \bar{y})$ of the beam. To be able to put the beam element into a frame, we have to establish an element relation where forces and displacements are expressed in the global coordinate system (x, y)

of the frame (Figure 4.19). Previously, in Section 3.2.3, we have performed a transformation of the element relation for a bar, where we have gone from *one* degree of freedom at each node of the bar element in a local coordinate system to *two* degrees of freedom with new directions for the bar element in a global coordinate system. Here, we go from *three* degrees of freedom at each node of the beam element in a local coordinate system to *three* degrees of freedom with new directions for the beam element in the global coordinate system.

The transformation of displacements between the local and the global coordinate system is done separately for each degree of freedom. From (3.72), we know that the displacement $\bar{u}_1$ in the direction of the local $\bar{x}$-axis can be written as

$$\bar{u}_1 = n_{x\bar{x}}u_1 + n_{y\bar{x}}u_2 \tag{4.90}$$

In the corresponding manner, the displacement $\bar{u}_2$ in the direction of the local $\bar{y}$-axis can be written as

$$\bar{u}_2 = n_{x\bar{y}}u_1 + n_{y\bar{y}}u_2 \tag{4.91}$$

The third displacement degree of freedom, which is a rotation, is not affected by the orientation of the coordinate system, which gives

$$\bar{u}_3 = u_3 \tag{4.92}$$

For the node at the other end of the beam, the following corresponding relations can be established:

$$\bar{u}_4 = n_{x\bar{x}}u_4 + n_{y\bar{x}}u_5 \tag{4.93}$$

$$\bar{u}_5 = n_{x\bar{y}}u_4 + n_{y\bar{y}}u_5 \tag{4.94}$$

$$\bar{u}_6 = u_6 \tag{4.95}$$

In matrix form, this can be expressed as

$$\boxed{\bar{\mathbf{a}}^e = \mathbf{G}\mathbf{a}^e} \tag{4.96}$$

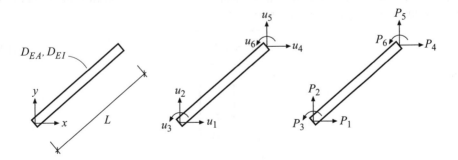

Figure 4.19   A beam element in a global coordinate system

where

$$
\bar{\mathbf{a}}^e = \begin{bmatrix} \bar{u}_1 \\ \bar{u}_2 \\ \bar{u}_3 \\ \bar{u}_4 \\ \bar{u}_5 \\ \bar{u}_6 \end{bmatrix}; \quad \mathbf{G} = \begin{bmatrix} n_{x\bar{x}} & n_{y\bar{x}} & 0 & 0 & 0 & 0 \\ n_{x\bar{y}} & n_{y\bar{y}} & 0 & 0 & 0 & 0 \\ 0 & 0 & 1 & 0 & 0 & 0 \\ 0 & 0 & 0 & n_{x\bar{x}} & n_{y\bar{x}} & 0 \\ 0 & 0 & 0 & n_{x\bar{y}} & n_{y\bar{y}} & 0 \\ 0 & 0 & 0 & 0 & 0 & 1 \end{bmatrix}; \quad \mathbf{a}^e = \begin{bmatrix} u_1 \\ u_2 \\ u_3 \\ u_4 \\ u_5 \\ u_6 \end{bmatrix} \tag{4.97}
$$

Using the relations (3.69), P_1 and P_2 can be expressed as

$$
P_1 = n_{x\bar{x}}\bar{P}_1 + n_{x\bar{y}}\bar{P}_2 \tag{4.98}
$$

$$
P_2 = n_{y\bar{x}}\bar{P}_1 + n_{y\bar{y}}\bar{P}_2 \tag{4.99}
$$

and P_4 and P_5 as

$$
P_4 = n_{x\bar{x}}\bar{P}_4 + n_{x\bar{y}}\bar{P}_5 \tag{4.100}
$$

$$
P_5 = n_{y\bar{x}}\bar{P}_4 + n_{y\bar{y}}\bar{P}_5 \tag{4.101}
$$

For the moments, we have

$$
P_3 = \bar{P}_3 \tag{4.102}
$$

$$
P_6 = \bar{P}_6 \tag{4.103}
$$

In matrix form, these relations can be written as

$$
\boxed{\mathbf{f}_b^e = \mathbf{G}^T \bar{\mathbf{f}}_b^e} \tag{4.104}
$$

where

$$
\mathbf{f}_b^e = \begin{bmatrix} P_1 \\ P_2 \\ P_3 \\ P_4 \\ P_5 \\ P_6 \end{bmatrix}; \quad \mathbf{G}^T = \begin{bmatrix} n_{x\bar{x}} & n_{x\bar{y}} & 0 & 0 & 0 & 0 \\ n_{y\bar{x}} & n_{y\bar{y}} & 0 & 0 & 0 & 0 \\ 0 & 0 & 1 & 0 & 0 & 0 \\ 0 & 0 & 0 & n_{x\bar{x}} & n_{x\bar{y}} & 0 \\ 0 & 0 & 0 & n_{y\bar{x}} & n_{y\bar{y}} & 0 \\ 0 & 0 & 0 & 0 & 0 & 1 \end{bmatrix}; \quad \bar{\mathbf{f}}_b^e = \begin{bmatrix} \bar{P}_1 \\ \bar{P}_2 \\ \bar{P}_3 \\ \bar{P}_4 \\ \bar{P}_5 \\ \bar{P}_6 \end{bmatrix} \tag{4.105}
$$

The relation between element loads $\mathbf{f}_l^e$ in a global system and element loads $\bar{\mathbf{f}}_l^e$ in a local system can, in a corresponding manner, be written as

$$
\boxed{\mathbf{f}_l^e = \mathbf{G}^T \bar{\mathbf{f}}_l^e} \tag{4.106}
$$

where

$$
\mathbf{f}_l^e = \begin{bmatrix} f_{l1}^e \\ f_{l2}^e \\ f_{l3}^e \\ f_{l4}^e \\ f_{l5}^e \\ f_{l6}^e \end{bmatrix} \tag{4.107}
$$

Substitution of the transformations (4.104), (4.96) and (4.106) into the element relation (4.86) gives an element relation with quantities expressed in the directions of the global coordinate system

$$\boxed{\mathbf{K}^e \mathbf{a}^e = \mathbf{f}^e}$$

(4.108)

where

$$\mathbf{K}^e = \mathbf{G}^T \bar{\mathbf{K}}^e \mathbf{G}; \quad \mathbf{f}^e = \mathbf{f}^e_b + \mathbf{f}^e_l$$

(4.109)

How transformations of displacements and forces between different coordinate systems lead to a relation for the beam element in global coordinates is shown in Figure 4.20.

$$
\left.
\begin{array}{ll}
\mathbf{f}^e_b = \mathbf{G}^T \bar{\mathbf{f}}^e_b & (4.104) \\
\mathbf{f}^e_l = \mathbf{G}^T \bar{\mathbf{f}}^e_l & (4.106) \\
\bar{\mathbf{K}}^e \bar{\mathbf{a}}^e = \bar{\mathbf{f}}^e & (4.86) \\
\bar{\mathbf{f}}^e = \bar{\mathbf{f}}^e_b + \bar{\mathbf{f}}^e_l & (4.88) \\
\bar{\mathbf{a}}^e = \mathbf{G} \mathbf{a}^e & (4.96)
\end{array}
\right\}
\Rightarrow
\begin{array}{l}
\mathbf{K}^e \mathbf{a}^e = \mathbf{f}^e \;\; (4.108) \\[4pt]
\text{where} \\[4pt]
\mathbf{K}^e = \mathbf{G}^T \bar{\mathbf{K}}^e \mathbf{G}; \quad \mathbf{f}^e = \mathbf{f}^e_b + \mathbf{f}^e_l
\end{array}
$$

Figure 4.20   From local coordinates to global coordinates

4.3   Frames

A frame consists of beam elements connected to each other. In the computational model for a frame, a beam element is represented by the system line for bar and beam action, and the joints between the beam elements by nodes (Figure 4.21). We assume that the nodes are located where the system lines cross. The beam element we have now formulated can be used to model frames (Figure 4.22).

In the same manner as for the truss, we introduce a global numbering for all displacement degrees of freedom in the frame. In the element relations for each of the beam elements, a local numbering of the degrees of freedom is used (u_1–u_6). Each one of the degrees of freedom at the element level is related to a degree of freedom at the global level by compatibility conditions.

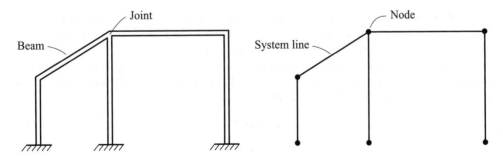

Figure 4.21   A frame and the associated computational model

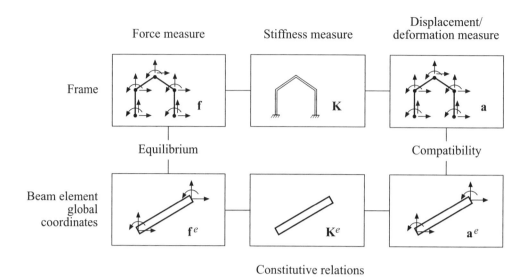

Figure 4.22　From beam element to frame

For an element related to the global degrees of freedom a_i–a_n, we obtain the compatibility conditions

$$u_1 = a_i \tag{4.110}$$

$$u_2 = a_j \tag{4.111}$$

$$u_3 = a_k \tag{4.112}$$

$$u_4 = a_l \tag{4.113}$$

$$u_5 = a_m \tag{4.114}$$

$$u_6 = a_n \tag{4.115}$$

The compatibility conditions can be written as

$$\mathbf{a}^e = \mathbf{H}\mathbf{a} \tag{4.116}$$

where $\mathbf{a}^e$ is the nodal displacements of the element in global directions (4.97), $\mathbf{a}$ is the displacement vector of the frame and $\mathbf{H}$ is a matrix where $H_{1,i} = 1$, $H_{2,j} = 1$, $H_{3,k} = 1$, $H_{4,l} = 1$, $H_{5,m} = 1$, $H_{6,n} = 1$ and all other elements are equal to 0.

To be able to establish equilibrium conditions for the nodes of the frame, we introduce an expanded element force vector $\hat{\mathbf{f}}_b^e$ and an expanded element load vector $\hat{\mathbf{f}}_l^e$. Both with equally many rows as there are degrees of freedom in the system,

$$\hat{\mathbf{f}}_b^e = \mathbf{H}^T \mathbf{f}_b^e \tag{4.117}$$

$$\hat{\mathbf{f}}_l^e = \mathbf{H}^T \mathbf{f}_l^e \tag{4.118}$$

with $\mathbf{H}$ as above. The vectors $\hat{\mathbf{f}}_l^e$ and $\hat{\mathbf{f}}_b^e$ contain the vector elements of $\mathbf{f}_l^e$ and $\mathbf{f}_b^e$, respectively, put on the rows which correspond to the global degrees of freedom number that the element should be connected to. Substituting the relations (4.116)–(4.118) into (4.108) gives

$$\hat{\mathbf{f}}_b^e = \hat{\mathbf{K}}^e \mathbf{a} - \hat{\mathbf{f}}_l^e \qquad (4.119)$$

where

$$\hat{\mathbf{K}}^e = \mathbf{H}^T \mathbf{K}^e \mathbf{H} \qquad (4.120)$$

Equation (4.119) is the expanded element relation for a beam element with six degrees of freedom, where $\hat{\mathbf{K}}^e$ contains the matrix elements of $\mathbf{K}^e$ located in positions corresponding to the global degrees of freedom.

An equilibrium equation for the forces acting in the direction of degree of freedom i at a node gives

$$\sum_{e=1}^{m} f_{b,i}^e = f_{ln,i} + f_{b,i} \qquad (4.121)$$

where e denotes the element number, $f_{ln,i}$ a possible point load acting on the node (a nodal load) and $f_{b,i}$ a possible support force. By establishing an equilibrium equation for each degree of freedom, we obtain for the entire frame

$$\sum_{e=1}^{m} \hat{\mathbf{f}}_b^e = \mathbf{f}_{ln} + \mathbf{f}_b \qquad (4.122)$$

Substituting the expanded element Equations (4.119) into (4.122), we obtain

$$\sum_{e=1}^{m} (\hat{\mathbf{K}}^e \mathbf{a} - \hat{\mathbf{f}}_l^e) = \mathbf{f}_{ln} + \mathbf{f}_b \qquad (4.123)$$

or

$$\boxed{\mathbf{Ka} = \mathbf{f}} \qquad (4.124)$$

where

$$\mathbf{K} = \sum_{e=1}^{m} \hat{\mathbf{K}}^e; \quad \mathbf{f} = \mathbf{f}_l + \mathbf{f}_b; \quad \mathbf{f}_l = \mathbf{f}_{ln} + \mathbf{f}_{lq}; \quad \mathbf{f}_{lq} = \sum_{e=1}^{m} \hat{\mathbf{f}}_l^e \qquad (4.125)$$

How compatibility conditions, element relations and equilibrium lead to a system of equations for a frame is shown in Figure 4.23.

For present boundary conditions, the nodal displacements and the support forces can be determined from (4.124). When the displacements $\mathbf{a}$ have been computed, the element displacements $\mathbf{a}^e$ can be determined. After that, displacements and rotations expressed in the local coordinate system can be computed from (4.96). Bending moment, shear force and normal force can then be determined using (4.71), (4.72) and (3.52).

The stiffness matrix $\mathbf{K}$ and the load vector $\mathbf{f}_l$ have been described as sums of expanded matrices $\hat{\mathbf{K}}^e$ and vectors $\hat{\mathbf{f}}_l^e$.

$\mathbf{K}$ and $\mathbf{f}_l$ are in the computational procedure introduced by defining them as matrices initially filled with zeros. The matrix elements of $\mathbf{K}^e$ and $\mathbf{f}_l^e$ are then added in the positions of the global degrees of freedom associated to each element; cf. Figure 2.14. The procedure is called assembling.

$$\begin{aligned}
\hat{\mathbf{f}}_b^e &= \mathbf{H}^T \mathbf{f}_b^e \quad (4.117) \\
\hat{\mathbf{f}}_l^e &= \mathbf{H}^T \mathbf{f}_l^e \quad (4.118) \\
\mathbf{K}^e \mathbf{a}^e &= \mathbf{f}^e \quad (4.108) \\
\mathbf{f}^e &= \mathbf{f}_b^e + \mathbf{f}_l^e \quad (4.109) \\
\mathbf{a}^e &= \mathbf{H}\mathbf{a} \quad (4.116)
\end{aligned}\right\} \Rightarrow \begin{array}{l} \hat{\mathbf{f}}_b^e = \hat{\mathbf{K}}^e \mathbf{a} - \hat{\mathbf{f}}_l^e \quad (4.119) \\[4pt] \text{where} \\[2pt] \hat{\mathbf{K}}^e = \mathbf{H}^T \mathbf{K}^e \mathbf{H} \end{array}$$

$$\sum_{e=1}^{m} \hat{\mathbf{f}}_b^e = \mathbf{f}_{ln} + \mathbf{f}_b \quad (4.122)$$

$$\Rightarrow \quad \mathbf{Ka} = \mathbf{f} \quad (4.124)$$

where

$$\mathbf{K} = \sum_{e=1}^{m} \hat{\mathbf{K}}^e$$

$$\mathbf{f} = \mathbf{f}_l + \mathbf{f}_b$$

$$\mathbf{f}_l = \mathbf{f}_{ln} + \sum_{e=1}^{m} \hat{\mathbf{f}}_l^e$$

Figure 4.23   From beam element to frame

Example 4.2 Frame

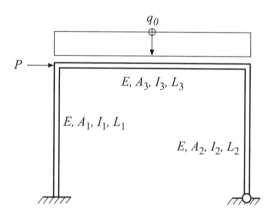

Figure 1   Frame

The frame in Figure 1 consists of three beams with lengths $L_1 = 4.0$ m, $L_2 = 4.0$ m and $L_3 = 6.0$ m. The cross-sectional areas of the beams are $A_1 = 2.0 \times 10^{-3}$ m^2, $A_2 = 2.0 \times 10^{-3}$ m^2 and $A_3 = 6.0 \times 10^{-3}$ m^2 and the moments of inertia are $I_1 = 1.6 \times 10^{-5}$ m^4, $I_2 = 1.6 \times 10^{-5}$ m^4 and $I_3 = 5.4 \times 10^{-5}$ m^4. The elastic modulus is $E = 200.0$ GPa for all the beams. The frame is loaded with a uniformly distributed load $q_0 = 10$ kN/m and a point load $P = 2$ kN. The structure is rigidly fixed at the left support and hinged at the right support.

The displacements, the support forces and the distribution of internal forces shall be determined.

Computational model

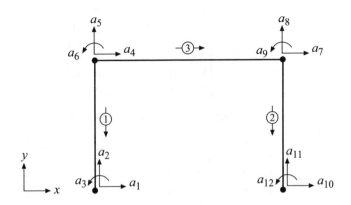

Figure 2   Computational model

The frame is modelled with three beam elements, denoted as 1, 2 and 3 (Figure 2). The system has the displacement degrees of freedom $a_1, a_2, \cdots, a_{12}$. The translations a_1, a_2, a_{10} and a_{11} and the rotation a_3 are prescribed to be zero.

Element matrices

For each beam element, an element relation $\mathbf{K}^e \mathbf{a}^e = \mathbf{f}_l^e + \mathbf{f}_b^e$ can be established. The element stiffness matrix in local coordinates $\bar{\mathbf{K}}^e$ is given by (4.87). The element stiffness matrix in global coordinates $\mathbf{K}^e$ is given by (4.109), where the transformation matrix being $\mathbf{G}$ is given by (4.97). For the three elements we have the following:

Element 1:
   With A_1, I_1, E and L_1 known, the element stiffness matrix in local coordinates can be computed,

$$
\bar{\mathbf{K}}^1 =
\begin{bmatrix}
100 & 0 & 0 & -100 & 0 & 0 \\
0 & 0.6 & 1.2 & 0 & -0.6 & 1.2 \\
0 & 1.2 & 3.2 & 0 & -1.2 & 1.6 \\
-100 & 0 & 0 & 100 & 0 & 0 \\
0 & -0.6 & -1.2 & 0 & 0.6 & -1.2 \\
0 & 1.2 & 1.6 & 0 & -1.2 & 3.2
\end{bmatrix} 10^6
\tag{1}
$$

The local $\bar{x}$-axis is oppositely directed compared with the global y-axis and the local $\bar{y}$-axis coincides with the global x-axis. The direction cosines for the angle between these are, therefore, $n_{x\bar{x}} = \cos(x, \bar{x}) = 0$, $n_{y\bar{x}} = \cos(y, \bar{x}) = -1$, $n_{x\bar{y}} = \cos(x, \bar{y}) = 1$ and

$n_{y\bar{y}} = \cos(y, \bar{y}) = 0$. For this element, the transformation matrix becomes

$$\mathbf{G}^1 = \begin{bmatrix} 0 & -1 & 0 & 0 & 0 & 0 \\ 1 & 0 & 0 & 0 & 0 & 0 \\ 0 & 0 & 1 & 0 & 0 & 0 \\ 0 & 0 & 0 & 0 & -1 & 0 \\ 0 & 0 & 0 & 1 & 0 & 0 \\ 0 & 0 & 0 & 0 & 0 & 1 \end{bmatrix} \tag{2}$$

which gives the element stiffness matrix in global coordinates,

$$\mathbf{K}^1 = \begin{bmatrix} 0.6 & 0 & 1.2 & -0.6 & 0 & 1.2 \\ 0 & 100 & 0 & 0 & -100 & 0 \\ 1.2 & 0 & 3.2 & -1.2 & 0 & 1.6 \\ -0.6 & 0 & -1.2 & 0.6 & 0 & -1.2 \\ 0 & -100 & 0 & 0 & 100 & 0 \\ 1.2 & 0 & 1.6 & -1.2 & 0 & 3.2 \end{bmatrix} 10^6 \tag{3}$$

Element 2:
   This element has the same properties and the same direction as Element 1. The element stiffness matrix in local coordinates as well as the transformation matrix therefore become identical for Elements 1 and 2, which gives $\mathbf{K}^2 = \mathbf{K}^1$.
Element 3:
   With A_3, I_3, E and L_3 known, the element stiffness matrix in local coordinates can be computed,

$$\bar{\mathbf{K}}^3 = \begin{bmatrix} 200 & 0 & 0 & -200 & 0 & 0 \\ 0 & 0.6 & 1.8 & 0 & -0.6 & 1.8 \\ 0 & 1.8 & 7.2 & 0 & -1.8 & 3.6 \\ -200 & 0 & 0 & 200 & 0 & 0 \\ 0 & -0.6 & -1.8 & 0 & 0.6 & -1.8 \\ 0 & 1.8 & 3.6 & 0 & -1.8 & 7.2 \end{bmatrix} 10^6 \tag{4}$$

This element is loaded with a uniformly distributed load $q_y = -10$ kN/m, which gives rise to an element load vector $\bar{\mathbf{f}}_l^e$ according to (4.89). From Example 4.1, we obtain $V_p(0) = q_{\bar{y}}\frac{L}{2}$, $M_p(0) = q_{\bar{y}}\frac{L^2}{12}$, $V_p(L) = q_{\bar{y}}\left(-\frac{L}{2}\right)$ and $M_p(L) = q_{\bar{y}}\frac{L^2}{12}$, which give

$$\bar{\mathbf{f}}_l^3 = \begin{bmatrix} 0 \\ -30 \\ -30 \\ 0 \\ -30 \\ 30 \end{bmatrix} 10^3 \tag{5}$$

The local coordinate system coincides with the global, which gives the direction cosines $n_{x\bar{x}} = \cos(x, \bar{x}) = 1.0$, $n_{y\bar{x}} = \cos(y, \bar{x}) = 0.0$, $n_{x\bar{y}} = \cos(x, \bar{y}) = 0.0$ and

$n_{y\bar{y}} = \cos(y, \bar{y}) = 1.0$, and the transformation matrix

$$\mathbf{G}^3 = \begin{bmatrix} 1 & 0 & 0 & 0 & 0 & 0 \\ 0 & 1 & 0 & 0 & 0 & 0 \\ 0 & 0 & 1 & 0 & 0 & 0 \\ 0 & 0 & 0 & 1 & 0 & 0 \\ 0 & 0 & 0 & 0 & 1 & 0 \\ 0 & 0 & 0 & 0 & 0 & 1 \end{bmatrix} \tag{6}$$

Since the transformation matrix is a unit matrix, the element stiffness matrix and the load vector in the global coordinate system become equal to the ones in the local coordinate system, that is $\mathbf{K}^3 = \bar{\mathbf{K}}^3$ and $\mathbf{f}_l^3 = \bar{\mathbf{f}}_l^3$.

Compatibility conditions

The relation between the local degrees of freedom and the global degrees of freedom is described by the topology matrix:

$$\text{topology} = \begin{bmatrix} 1 & 4 & 5 & 6 & 1 & 2 & 3 \\ 2 & 7 & 8 & 9 & 10 & 11 & 12 \\ 3 & 4 & 5 & 6 & 7 & 8 & 9 \end{bmatrix} \tag{7}$$

Assembling

Adding the coefficients of the element stiffness matrices to a global matrix using the topology information gives the global stiffness matrix

$$\mathbf{K} = \begin{bmatrix}
0.6 & 0 & -1.2 & -0.6 & 0 & -1.2 & 0 & 0 & 0 & 0 & 0 & 0 \\
0 & 100 & 0 & 0 & -100 & 0 & 0 & 0 & 0 & 0 & 0 & 0 \\
-1.2 & 0 & 3.2 & 1.2 & 0 & 1.6 & 0 & 0 & 0 & 0 & 0 & 0 \\
-0.6 & 0 & 1.2 & 200.6 & 0 & 1.2 & -200 & 0 & 0 & 0 & 0 & 0 \\
0 & -100 & 0 & 0 & 100.6 & 1.8 & 0 & -0.6 & 1.8 & 0 & 0 & 0 \\
-1.2 & 0 & 1.6 & 1.2 & 1.8 & 10.4 & 0 & -1.8 & 3.6 & 0 & 0 & 0 \\
0 & 0 & 0 & -200 & 0 & 0 & 200.6 & 0 & 1.2 & -0.6 & 0 & 1.2 \\
0 & 0 & 0 & 0 & -0.6 & -1.8 & 0 & 100.6 & -1.8 & 0 & -100 & 0 \\
0 & 0 & 0 & 0 & 1.8 & 3.6 & 1.2 & -1.8 & 10.4 & -1.2 & 0 & 1.6 \\
0 & 0 & 0 & 0 & 0 & 0 & -0.6 & 0 & -1.2 & 0.6 & 0 & -1.2 \\
0 & 0 & 0 & 0 & 0 & 0 & 0 & -100 & 0 & 0 & 100 & 0 \\
0 & 0 & 0 & 0 & 0 & 0 & 1.2 & 0 & 1.6 & -1.2 & 0 & 3.2
\end{bmatrix} 10^6 \tag{8}$$

The element load vector for Element 3 is added to in a global load vector using the topology information. The nodal load of 2 kN acting at the upper left corner of the frame is also placed in the global load vector. Altogether, this gives

$$
\mathbf{f}_l = \begin{bmatrix} 0 \\ 0 \\ 0 \\ 2 \\ -30 \\ -30 \\ 0 \\ -30 \\ 30 \\ 0 \\ 0 \\ 0 \end{bmatrix} 10^3 \tag{9}
$$

Boundary conditions

The displacement is prescribed to zero in the degrees of freedom, where the structure is fixed, that is $a_1 = 0$, $a_2 = 0$, $a_3 = 0$, $a_{10} = 0$ and $a_{11} = 0$. This is described by the boundary condition matrix

$$
\text{boundary conditions} = \begin{bmatrix} 1 & 0 \\ 2 & 0 \\ 3 & 0 \\ 10 & 0 \\ 11 & 0 \end{bmatrix} \tag{10}
$$

The degrees of freedom where the displacement is not prescribed are a_4, a_5, a_6, a_7, a_8, a_9 and a_{12}. Note that a_{12} shall not be prescribed, since the hinge allows rotation. In the degrees of freedom where the displacement is prescribed, support forces arise. These are unknown for now and denoted as $f_{b,1}$, $f_{b,2}$, $f_{b,3}$, $f_{b,10}$ and $f_{b,11}$. The displacement vector $\mathbf{a}$ and the boundary force vector $\mathbf{f}_b$ can now be written as

$$
\mathbf{a} = \begin{bmatrix} 0 \\ 0 \\ 0 \\ a_4 \\ a_5 \\ a_6 \\ a_7 \\ a_8 \\ a_9 \\ 0 \\ 0 \\ a_{12} \end{bmatrix} ; \quad \mathbf{f}_b = \begin{bmatrix} f_{b,1} \\ f_{b,2} \\ f_{b,3} \\ 0 \\ 0 \\ 0 \\ 0 \\ 0 \\ 0 \\ f_{b,10} \\ f_{b,11} \\ 0 \end{bmatrix} \tag{11}
$$

Solving the system of equations

By solving the system of equations, we obtain the result

$$\begin{bmatrix} a_4 \\ a_5 \\ a_6 \\ a_7 \\ a_8 \\ a_9 \\ a_{12} \end{bmatrix} = \begin{bmatrix} 7.5357 \\ -0.2874 \\ -5.3735 \\ 7.5161 \\ -0.3126 \\ 4.6656 \\ -5.1513 \end{bmatrix} 10^{-3} \tag{12}$$

which means that the horizontal part of the frame is displaced 7.5 mm to the right and 0.3 mm downwards. The upper joints rotate 5.4×10^{-3} clockwise and 4.7×10^{-3} counterclockwise, respectively. At the hinge, the rotation becomes 5.2×10^{-3} clockwise. The computed nodal displacements are shown in Figure 3.

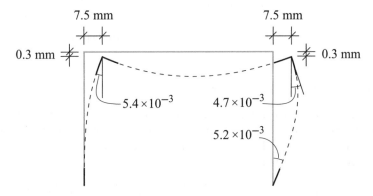

Figure 3 Computed nodal displacements (translations and rotations) drawn in an exaggerated scale

We also obtain the support forces

$$\begin{bmatrix} f_{b,1} \\ f_{b,2} \\ f_{b,3} \\ f_{b,10} \\ f_{b,11} \end{bmatrix} = \begin{bmatrix} 1.9268 \\ 28.7409 \\ 0.4453 \\ -3.9268 \\ 31.2591 \end{bmatrix} 10^{3} \tag{13}$$

This means that the horizontal support forces are 1.93 kN directed to the right and 3.93 kN directed to the left. Since the frame is loaded with a horizontal force of 2 kN directed to the right, the horizontal equilibrium is satisfied. The vertical support forces are 28.74 kN and 31.26 kN, both directed upwards. The sum of these is 60 kN, which is equal to the distributed load on Element 3. By establishing a moment equation, we can also conclude that the external moment equilibrium is satisfied. Figure 4 shows the external load and the computed support forces.

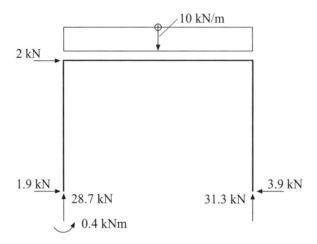

Figure 4   External load and computed support forces

Displacements and internal forces

With the global node displacements **a** known, we can, with use of compatibility relations, find the node displacements for each element. After that, the node displacements for each element can be transformed to local coordinates using (4.96).

For Element 1, we obtain

$$
\bar{\mathbf{a}}^1 = \mathbf{G}^1 \mathbf{a}^1 =
\begin{bmatrix}
0 & -1 & 0 & 0 & 0 & 0 \\
1 & 0 & 0 & 0 & 0 & 0 \\
0 & 0 & 1 & 0 & 0 & 0 \\
0 & 0 & 0 & 0 & -1 & 0 \\
0 & 0 & 0 & 1 & 0 & 0 \\
0 & 0 & 0 & 0 & 0 & 1
\end{bmatrix}
\begin{bmatrix}
7.5357 \\
-0.2874 \\
-5.3735 \\
0 \\
0 \\
0
\end{bmatrix} 10^{-3} =
\begin{bmatrix}
0.2874 \\
7.5357 \\
-5.3735 \\
0 \\
0 \\
0
\end{bmatrix} 10^{-3}
\tag{14}
$$

The axial and the transversal displacements along the element are determined by (3.46) and (4.55) giving

$$
u^{(1)}(\bar{x}) = \begin{bmatrix} 1 - \dfrac{\bar{x}}{4.0} & \dfrac{\bar{x}}{4.0} \end{bmatrix}
\begin{bmatrix} 0.2874 \\ 0 \end{bmatrix} 10^{-3} = (0.2874 - 0.0718\bar{x}) \times 10^{-3}
\tag{15}
$$

$$v^{(1)}(\bar{x}) = \begin{bmatrix} 1 - 3\frac{\bar{x}^2}{4.0^2} + 2\frac{\bar{x}^3}{4.0^3} \\ \bar{x} - 2\frac{\bar{x}^2}{4.0} + \frac{\bar{x}^3}{4.0^2} \\ 3\frac{\bar{x}^2}{4.0^2} - 2\frac{\bar{x}^3}{4.0^3} \\ -\frac{\bar{x}^2}{4.0} + \frac{\bar{x}^3}{4.0^2} \end{bmatrix}^T \begin{bmatrix} 7.5357 \\ -5.3735 \\ 0 \\ 0 \end{bmatrix} 10^{-3}$$

$$= (7.5357 - 5.3735\bar{x} + 1.2738\bar{x}^2 - 0.1004\bar{x}^3) \times 10^{-3} \tag{16}$$

Next, we compute the section forces along the beam. For the normal force, we substitute the displacements directed along the element into (3.52). For Element 1, we then obtain

$$N^{(1)} = 400.0 \times 10^6 \begin{bmatrix} -\frac{1}{4.0} & \frac{1}{4.0} \end{bmatrix} \begin{bmatrix} 0.2874 \\ 0 \end{bmatrix} 10^{-3} = -28.740 \times 10^3 \tag{17}$$

For the moment and the shear force, we substitute the displacements directed perpendicular to the element and the rotations into (4.71) and (4.72), respectively. Since no load acts along the element, $M_p(\bar{x}) = 0$ and $V_p(\bar{x}) = 0$, we obtain

$$V^{(1)} = -3.2 \times 10^6 \begin{bmatrix} \frac{12}{4.0^3} \\ \frac{6}{4.0^2} \\ -\frac{12}{4.0^3} \\ \frac{6}{4.0^2} \end{bmatrix}^T \begin{bmatrix} 7.5357 \\ -5.3735 \\ 0 \\ 0 \end{bmatrix} 10^{-3} = 1.927 \times 10^3 \tag{18}$$

$$M^{(1)}(\bar{x}) = 3.2 \times 10^6 \begin{bmatrix} -\frac{6}{4.0^2} + 12\frac{\bar{x}}{4.0^3} \\ -\frac{4}{4.0} + 6\frac{\bar{x}}{4.0^2} \\ \frac{6}{4.0^2} - 12\frac{\bar{x}}{4.0^3} \\ -\frac{2}{4.0} + 6\frac{\bar{x}}{4.0^2} \end{bmatrix}^T \begin{bmatrix} 7.5357 \\ -5.3735 \\ 0 \\ 0 \end{bmatrix} 10^{-3}$$

$$= (8.152 - 1.927\bar{x}) \times 10^3 \tag{19}$$

At the end points of the element, the moment is

$$M^{(1)}(0) = 8.152 \times 10^3 \tag{20}$$

$$M^{(1)}(4.0) = 0.445 \times 10^3 \tag{21}$$

For Element 2, we obtain

$$\bar{\mathbf{a}}^2 = \mathbf{G}^2 \mathbf{a}^2 = \begin{bmatrix} 0 & -1 & 0 & 0 & 0 & 0 \\ 1 & 0 & 0 & 0 & 0 & 0 \\ 0 & 0 & 1 & 0 & 0 & 0 \\ 0 & 0 & 0 & 0 & -1 & 0 \\ 0 & 0 & 0 & 1 & 0 & 0 \\ 0 & 0 & 0 & 0 & 0 & 1 \end{bmatrix} \begin{bmatrix} 7.5161 \\ -0.3126 \\ 4.6656 \\ 0 \\ 0 \\ -5.1513 \end{bmatrix} 10^{-3} = \begin{bmatrix} 0.3126 \\ 7.5161 \\ 4.6656 \\ 0 \\ 0 \\ -5.1513 \end{bmatrix} 10^{-3} \tag{22}$$

$$u^{(2)}(\bar{x}) = \begin{bmatrix} 1 - \frac{\bar{x}}{4.0} & \frac{\bar{x}}{4.0} \end{bmatrix} \begin{bmatrix} 0.3126 \\ 0 \end{bmatrix} 10^{-3} = (0.3126 - 0.0782\bar{x}) \times 10^{-3} \tag{23}$$

$$v^{(2)}(\bar{x}) = \begin{bmatrix} 1 - 3\frac{\bar{x}^2}{4.0^2} + 2\frac{\bar{x}^3}{4.0^3} \\ \bar{x} - 2\frac{\bar{x}^2}{4.0} + \frac{\bar{x}^3}{4.0^2} \\ 3\frac{\bar{x}^2}{4.0^2} - 2\frac{\bar{x}^3}{4.0^3} \\ -\frac{\bar{x}^2}{4.0} + \frac{\bar{x}^3}{4.0^2} \end{bmatrix}^T \begin{bmatrix} 7.5161 \\ 4.6656 \\ 0 \\ -5.1513 \end{bmatrix} 10^{-3}$$

$$= (7.5161 + 4.6656\bar{x} - 2.4542\bar{x}^2 + 0.2045\bar{x}^3) \times 10^{-3} \tag{24}$$

$$N^{(2)} = 400.0 \times 10^6 \begin{bmatrix} -\frac{1}{4.0} & \frac{1}{4.0} \end{bmatrix} \begin{bmatrix} 0.3126 \\ 0 \end{bmatrix} 10^{-3} = -31.26 \times 10^3 \tag{25}$$

$$V^{(2)} = -3.2 \times 10^6 \begin{bmatrix} \frac{12}{4.0^3} \\ \frac{6}{4.0^2} \\ -\frac{12}{4.0^3} \\ \frac{6}{4.0^2} \end{bmatrix}^T \begin{bmatrix} 7.5161 \\ 4.6656 \\ 0 \\ -5.1513 \end{bmatrix} 10^{-3} = -3.927 \times 10^3 \tag{26}$$

$$M^{(2)}(\bar{x}) = 3.2 \times 10^6 \begin{bmatrix} -\frac{6}{4.0^2} + 12\frac{\bar{x}}{4.0^3} \\ -\frac{4}{4.0} + 6\frac{\bar{x}}{4.0^2} \\ \frac{6}{4.0^2} - 12\frac{\bar{x}}{4.0^3} \\ -\frac{2}{4.0} + 6\frac{\bar{x}}{4.0^2} \end{bmatrix}^T \begin{bmatrix} 7.5161 \\ 4.6656 \\ 0 \\ -5.1513 \end{bmatrix} 10^{-3}$$

$$= (-15.707 + 3.927\bar{x}) \times 10^3 \tag{27}$$

At the end points of the element, the moment is

$$M^{(2)}(0) = -15.707 \times 10^3 \tag{28}$$

$$M^{(2)}(4.0) = 0 \tag{29}$$

For Element 3, we obtain

$$\bar{\mathbf{a}}^3 = \mathbf{G}^3 \mathbf{a}^3 = \begin{bmatrix} 1 & 0 & 0 & 0 & 0 & 0 \\ 0 & 1 & 0 & 0 & 0 & 0 \\ 0 & 0 & 1 & 0 & 0 & 0 \\ 0 & 0 & 0 & 1 & 0 & 0 \\ 0 & 0 & 0 & 0 & 1 & 0 \\ 0 & 0 & 0 & 0 & 0 & 1 \end{bmatrix} \begin{bmatrix} 7.5357 \\ -0.2874 \\ -5.3735 \\ 7.5161 \\ -0.3126 \\ 4.6656 \end{bmatrix} 10^{-3} = \begin{bmatrix} 7.5357 \\ -0.2874 \\ -5.3735 \\ 7.5161 \\ -0.3126 \\ 4.6656 \end{bmatrix} 10^{-3} \tag{30}$$

$$u^{(3)}(\bar{x}) = \begin{bmatrix} 1 - \frac{\bar{x}}{6.0} & \frac{\bar{x}}{6.0} \end{bmatrix} \begin{bmatrix} 7.5357 \\ 7.5161 \end{bmatrix} 10^{-3} = (7.5357 - 0.0033\bar{x}) \times 10^{-3} \tag{31}$$

$$v^{(3)}(\bar{x}) = \begin{bmatrix} 1 - 3\frac{\bar{x}^2}{6.0^2} + 2\frac{\bar{x}^3}{6.0^3} \\ \bar{x} - 2\frac{\bar{x}^2}{6.0} + \frac{\bar{x}^3}{6.0^2} \\ 3\frac{\bar{x}^2}{6.0^2} - 2\frac{\bar{x}^3}{6.0^3} \\ -\frac{\bar{x}^2}{6.0} + \frac{\bar{x}^3}{6.0^2} \end{bmatrix}^T \begin{bmatrix} -0.2874 \\ -5.3735 \\ -0.3126 \\ 4.6656 \end{bmatrix} 10^{-3} + v_p(\bar{x}) \tag{32}$$

$$N^{(3)} = 1200.0 \times 10^6 \begin{bmatrix} -\frac{1}{6.0} & \frac{1}{6.0} \end{bmatrix} \begin{bmatrix} 7.5357 \\ 7.5161 \end{bmatrix} 10^{-3} = -3.927 \times 10^3 \tag{33}$$

$$V^{(3)}(\bar{x}) = -10.8 \times 10^6 \begin{bmatrix} \frac{12}{6.0^3} \\ \frac{6}{6.0^2} \\ -\frac{12}{6.0^3} \\ \frac{6}{6.0^2} \end{bmatrix}^T \begin{bmatrix} -0.2874 \\ -5.3735 \\ -0.3126 \\ 4.6656 \end{bmatrix} 10^{-3} + V_p(\bar{x}) \tag{34}$$

$$M^{(3)}(\bar{x}) = 10.8 \times 10^6 \begin{bmatrix} -\frac{6}{6.0^2} + 12\frac{\bar{x}}{6.0^3} \\ -\frac{4}{6.0} + 6\frac{\bar{x}}{6.0^2} \\ \frac{6}{6.0^2} - 12\frac{\bar{x}}{6.0^3} \\ -\frac{2}{6.0} + 6\frac{\bar{x}}{6.0^2} \end{bmatrix}^T \begin{bmatrix} -0.2874 \\ -5.3735 \\ -0.3126 \\ 4.6656 \end{bmatrix} 10^{-3} + M_p(\bar{x}) \tag{35}$$

with

$$v_p(\bar{x}) = \frac{-10 \times 10^3}{10.8 \times 10^6} \left(\frac{\bar{x}^4}{24} - \frac{\bar{x}^3 \cdot 6.0}{12} + \frac{\bar{x}^2 \cdot 6.0^2}{24} \right) \tag{36}$$

$$V_p(\bar{x}) = -10 \times 10^3 \left(-\bar{x} + \frac{6.0}{2} \right) \tag{37}$$

$$M_p(\bar{x}) = -10 \times 10^3 \left(\frac{\bar{x}^2}{2} - \frac{\bar{x} \cdot 6.0}{2} + \frac{6.0^2}{12} \right) \tag{38}$$

that is

$$v^{(3)}(\bar{x}) = (-0.2874 - 5.3735\bar{x} - 0.3774\bar{x}^2 + 0.4435\bar{x}^3 - 0.0386\bar{x}^4) \times 10^{-3} \tag{39}$$

$$V^{(3)}(\bar{x}) = (-28.74 + 10.0\bar{x}) \times 10^3 \tag{40}$$

$$M^{(3)}(\bar{x}) = (-8.152 + 28.741\bar{x} - 5.0\bar{x}^2) \times 10^3 \tag{41}$$

At the end points of the element, the shear force is

$$V^{(3)}(0) = -28.740 \times 10^3 \tag{42}$$

$$V^{(3)}(L) = 31.260 \times 10^3 \tag{43}$$

At the end points of the element and at the midpoint, the moment is

$$M^{(3)}(0) = -8.152 \times 10^3 \tag{44}$$

$$M^{(3)}(3.0) = 33.070 \times 10^3 \tag{45}$$

$$M^{(3)}(6.0) = -15.707 \times 10^3 \tag{46}$$

The displacements of the frame are shown in Figure 5.

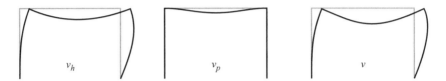

Figure 5   Displacements drawn in an exaggerated scale

The normal force, shear force and moment distributions in the frame are shown in Figure 6, where the moment diagram has been drawn at the side of the beam exposed to tension.

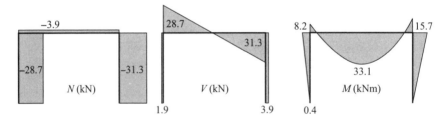

Figure 6   The normal force, shear force and moment distributions

An alternative way to find the section forces at the end points of the element is to determine the nodal forces $\bar{\mathbf{f}}_b^e$ using (4.88) and (4.86) and then compare the sign definitions we have introduced for nodal forces with the ones for section forces. For Element 3, we then obtain

$$\bar{\mathbf{f}}_b^3 = \bar{\mathbf{K}}^3 \bar{\mathbf{a}}^3 - \bar{\mathbf{f}}_l^3$$

$$= \begin{bmatrix} 200 & 0 & 0 & -200 & 0 & 0 \\ 0 & 0.6 & 1.8 & 0 & -0.6 & 1.8 \\ 0 & 1.8 & 7.2 & 0 & -1.8 & 3.6 \\ -200 & 0 & 0 & 200 & 0 & 0 \\ 0 & -0.6 & -1.8 & 0 & 0.6 & -1.8 \\ 0 & 1.8 & 3.6 & 0 & -1.8 & 7.2 \end{bmatrix} \begin{bmatrix} 7.5357 \\ -0.2874 \\ -5.3735 \\ 7.5161 \\ -0.3126 \\ 4.6656 \end{bmatrix} 10^3 - \begin{bmatrix} 0 \\ -30 \\ -30 \\ 0 \\ -30 \\ 30 \end{bmatrix} 10^3 \tag{47}$$

that is

$$\bar{\mathbf{f}}_b^3 = \begin{bmatrix} 3.927 \\ 28.740 \\ 8.152 \\ -3.927 \\ 31.260 \\ -15.770 \end{bmatrix} 10^3 \tag{48}$$

This corresponds to the section forces at the end points of the element (Figure 7):

$$N^{(3)}(0) = -\bar{P}_1^{(3)} = -3.927 \times 10^3 \tag{49}$$

$$V^{(3)}(0) = -\bar{P}_2^{(3)} = -28.740 \times 10^3 \tag{50}$$

$$M^{(3)}(0) = -\bar{P}_3^{(3)} = -8.152 \times 10^3 \tag{51}$$

$$N^{(3)}(L) = \bar{P}_4^{(3)} = -3.927 \times 10^3 \tag{52}$$

$$V^{(3)}(L) = \bar{P}_5^{(3)} = 31.260 \times 10^3 \tag{53}$$

$$M^{(3)}(L) = \bar{P}_6^{(3)} = -15.707 \times 10^3 \tag{54}$$

Figure 7   The section forces at the end points of Element 3

Exercises

4.1 Consider the cross-sections in Exercise 3.1. For the locations of the local $\bar{x}$-axis found in that exercise,
(a) determine the stiffness D_{EI} of the cross-section.
(b) determine the bending moment M for the curvature $\kappa_{\bar{x}} = 0.001$.

4.2 Consider a non-loaded beam of length $L = 1$. Start from (4.50) and let the displacements at the ends of the beam be $\bar{u}_1 = 0.001$, $\bar{u}_2 = 0.001$, $\bar{u}_3 = 0.002$ and $\bar{u}_4 = -0.002$.
(a) Draw the shape functions N_1, N_2, N_3 and N_4 as functions of $\bar{x}$.
(b) Draw $N_1\bar{u}_1, N_2\bar{u}_2, N_3\bar{u}_3$ and $N_4\bar{u}_4$.
(c) Draw $v_h(\bar{x}) = \mathbf{N}\bar{\mathbf{a}}^e = N_1\bar{u}_1 + N_2\bar{u}_2 + N_3\bar{u}_3 + N_4\bar{u}_4$. Compare with Figure 4.13.
(d) For $D_{EI} = 1.0 \times 10^7$, determine shear force $V(\bar{x})$ and bending moment $M(\bar{x})$.
(e) Determine the section forces $V(0), M(0), V(L)$ and $M(L)$ at both ends of the beam.

4.3 Consider the differential equation (4.27) for a linearly varying load $q_{\bar{y}}(\bar{x}) = q_0 \frac{\bar{x}}{L}$. With the method applied in Example 4.1,
(a) determine the particular solution $v_p(\bar{x})$.
(b) draw $v_p(\bar{x})$ and compare with Figure 4.13.
(c) determine and draw $V_p(\bar{x})$ and $M_p(\bar{x})$.
(d) determine element loads $\bar{\mathbf{f}}_l^e$.

4.4

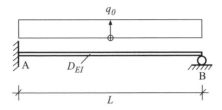

Consider a beam rigidly fixed at its left end and with a roller support at its right. The beam is loaded with a constant load q_0. With $\bar{\mathbf{f}}_l^e$ obtained in Example 4.1 and using element equation (4.81), determine the rotation of the beam at point B.

4.5 For the beam in Exercise 4.4 and with $D_{EI} = 1.0 \times 10^6$, $L = 1$, and $q_0 = 1.0 \times 10^3$, determine the displacement, shear force and moment distributions $v(\bar{x})$, $V(\bar{x})$ and $M(\bar{x})$.

4.6

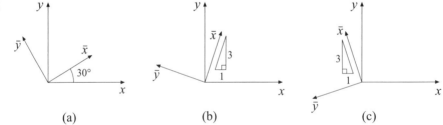

(a)   (b)   (c)

Determine the direction cosines $n_{x\bar{x}}$, $n_{y\bar{x}}$, $n_{x\bar{y}}$ and $n_{y\bar{y}}$ for the three cases in the figure.

4.7

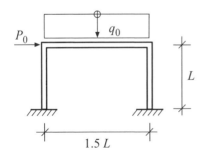

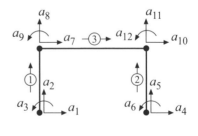

The frame in the figure to the left is loaded by a uniformly distributed load q_0 and by a point load P_0. To the right is shown a computational model for the frame. An analysis of the frame yields the displacements $a_7 = 8.7296$ mm, $a_8 = -0.6960$ mm, $a_9 = -13.8747 \times 10^{-3}$, $a_{10} = 8.6325$ mm, $a_{11} = -0.7440$ mm and $a_{12} = -11.8603 \times 10^{-3}$. The element stiffness matrix for Element 1 in local coordinates is

$$
\bar{\mathbf{K}}^1 = \begin{bmatrix}
50.0 & 0 & 0 & -50.0 & 0 & 0 \\
0 & 0.3 & 0.6 & 0 & -0.3 & 0.6 \\
0 & 0.6 & 1.6 & 0 & -0.6 & 0.8 \\
-50.0 & 0 & 0 & 50.0 & 0 & 0 \\
0 & -0.3 & -0.6 & 0 & 0.3 & -0.6 \\
0 & 0.6 & 0.8 & 0 & -0.6 & 1.6
\end{bmatrix} 10^6
$$

(a) Establish a topology matrix for the frame and using this, determine the element displacement vectors $\mathbf{a}^1$, $\mathbf{a}^2$ and $\mathbf{a}^3$.
(b) Determine for Element 1 the element displacement vector $\bar{\mathbf{a}}^1$ in local coordinates.
(c) Determine the section forces at the ends of the element.

4.8

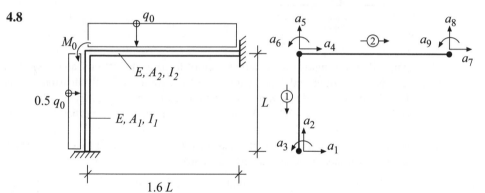

For the frame in the figure, $L = 3.0$ m, $E = 210.0$ GPa, $A_1 = 3.0 \times 10^{-3}$ m^2, $I_1 = 9.6 \times 10^{-6}$ m^4, $A_2 = 4.8 \times 10^{-3}$ m^2, $I_2 = 19.2 \times 10^{-6}$ m^4, $M_0 = 15.0$ kNm, $q_0 = 20.0$ kN/m.
Perform the following exercises manually.
(a) Establish element relations in a local coordinate system for both elements according to (4.86). Determine $\bar{\mathbf{K}}^e$ and $\bar{\mathbf{f}}_l^e$.
(b) Express the element stiffness matrix and the element load vector in the global coordinate system, $\mathbf{K}^e$ and $\mathbf{f}_l^e$.
(c) Assemble the element relations so that a system of equations $\mathbf{Ka} = \mathbf{f}_l + \mathbf{f}_b$ is obtained.
(d) Define the boundary conditions.
(e) Determine the unknown nodal displacements $\mathbf{a}$ and support forces $\mathbf{f}_b$ by solving the system of equations.
(f) Check that the equilibrium is satisfied. Draw the frame in its deformed state and check whether the result is reasonable.
(g) Determine the element displacement vector $\bar{\mathbf{a}}^e$ using (4.96) and the nodal force vector $\bar{\mathbf{f}}_b^e$ using (4.88) and (4.86). (Local coordinates.)
(h) Determine the section force distributions and draw a moment diagram and a shear force diagram.

4.9   Follow the method of calculation for frames in the example section in the CALFEM manual and analyse the frame in Exercise 4.8. Print out the matrices and compare with the corresponding matrices in the calculations done manually.

4.10

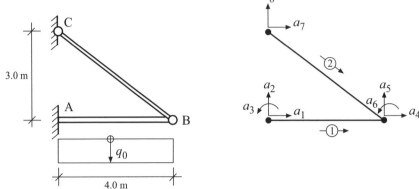

The beam AB in the figure to the left is rigidly fixed at its left end and suspended by a wire BC at its right end. The beam is loaded with a uniformly distributed load $q_0 = 40.0$ kN/m directed downwards. The beam has the properties $E_1 = 200.0$ GPa, $A_1 = 8.00 \times 10^{-3}$ m^2, $I_1 = 4.80 \times 10^{-5}$ m^4 and the wire has the properties $E_2 = 200.0$ GPa, $A_2 = 1.25 \times 10^{-3}$ m^2. In the figure to the right, a computational model is shown were the wire is modelled as a bar (it is assumed that it is exposed to tension). Using this computational model yields the stiffness matrix

$$
\mathbf{K} =
\begin{bmatrix}
400.0 & 0 & 0 & -400.0 & 0 & 0 & 0 & 0 \\
0 & 1.8 & 3.6 & 0 & -1.8 & 3.6 & 0 & 0 \\
0 & 3.6 & 9.6 & 0 & -3.6 & 4.8 & 0 & 0 \\
-400.0 & 0 & 0 & 432.0 & -24.0 & 0 & -32.0 & 24.0 \\
0 & -1.8 & -3.6 & -24.0 & 19.8 & -3.6 & 24.0 & -18.0 \\
0 & 3.6 & 4.8 & 0 & -3.6 & 9.6 & 0 & 0 \\
0 & 0 & 0 & -32.0 & 24.0 & 0 & 32.0 & -24.0 \\
0 & 0 & 0 & 24.0 & -18.0 & 0 & -24.0 & 18.0
\end{bmatrix} 10^6
$$

Compute
(a) the displacements at B (two translations and one rotation)
(b) the moment at A
(c) the tensile stress in the wire BC

4.11

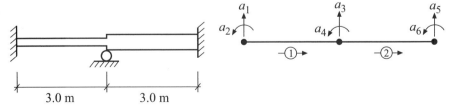

The beam in the figure to the left is clamped at both its ends and has a roller support at the midpoint. The figure to the right shows a computational model where only beam-action is considered. Using this computational model and $D_{EI} = 4.5$ MNm^2 for Element 1 and $D_{EI} = 18.0$ MNm^2 for Element 2, the following global stiffness matrix is obtained:

$$\mathbf{K} = \begin{bmatrix} 2.0 & 3.0 & -2.0 & 3.0 & 0 & 0 \\ 3.0 & 6.0 & -3.0 & 3.0 & 0 & 0 \\ -2.0 & -3.0 & 10.0 & 9.0 & -8.0 & 12.0 \\ 3.0 & 3.0 & 9.0 & 30.0 & -12.0 & 12.0 \\ 0 & 0 & -8.0 & -12.0 & 8.0 & -12.0 \\ 0 & 0 & 12.0 & 12.0 & -12.0 & 24.0 \end{bmatrix} 10^6$$

Determine all vertical support forces for the following cases:
(a) the left beam part is loaded by a uniformly distributed load $q = 40.0$ kN/m directed downwards.
(b) the roller support at the midpoint of the beam is imposed a displacement $\delta = 10.0$ mm directed downwards.
(c) a combination of the uniformly distributed load in task (a) and the support settlement in task (b).

4.12

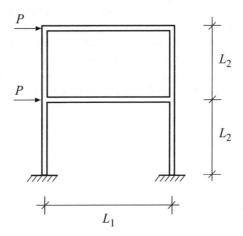

The frame in the figure is loaded with forces $P = 5.0$ kN. For the horizontal parts $E_1 = 210.0$ GPa, $A_1 = 4.5 \times 10^{-3}$ m^2, $I_1 = 25.0 \times 10^{-6}$ m^4, $L_1 = 5.6$ m and for the vertical parts $E_2 = 210.0$ GPa, $A_2 = 2.5 \times 10^{-3}$ m^2, $I_2 = 6.0 \times 10^{-6}$ m^4, $L_2 = 3.2$ m. Determine the moment diagram for the frame using CALFEM.

4.13

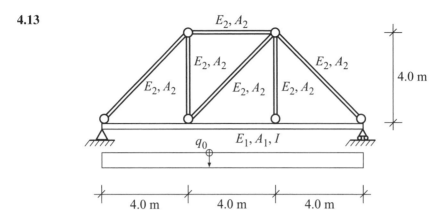

A beam with the length 12 m is strengthened by a system of bars according to the figure. The beam has the properties $E_1 = 210.0$ GPa, $A_1 = 10.0 \times 10^{-3}$ m^2 and $I_1 = 2.0 \times 10^{-4}$ m^4 and all bars have the properties $E_2 = 210.0$ GPa and $A_2 = 1.0 \times 10^{-3}$ m^2. The beam is loaded by a uniformly distributed load $q = 12.0$ kN/m directed downwards. The system of bars is designed in such a way that the system lines of the bars meet the system line of the beam at the common nodes.

(a) Compute the maximal normal stress occurring in the system of bars.

(b) Draw a moment diagram for the beam and determine the maximal bending moment.

(c) Repeat (b) for the case when the stiffness of the bars is negligible.

(d) Repeat (b) for the case when the stiffness of the bars is infinitely large.

5

Modelling at the System Level

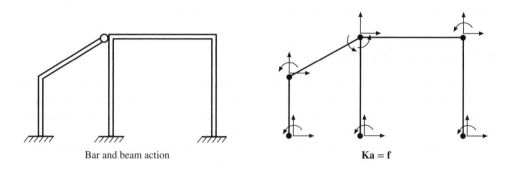

Bar and beam action   Ka = f

Figure 5.1 A computational model

In Chapters 2–4, we have created computational models for different mechanical systems. With assumptions on the motion patterns of the bodies, assumptions on the characteristics of the material and with the use of equilibrium, a *physical model* of the system has been defined. Simultaneously, a *mathematical formulation* (a system of equations) has been established by choosing the locations of local reference axes (system lines), introducing discrete degrees of freedom, defining the positive directions of quantities, choosing the directions of global coordinate systems and using the compact way of writing of matrix algebra. Hence, the computational model describing the mechanical system has two sides: the physical model and the mathematical formulation (Figure 5.1).

While the physical model is general in the sense that its mathematical definition is unspecified, the computational model is specific and directly connected to a mathematical formulation. The formulation described here is in its general form referred to as *the finite element method* (FEM). A strength of this method is that matrix formulation clearly distinguishes the basics of the physics: force, deformation and stiffness.

In Chapter 5, we broaden our perspective and discuss how the physical model (the mechanical system) can be interpreted and understood in relation to the mathematical representation

Structural Mechanics: Modelling and Analysis of Frames and Trusses, First Edition.
Karl-Gunnar Olsson and Ola Dahlblom.
© 2016 John Wiley & Sons, Ltd. Published 2016 by John Wiley & Sons, Ltd.

(the system of equations). The purpose is to give an increased understanding of the system and at the same time introduce some practical modelling aids. In Section 5.1, we discuss the concept of symmetry and how symmetry at the system level can simplify the system of equations and reduce the computational effort. Section 5.2 explains the different components of the system: the displacements and deformations, forces and stiffnesses. Here, we emphasize different patterns within the system – motion patterns, force patterns and stiffness patterns – and relate these to manipulations that can be performed on the system of equations, such as static condensation and the introduction of constraints. In Section 5.3, the general properties of the mechanical system and its behaviour are discussed. Here, it is discussed how the design of load-carrying structures affect their mode of action and efficiency. What is necessary to enable a structure to carry load at all, and how should it be designed to efficiently perform its tasks? Based on this type of system understanding, it is shown how approximations and simplifications can be introduced to reduce the computational effort.

5.1 Symmetry Properties

Mechanical systems can be described based on the presence of symmetries. The concept *symmetry*, which means reflection, exists at all scale levels. If a material is completely symmetric, it is referred to as *isotropic*. No matter how a section surface through a material point is oriented, a mirroring about this surface gives equal material properties. This means that the material has the same properties in all directions. A symmetry that is not equally complete is described by the concept *orthotropy*. Here, the material has three principal planes, *symmetry planes*. A mirroring about any of these planes gives equal properties. For isotropic and orthotropic materials, the number of coefficients that are necessary to describe material properties, such as stiffness and strength, is reduced compared with anisotropic materials that have no symmetry planes (Figure 5.2). For the applications we have studied so far, only one material property has been of interest, the elastic modulus E for a material fibre oriented perpendicular to the cross-sectional plane of the bar and the beam. It means that structures of both isotropic and orthotropic materials can be described, but note that for the latter it is required that the cross-sectional plane is one of the symmetry planes.

Moreover, it is often possible to identify symmetry planes at the system level. For two-dimensional (plane) applications, these symmetry planes become *symmetry lines*. Here,

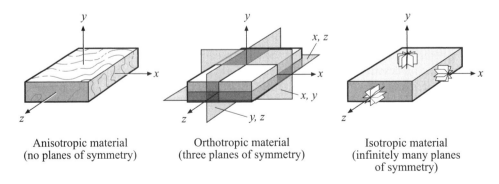

| Anisotropic material (no planes of symmetry) | Orthotropic material (three planes of symmetry) | Isotropic material (infinitely many planes of symmetry) |

Figure 5.2 Material symmetries

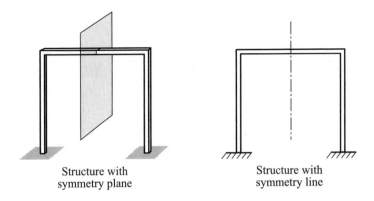

Structure with
symmetry plane

Structure with
symmetry line

Figure 5.3   Symmetry plane and symmetry line

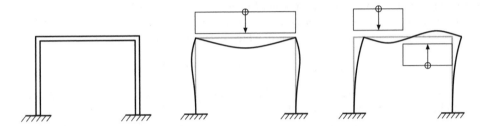

Figure 5.4   Symmetric structure, symmetric and anti-symmetric load

symmetry means that geometry as well as material properties and boundary conditions, that is properties that in some sense affect the stiffness matrix, can be mirrored in the symmetry line (Figure 5.3).

At the system level, the concept of symmetry is basically related to the material and shape of the structure. However, for load-carrying structures, there are two additional concepts of symmetry connected to the loading of the structure and the associated modes of action. A symmetric structure can be loaded with a *symmetric load* resulting in a symmetric mode of action, or with an *anti-symmetric load* resulting in an anti-symmetric mode of action (Figure 5.4). For both cases, an analysis of half the structure gives enough information to determine the mode of action of the entire structure.

A symmetric load gives symmetric deformations and section forces. The results obtained from an analysis of half the structure can thus, by mirroring, be applied also to the other half. A symmetric mirroring means that the sign of translations normal to the symmetry line a_n, rotations a_θ and shear forces V is reversed, while the sign of translations tangential to the symmetry line a_t, normal forces N and moments M remains unchanged. To achieve continuity across the boundary, the boundary conditions of which the sign is reversed have to be zero at the symmetry line (Figure 5.5), which gives the boundary conditions[1]

$$a_n = 0; \qquad a_\theta = 0; \qquad V = 0 \tag{5.1}$$

---

[1] Here, the notation boundary condition is used in a general sense and includes also conditions at symmetry lines.

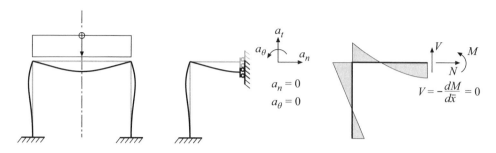

Figure 5.5 Symmetric load gives a symmetric mode of action

An anti-symmetric load gives anti-symmetric deformations and section forces. For an anti-symmetric mirroring, the sign remains unchanged for translations a_n normal to the symmetry line, for rotations a_θ and for section forces V, while the sign of translations tangential to the symmetry line a_t, normal forces N and moments M is reversed. To achieve continuity across the boundary, the boundary conditions of which the sign is reversed have to be zero on the symmetry line (Figure 5.6), which gives the boundary conditions

$$a_t = 0; \qquad N = 0; \qquad M = 0 \qquad (5.2)$$

For elements located along a symmetry line (Figure 5.7), it is not obvious which properties should be given and how the results from the computations should be interpreted. The issue is how bar action and beam action behave at the symmetry line. In the symmetric case, we

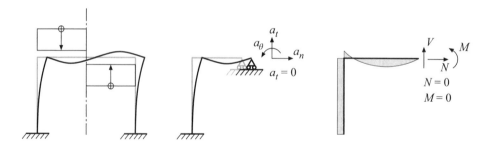

Figure 5.6 Anti-symmetric load gives an anti-symmetric mode of action

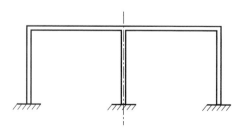

Figure 5.7 An element at the symmetry line

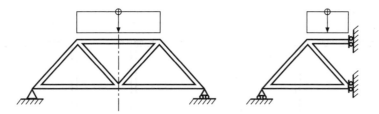

Figure 5.8 A simply supported structure with the corresponding boundary conditions at the symmetry line

have $a_n = a_\theta = 0$, which means that there will be no displacements related to beam action. Consequently, the stiffness properties for beam action may be arbitrary values, for example $D_{EI} = 0$. For bar action, we consider the bar as divided into two halves, one on each side of the symmetry line. The half included in the computation is modelled with its total stiffness E and with half its area $A/2$. As a result, we get half the normal force $N/2$ acting on half the area of the bar. For the anti-symmetric case, we have $a_t = 0$, which means that we will not have any displacements related to bar action. Consequently, the properties specifically related to bar action can be given arbitrary values, for example $D_{EA} = 0$. For beam action, we consider the whole element with stiffness E but with a halved moment of inertia $I/2$.[2] As a result, we obtain half the moment $M/2$ acting on a cross-section with a halved moment of inertia.

Another case where it is not obvious how to handle the boundary conditions is the modelling of a symmetric and 'simply supported' structure. 'Simply supported' means that the structure is given two vertical supports and one horizontal. By prescribing an arbitrarily located horizontal support, we avoid creating a computational model that is an external mechanism. If more horizontal boundary conditions are prescribed, the structure is no longer 'simply supported'. We have then instead introduced a horizontal restraint in the model, which is rather rare in real structures. For a simply supported structure combined with symmetry, this means that the boundary conditions of the symmetry line $a_n = 0$ are sufficient to obtain a 'simply supported' structure (Figure 5.8). With additional horizontal supports, the structure is no longer 'simply supported'. In the corresponding manner, a 'simply supported' structure combined with anti-symmetry has to be given one horizontal support at some point. This single point is arbitrary and can be chosen to lay on the symmetry line. For both symmetry and anti-symmetry, the horizontal support constitutes a reference point for the horizontal displacements. For the anti-symmetric case, it can therefore be appropriate to locate the horizontal fixing at one of the points on the symmetry line.

For structures with a linear mode of action,[3] different load cases can be analysed separately, after which the load cases and the results can be superposed (added to each other). With the previously made assumptions, primarily the assumptions of linear elastic material and of equilibrium in the undeformed state of the structure, this linear mode of action will apply here. For a symmetric structure with an arbitrary load, it is always possible to divide the load into two parts – one symmetric and one anti-symmetric (Figure 5.9). In this way, an arbitrary load on a symmetric structure can be analysed with a computational model describing only half the structure.

---

[2] Note that it is *not* the height of the cross-section, but the moment of inertia that should be halved.

[3] The load and the deformations are proportional in the sense that a doubled load gives doubled deformations.

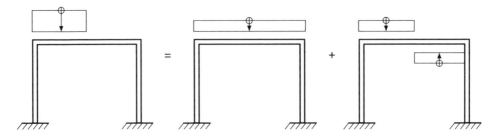

Figure 5.9   Division of an arbitrary load into a symmetric and an anti-symmetric part

5.2   The Structure and the System of Equations

The systematic method we have used to create computational models for load-carrying structures has resulted in a description of the load-carrying structure in terms of a system of equations. The system of equations organises the quantities and fundamental relations of the load-carrying structure in a way which makes them distinguishable and possible to interpret directly from the assumed for numbers in the system of equations (Figure 5.10).

We now in a systematic manner consider the different parts of the system of equations, both to review the contribution of the different parts and to show ways to manipulate the system, which can affect and control a computational model in a desired way. One purpose may be to

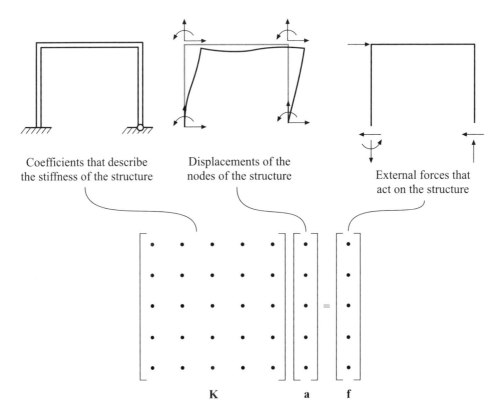

Figure 5.10   The structure and the system of equations

more carefully depict the described structure, another to reduce the computation effort with maintained exactness. Sometimes errors arise, either because the load-carrying structure has an inappropriate design or because we have made some mistake when constructing the computational model. A common mistake is that a local degree of freedom is connected to the wrong global one in the topology matrix. Troubleshooting the system of equations can be made considerably easier when the inner structure of it is understood.

5.2.1   The Deformations and Displacements of the System

In the derivations of the systems of equations for trusses and frames, we have worked ourselves from a three-dimensional description at the material level, where each material point has its own displacement, for example $u = u(\bar{x}, \bar{y}, \bar{z})$, via displacements $u(\bar{x})$ and $v(\bar{x})$ of the system line of the element, to a discrete model of the structure with a displacement vector **a** containing the displacements in a certain set of degrees of freedom (Figure 5.11).

On the way, we have made kinematic assumptions implying that we have prescribed deformation patterns (Figure 5.12). At the cross-section level, we have introduced deformation modes, which limit the possibilities of motion for a cross-section lamella. In Section 4.1.3, this was formulated by the assumption: 'plane cross-sections remain plane and perpendicular to the system line'.

Along the local $\bar{x}$-axis, we have introduced displacement assumptions. So far, we have found the assumptions $u = u_h + u_p$ and $v = v_h + v_p$, which solve the differential equation exactly; thus, at this level, no kinematic restriction is introduced. Later on, when the solutions of the differential equation become more complex, we introduce approximations.

At the system level, we have, by the compatibility requirements, made the choice that all elements connected to a certain node share the same set of degrees of freedom. This means that we fix the translations as well as rotations of adjacent elements to each other. When an equal rotation is assumed for all elements connected at a node, the connection is referred to as a *rigid connection*.

> *For each kinematic restriction we have introduced, the system has become slightly stiffer.*

So far, the computational models that we have constructed have been slightly too stiff. The computed deformations slightly are too small. The assumption of plane deformation of the cross-sectional surface gives a negligible increase of the stiffness as long as h/L is less

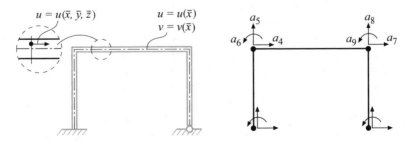

Figure 5.11   The displacement of a material point and the displacement degrees of freedom of the system

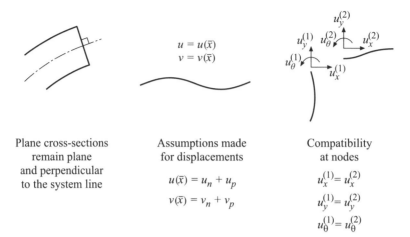

Plane cross-sections
remain plane
and perpendicular
to the system line

Assumptions made
for displacements

$$u(\bar{x}) = u_n + u_p$$

$$v(\bar{x}) = v_n + v_p$$

Compatibility
at nodes

$$u_x^{(1)} = u_x^{(2)}$$

$$u_y^{(1)} = u_y^{(2)}$$

$$u_\theta^{(1)} = u_\theta^{(2)}$$

Figure 5.12   Kinematic assumptions

than 1/5–1/10, depending on the shape of the cross-section, but the assumption of rigid connections can give an estimated stiffness that is considerably larger than the actual stiffness of the structure.

Adding Displacement Degrees of Freedom

In the method of computation that we are using, the degrees of freedom have an essential importance. The introduction of degrees of freedom implies that we have chosen to represent a more or less continuous mechanical system with quantities that can be associated with discrete nodes. A direct interpretation of the degrees of freedom is that they define the possibility to introduce loading of the system into the computational model and to read the displacements. But at a more fundamental level it is a matter of two other things: it is in the degrees of freedom that the equilibrium of the system is satisfied and it is through the degrees of freedom that the deformation pattern (kinematics) of the system is described. We here discuss the latter.

Displacement degrees of freedom can be understood as a set of possible displacements in a number of reference points in the system. In each point, several displacements may appear. It can be translations and rotations, whose number depends on the number of bodies that are connected at the point and whether these bodies are rigidly connected.

In Figure 5.13a, all elements are modelled as rigidly connected in the node. This means that one rotational degree of freedom is sufficient to describe the rotation of all the element ends at the node. By associating two or more element degrees of freedom to the same global degree of freedom, we have decided that their displacements are equal. This can be difficult to achieve in practice nor is it always beneficial for an effective mode of action of the system.

A better model of the connection at the node may be to let the translations of the element interact while their rotations are independent of each other. This is achieved by introducing extra rotational degrees of freedom at the node and associating these with different elements. In Figure 5.13b, the two horizontal elements are rigidly connected, while the two

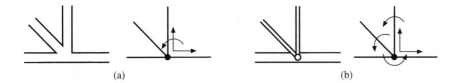

Figure 5.13   Nodes with different numbers of displacement degrees of freedom

connecting from above are hinged and consequently have their own rotations. Three rotational degrees of freedom are therefore required in this model. Something in between rigidly connected and completely independent displacements is accomplished by assembling a rotational spring between two of the rotational degrees of freedom at the node. This case is discussed in Chapter 6.

The major kinematic approximations in a frame are rarely those belonging to cross-sections or elements, but those done when compatibility conditions are formulated for elements to be linked into a global system.

To introduce extra degrees of freedom at a node makes the system more flexible.

Constraints

By formulating constraints and introducing them into a computational model, it is possible to link parts in the model together when differing geometries or directions prevent them from directly being assembled into the chosen frame of reference (chosen global coordinate system, chosen system lines and chosen positions of nodes). The constraints then become a method for translating differing locations and directions to common ones.

For example, consider a structure where the connection between two beams is such that the two system lines do not intersect. Then, the problem of where to place the node of the global system arises (Figure 5.14). By choosing one of the sets of degrees of freedom $\mathbf{a}_m$ as the superior (main variables) and the other $\mathbf{a}_s$ as the subordinated (sub-variables) and thereafter with use of kinematic conditions expressing the latter in terms of the former, the problem is solved. When both system lines have a cross-sectional surface at their ends perpendicular to the system line, this surface can be considered to be common. With the kinematic assumption of the motion pattern of a cross-section, plane cross-sections remain plane and perpendicular to the system line (Figure 5.15), three kinematic relations can be formulated to connect the

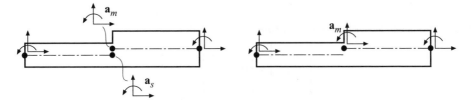

Figure 5.14   Constraints – translation of degrees of freedom

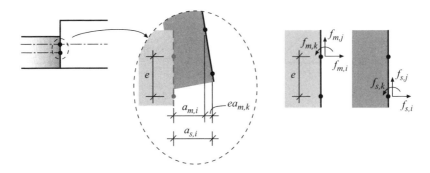

Figure 5.15   Kinematic and static equivalence in the common cross-section

two elements to each other. These kinematic relations between the degrees of freedom in the nodes to be connected read

$$a_{s,i} = a_{m,i} + e\, a_{m,k} \tag{5.3}$$

$$a_{s,j} = a_{m,j} \tag{5.4}$$

$$a_{s,k} = a_{m,k} \tag{5.5}$$

The relations are referred to as *constraints* and can be written in matrix form as

$$\begin{bmatrix} a_{s,i} \\ a_{s,j} \\ a_{s,k} \end{bmatrix} = \begin{bmatrix} 1 & 0 & e \\ 0 & 1 & 0 \\ 0 & 0 & 1 \end{bmatrix} \begin{bmatrix} a_{m,i} \\ a_{m,j} \\ a_{m,k} \end{bmatrix} \tag{5.6}$$

or

$$\mathbf{a}_s = \mathbf{C}\, \mathbf{a}_m \tag{5.7}$$

A constraint can either, as in the present case, follow a previously made kinematic assumption, which does not affect the stiffness of the system, or it may add another kinematic restraint to the structure, which then becomes stiffer.

A set of relations formulating static equivalence between forces in the two sets of degrees of freedom is also associated with the constraint (Figure 5.15).

$$f_{m,i} = f_{s,i} \tag{5.8}$$

$$f_{m,j} = f_{s,j} \tag{5.9}$$

$$f_{m,k} = e\, f_{s,i} + f_{s,k} \tag{5.10}$$

The relations can be written in matrix form as

$$\begin{bmatrix} f_{m,i} \\ f_{m,j} \\ f_{m,k} \end{bmatrix} = \begin{bmatrix} 1 & 0 & 0 \\ 0 & 1 & 0 \\ e & 0 & 1 \end{bmatrix} \begin{bmatrix} f_{s,i} \\ f_{s,j} \\ f_{s,k} \end{bmatrix} \tag{5.11}$$

or

$$\mathbf{f}_m = \mathbf{C}^T \mathbf{f}_s \tag{5.12}$$

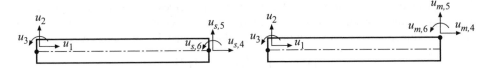

Figure 5.16 Beam element before and after translation of degrees of freedom

The constraints can now be introduced into the element equation to relate the degrees of freedom of the element to the degrees of freedom of a global system (Figure 5.16). If we partition the system of equations of the element, we obtain

$$
\begin{bmatrix} \mathbf{K}^e_l & \mathbf{K}^e_{ls} \\ (\mathbf{K}^e_{ls})^T & \mathbf{K}^e_s \end{bmatrix} \begin{bmatrix} \mathbf{a}^e_l \\ \mathbf{a}^e_s \end{bmatrix} = \begin{bmatrix} \mathbf{f}^e_l \\ \mathbf{f}^e_s \end{bmatrix}
\tag{5.13}
$$

where $\mathbf{a}^e_s$ is the displacement degrees of freedom to be replaced by new degrees of freedom $\mathbf{a}^e_m$. Substitution of the kinematic relation (5.7) into (5.13) gives

$$
\begin{bmatrix} \mathbf{K}^e_l & \mathbf{K}^e_{ls}\mathbf{C} \\ (\mathbf{K}^e_{ls})^T & \mathbf{K}^e_s\mathbf{C} \end{bmatrix} \begin{bmatrix} \mathbf{a}^e_l \\ \mathbf{a}^e_m \end{bmatrix} = \begin{bmatrix} \mathbf{f}^e_l \\ \mathbf{f}^e_s \end{bmatrix}
\tag{5.14}
$$

Left-multiplying the lower part of the system of equations by $\mathbf{C}^T$ and using the force relation (5.12) give

$$
\begin{bmatrix} \mathbf{K}^e_l & \mathbf{K}^e_{ls}\mathbf{C} \\ (\mathbf{K}^e_{ls}\mathbf{C})^T & \mathbf{C}^T\mathbf{K}^e_s\mathbf{C} \end{bmatrix} \begin{bmatrix} \mathbf{a}^e_l \\ \mathbf{a}^e_m \end{bmatrix} = \begin{bmatrix} \mathbf{f}^e_l \\ \mathbf{f}^e_m \end{bmatrix}
\tag{5.15}
$$

which is the system of equations of the element expressed in the new degrees of freedom. If we solve the global system of equations, we can obtain the displacements $\mathbf{a}^e_m$. If we want to, we can thereafter, with use of (5.7), determine the displacements for the original set of degrees of freedom $\mathbf{a}^e_s$.

Alternatively, the constraints can be introduced directly at the system level (Figure 5.17). The constraints (5.7) can then be formulated as

$$
\begin{bmatrix} \mathbf{a}_m \\ \mathbf{a}_s \end{bmatrix} = \begin{bmatrix} \mathbf{I} \\ \mathbf{C} \end{bmatrix} \mathbf{a}; \qquad \mathbf{f} = \begin{bmatrix} \mathbf{I} & \mathbf{C}^T \end{bmatrix} \begin{bmatrix} \mathbf{f}_m \\ \mathbf{f}_s \end{bmatrix}
\tag{5.16}
$$

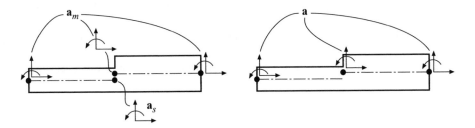

Figure 5.17 Constraints introduced at the system level

where $\mathbf{a}_m$ with dimensions ($m \times 1$) are the displacement degrees of freedom of the global system, $\mathbf{a}_s$ with dimensions ($s \times 1$) are auxiliary degrees of freedom and $\mathbf{C}$ is a transformation matrix with dimensions ($s \times m$). Partitioning the global system of equations gives

$$\begin{bmatrix} \mathbf{K}_{mm} & \mathbf{K}_{ms} \\ \mathbf{K}_{ms}^T & \mathbf{K}_{ss} \end{bmatrix} \begin{bmatrix} \mathbf{a}_m \\ \mathbf{a}_s \end{bmatrix} = \begin{bmatrix} \mathbf{f}_m \\ \mathbf{f}_s \end{bmatrix} \tag{5.17}$$

Left-multiplication of the left and the right-hand sides in (5.17) with the matrix $\begin{bmatrix} \mathbf{I} & \mathbf{C}^T \end{bmatrix}$ gives

$$\begin{bmatrix} \mathbf{I} & \mathbf{C}^T \end{bmatrix} \begin{bmatrix} \mathbf{K}_{mm} & \mathbf{K}_{ms} \\ \mathbf{K}_{ms}^T & \mathbf{K}_{ss} \end{bmatrix} \begin{bmatrix} \mathbf{a}_m \\ \mathbf{a}_s \end{bmatrix} = \begin{bmatrix} \mathbf{I} & \mathbf{C}^T \end{bmatrix} \begin{bmatrix} \mathbf{f}_m \\ \mathbf{f}_s \end{bmatrix} \tag{5.18}$$

Substituting (5.16) into (5.18), the following reduced system of equations is obtained

$$\mathbf{Ka} = \mathbf{f} \tag{5.19}$$

where

$$\mathbf{K} = \mathbf{K}_{mm} + \left(\mathbf{K}_{ms} \mathbf{C} \right)^T + \mathbf{K}_{ms} \mathbf{C} + \mathbf{C}^T \mathbf{K}_{ss} \mathbf{C} \tag{5.20}$$

Figure 5.18 shows three examples of common situations where constraints are useful: In panel (a), a structure with an inclined roller support is shown. In panel (b), the case studied earlier with translation of degrees of freedom along a rigid line is shown, and in panel (c), a translation of degrees of freedom in connection to a rigid body displacement is shown.

If a constraint follows the kinematic approximations already done, the stiffness of the system is maintained. If the constraint introduces a rigid connection between two points, the stiffness increases.

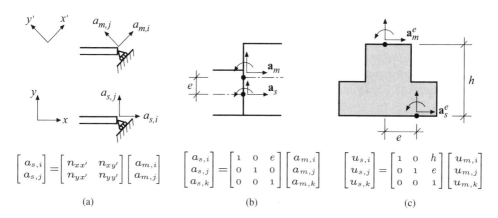

(a)   (b)   (c)

Figure 5.18   Constraints – rotation of degrees of freedom, of a rigid line and of a rigid body

Example 5.1 Constraints – translation of degrees of freedom

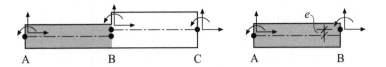

Figure 1 Beam with discontinuous system line

Consider a beam ABC whose cross-section varies so that the system line of the beam is discontinuous at B (Figure 1). An element relation for the beam part AB with degrees of freedom that connect to the system line of the right beam part is to be determined. The left beam part has the modulus of elasticity E, the cross-sectional area A, the moment of inertia I and the length L. The distance between the system lines of the beam parts is e.

Computational model

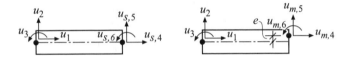

Figure 2 Original beam element and beam element with translated degrees of freedom

We begin with a beam element with six degrees of freedom and with element relations according to (4.86). The degrees of freedom to move are given index s (sub) and the degrees of freedom to be established are given index m (main) (Figure 1).

Establishment of element relation

The system of equations (5.15) gives the sought element relation as

$$\begin{bmatrix} \mathbf{K}_1^e & \mathbf{K}_{1s}^e\mathbf{C} \\ (\mathbf{K}_{1s}^e\mathbf{C})^T & \mathbf{C}^T\mathbf{K}_s^e\mathbf{C} \end{bmatrix}\begin{bmatrix} \mathbf{a}_1^e \\ \mathbf{a}_m^e \end{bmatrix} = \begin{bmatrix} \mathbf{f}_1^e \\ \mathbf{f}_m^e \end{bmatrix} \tag{1}$$

We therefore need to determine the matrices $\mathbf{K}_1^e$, $\mathbf{K}_{1s}^e$, $\mathbf{K}_s^e$ and $\mathbf{C}$. Partitioning of (4.86) with respect to the degrees of freedom to move gives

$$
\begin{bmatrix}
\dfrac{D_{EA}}{L} & 0 & 0 & -\dfrac{D_{EA}}{L} & 0 & 0 \\
0 & \dfrac{12D_{EI}}{L^3} & \dfrac{6D_{EI}}{L^2} & 0 & -\dfrac{12D_{EI}}{L^3} & \dfrac{6D_{EI}}{L^2} \\
0 & \dfrac{6D_{EI}}{L^2} & \dfrac{4D_{EI}}{L} & 0 & -\dfrac{6D_{EI}}{L^2} & \dfrac{2D_{EI}}{L} \\
\hline
-\dfrac{D_{EA}}{L} & 0 & 0 & \dfrac{D_{EA}}{L} & 0 & 0 \\
0 & -\dfrac{12D_{EI}}{L^3} & -\dfrac{6D_{EI}}{L^2} & 0 & \dfrac{12D_{EI}}{L^3} & -\dfrac{6D_{EI}}{L^2} \\
0 & \dfrac{6D_{EI}}{L^2} & \dfrac{2D_{EI}}{L} & 0 & -\dfrac{6D_{EI}}{L^2} & \dfrac{4D_{EI}}{L}
\end{bmatrix}
\begin{bmatrix}
u_1 \\ u_2 \\ u_3 \\ \hline u_{s,4} \\ u_{s,5} \\ u_{s,6}
\end{bmatrix}
=
\begin{bmatrix}
P_1 \\ P_2 \\ P_3 \\ \hline P_{s,4} \\ P_{s,5} \\ P_{s,6}
\end{bmatrix}
\tag{2}
$$

or

$$
\begin{bmatrix}
\mathbf{K}_1^e & \mathbf{K}_{1s}^e \\
(\mathbf{K}_{1s}^e)^T & \mathbf{K}_s^e
\end{bmatrix}
\begin{bmatrix}
\mathbf{a}_1^e \\ \mathbf{a}_s^e
\end{bmatrix}
=
\begin{bmatrix}
\mathbf{f}_1^e \\ \mathbf{f}_s^e
\end{bmatrix}
\tag{3}
$$

that is

$$
\mathbf{K}_1^e =
\begin{bmatrix}
\dfrac{D_{EA}}{L} & 0 & 0 \\
0 & \dfrac{12D_{EI}}{L^3} & \dfrac{6D_{EI}}{L^2} \\
0 & \dfrac{6D_{EI}}{L^2} & \dfrac{4D_{EI}}{L}
\end{bmatrix}
\tag{4}
$$

$$
\mathbf{K}_{1s}^e =
\begin{bmatrix}
-\dfrac{D_{EA}}{L} & 0 & 0 \\
0 & -\dfrac{12D_{EI}}{L^3} & \dfrac{6D_{EI}}{L^2} \\
0 & -\dfrac{6D_{EI}}{L^2} & \dfrac{2D_{EI}}{L}
\end{bmatrix}
\tag{5}
$$

$$
\mathbf{K}_s^e =
\begin{bmatrix}
\dfrac{D_{EA}}{L} & 0 & 0 \\
0 & \dfrac{12D_{EI}}{L^3} & -\dfrac{6D_{EI}}{L^2} \\
0 & -\dfrac{6D_{EI}}{L^2} & \dfrac{4D_{EI}}{L}
\end{bmatrix}
\tag{6}
$$

The matrix $\mathbf{C}$ is determined by constraints and/or static equivalence. When the beam deforms we have the kinematic condition that plane cross-sections remain plane; see Figure 3a. At the interface B, these conditions can be reformulated as the constraints

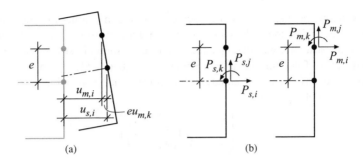

Figure 3   Constraints and static equivalence

$$\begin{bmatrix} u_{s,4} \\ u_{s,5} \\ u_{s,6} \end{bmatrix} = \begin{bmatrix} 1 & 0 & e \\ 0 & 1 & 0 \\ 0 & 0 & 1 \end{bmatrix} \begin{bmatrix} u_{m,4} \\ u_{m,5} \\ u_{m,6} \end{bmatrix} \tag{7}$$

or

$$\mathbf{a}_s^e = \mathbf{C}\,\mathbf{a}_m^e \tag{8}$$

The forces acting at node s can also be expressed as statically equivalent forces at node m

$$\begin{bmatrix} P_{m,4} \\ P_{m,5} \\ P_{m,6} \end{bmatrix} = \begin{bmatrix} 1 & 0 & 0 \\ 0 & 1 & 0 \\ e & 0 & 1 \end{bmatrix} \begin{bmatrix} P_{s,4} \\ P_{s,5} \\ P_{s,6} \end{bmatrix} \tag{9}$$

or

$$\mathbf{f}_m^e = \mathbf{C}^T\,\mathbf{f}_s^e \tag{10}$$

As the matrix $\mathbf{C}$ is obtained both from kinematic constraints and static equivalence, there is an opportunity to compare and check. Substitution of the matrices $\mathbf{K}_1^e$, $\mathbf{K}_{1s}^e$, $\mathbf{K}_s^e$ and $\mathbf{C}$ in (1) eventually gives the element relation

$$\begin{bmatrix} \dfrac{D_{EA}}{L} & 0 & 0 & -\dfrac{D_{EA}}{L} & 0 & -\dfrac{D_{EA}}{L}e \\ 0 & \dfrac{12D_{EI}}{L^3} & \dfrac{6D_{EI}}{L^2} & 0 & -\dfrac{12D_{EI}}{L^3} & \dfrac{6D_{EI}}{L^2} \\ 0 & \dfrac{6D_{EI}}{L^2} & \dfrac{4D_{EI}}{L} & 0 & -\dfrac{6D_{EI}}{L^2} & \dfrac{2D_{EI}}{L} \\ -\dfrac{D_{EA}}{L} & 0 & 0 & \dfrac{D_{EA}}{L} & 0 & \dfrac{D_{EA}}{L}e \\ 0 & -\dfrac{12D_{EI}}{L^3} & -\dfrac{6D_{EI}}{L^2} & 0 & \dfrac{12D_{EI}}{L^3} & -\dfrac{6D_{EI}}{L^2} \\ -\dfrac{D_{EA}}{L}e & \dfrac{6D_{EI}}{L^2} & \dfrac{2D_{EI}}{L} & \dfrac{D_{EA}}{L}e & -\dfrac{6D_{EI}}{L^2} & \dfrac{4D_{EI}}{L}+\dfrac{D_{EA}}{L}e^2 \end{bmatrix} \begin{bmatrix} u_1 \\ u_2 \\ u_3 \\ u_{m,4} \\ u_{m,5} \\ u_{m,6} \end{bmatrix} = \begin{bmatrix} P_1 \\ P_2 \\ P_3 \\ P_{m,4} \\ P_{m,5} \\ P_{m,6} \end{bmatrix} \tag{11}$$

Prescribed Displacements

When we prescribe the displacement of a degree of freedom – either to zero or to a non-zero value – the degree of freedom will be prevented from further displacements. An external loading acting on the structure will perceive the structure as infinitely stiff in the direction of the prescribed degree of freedom. We have previously, in Equation (1.40), observed that by using partition we do not have to use the equations that correspond to prescribed degrees of freedom to solve the system of equations

$$\begin{bmatrix} \mathbf{A}_1 & \mathbf{A}_2 \\ \mathbf{A}_3 & \tilde{\mathbf{K}} \end{bmatrix} \begin{bmatrix} \mathbf{g} \\ \tilde{\mathbf{a}} \end{bmatrix} = \begin{bmatrix} \mathbf{r} \\ \tilde{\mathbf{f}} \end{bmatrix} \tag{5.21}$$

If the displacements $\mathbf{g}$ are prescribed to zero, $\mathbf{g} = \mathbf{0}$, (5.21) can be divided into two parts and written as

$$\tilde{\mathbf{K}}\,\tilde{\mathbf{a}} = \tilde{\mathbf{f}} \tag{5.22}$$

$$\mathbf{r} = \mathbf{A}_2\,\tilde{\mathbf{a}} \tag{5.23}$$

that is we get the reduced stiffness matrix $\tilde{\mathbf{K}}$ directly by deleting the rows and columns in $\mathbf{K}$ associated with the prescribed displacements $\mathbf{g}$. The support forces $\mathbf{r}$ can be determined by multiplication when the unknown displacements $\tilde{\mathbf{a}}$ have been determined. For prescribed non-zero displacements, $\mathbf{g} \neq \mathbf{0}$, we, instead of (5.22) and (5.23), get

$$\tilde{\mathbf{K}}\,\tilde{\mathbf{a}} = \tilde{\mathbf{f}} - \mathbf{A}_3\,\mathbf{g} \tag{5.24}$$

$$\mathbf{r} = \mathbf{A}_1\,\mathbf{g} + \mathbf{A}_2\,\tilde{\mathbf{a}} \tag{5.25}$$

where $\mathbf{A}_3\,\mathbf{g}$ corresponds to the values of the support forces necessary to keep the displacements at the prescribed values.

> *When a displacement is prescribed, a global restraint is introduced, which increases the stiffness of the system.*

5.2.2   The Forces and Equilibria of the System

The degrees of freedom of a system are references for the displacements of the structure, but also references for the *internal equilibria* of the structure. For each degree of freedom an internal equilibrium equation can be established (Figure 5.19) and for each node in a plane frame at least three equilibrium equations can then be established. Rigidly connected elements contribute with forces to the same equilibria, while bodies with independent displacements contribute to different equilibrium conditions. The latter implies that two bodies can meet at a point without transmitting forces between each other.

Figure 5.20 shows a node where two elements are hinged to an otherwise rigidly connected node (see also Figure 5.13). This means that to this node five equilibrium equations are associated

$$f_i = P_4^{(\alpha)} + P_4^{(\beta)} + P_1^{(\gamma)} + P_1^{(\delta)} \tag{5.26}$$

$$f_j = P_5^{(\alpha)} + P_5^{(\beta)} + P_2^{(\gamma)} + P_2^{(\delta)} \tag{5.27}$$

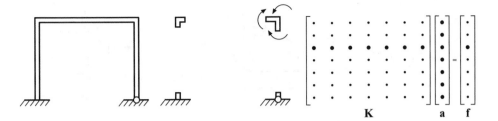

Figure 5.19   To each degree of freedom an internal equilibrium equation is associated

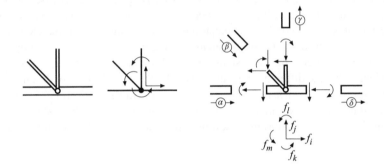

Figure 5.20   A node with two extra equilibrium equations

$$f_k = P_6^{(\alpha)} + P_3^{(\delta)} \tag{5.28}$$

$$f_l = P_3^{(\gamma)} \tag{5.29}$$

$$f_m = P_6^{(\beta)} \tag{5.30}$$

where α, β, γ and δ denote the elements in Figure 5.20. Each equilibrium equation constitutes one of the rows in the global system of equations.

Besides these internal equilibria, a set of *external equilibria* have to be fulfilled. For plane trusses and frames, there are three external equilibria,

$$\sum_i f_{ix} = 0 \tag{5.31}$$

$$\sum_i f_{iy} = 0 \tag{5.32}$$

$$\sum_i f_{ir} + x f_{iy} - y f_{ix} = 0 \tag{5.33}$$

where f_{ix} and f_{iy} are the force components and f_{ir} is the moment component of the system force vector **f**.

In the derivations of the systems of equations for trusses and frames, we have moved from a three-dimensional description at the material level, where the stresses σ_{ij} at each local material point balance each other, to a discrete model of the structure based on equilibrium equations in

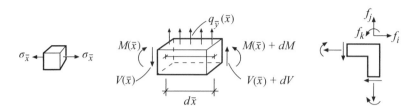

Figure 5.21   Equilibria at different levels

a finite number of nodes (Figure 5.21). On the way, we have gathered the stresses of the material and obtained resulting section forces. Between these section forces and the external loads, we have established exact equilibrium equations for small slices $d\bar{x}$ of the elements in their undeformed state. In the derivation of the element relations, we introduced assumptions which describe the displacements of the system line. With the assumptions, $u = u_h + u_p$ and $v = v_h + v_p$, which solve the differential equation exactly, the equilibrium conditions are fulfilled exactly[4] for each part $d\bar{x}$, that is for the entire element. In the derivation of the systems of equations, the equilibria are satisfied in the degrees of freedom at the nodes. Also here we have assumed that these equilibrium conditions are established in the undeformed state of the structure.

5.2.3   The Stiffness of the System

Designing a structure involves creating stiffness. At the element assembly, stiffness is added to the structure between the degrees of freedom to which the element to be assembled is connected. More elements, stiffer cross-sectional shape, stiffer material, stiffer connections and more external supports – all give local contributions to the creation of a stiffer structure. But the most important aspect of stiffness is not these local contributions, but how the topology of the system organises the elements of the structure. By an ingeniously organised structure, the stiffness can be increased significantly.

In the system of equations, the stiffness is the components of the stiffness matrix **K**. By examining the stiffness matrix, we can in different ways get it to reveal properties of the modelled structure. It may be anything from examining whether the structure has appropriate properties for the tasks it should carry out to checking whether the computational model is constructed correctly. We later study three aspects of a structure, which can be read from the stiffness matrix:

- how stiffly a degree of freedom is locally connected to its surroundings;
- whether a structure is unstable or stable;
- the inherent flexibility of a structure to different loads.

---

[4] In FEM analysis, there exist derivations where also the assumption made for the particular solution is a cubic equation for beam action and a linear equation for bar action. This leads to that the equilibrium equations are only approximatively satisfied within the element.

The Diagonal of the Stiffness Matrix

The assembly process means that we build up our structure by putting stiffness coefficients into the global stiffness matrix. There are two types of stiffness coefficients; diagonal elements K_{ii} and the other elements K_{ij}, where $i \neq j$. When assembling an element between given degrees of freedom in a structure, we create potential force paths between these degrees of freedom by the stiffness we introduce. The diagonal elements K_{ii} show how different elements contribute locally to the stiffness in the direction of a particular degree of freedom, while the rest of the stiffness coefficients $K_{ij} \neq 0$ show that there is a stiffness connected between two degrees of freedom i and j.

In the diagonal matrix elements, stiffnesses are gathered from different elements, all potentially contribute to counteract an external load in the direction of the degree of freedom in question. One can consider the magnitude of the diagonal element K_{ii} as a measure of how stiffly connected degree of freedom i is to its local surroundings (Figure 5.22). One can also, during the assembly, see how different elements contribute to creating this stiffness. However, the matrix element K_{ii} does not give an absolute stiffness value for forces acting in the direction of the degree of freedom since this stiffness is built up from stiffness chains including all the elements between degree of freedom i and the supports of the structure. To determine such an absolute stiffness, *static condensation* can be used. With static condensation all stiffnesses along an internal force path can be gathered to a total stiffness (Figure 5.23). In this manner, stiffness measures for different force paths through the structure can be determined and compared.

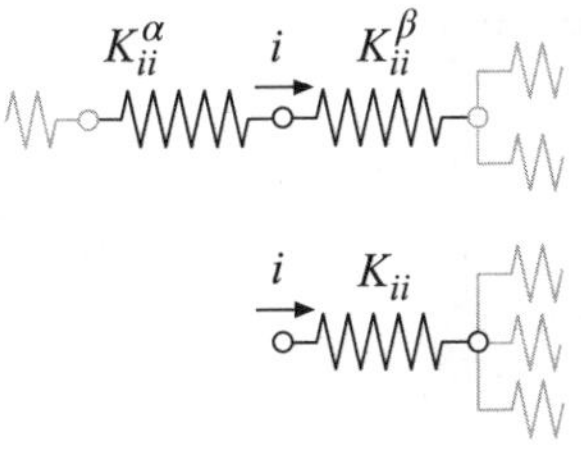

Figure 5.22 Local stiffness $K_{ii} = \sum_e K_{ii}^e$ in the direction of degree of freedom i

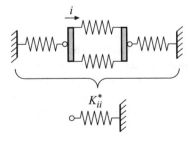

Figure 5.23 By static condensation, all stiffnesses along an internal force path can be gathered to a total stiffness K_{ii}^*

If any of the diagonal matrix elements K_{ii} of the system is equal to zero and degree of freedom i is not fixed with a boundary condition, then this degree of freedom has no stiffness. The structure is unstable and the system of equations has no unique solution. There are infinitely many solutions since the displacement of the degree of freedom cannot be given a unique value. This may arise if one makes a mistake in the construction of the computational model or if elements are assembled between the wrong degrees of freedom. In both cases, a check of the diagonal elements of the global stiffness matrix is a good diagnosis method. One gets to know if and where in the structure stiffness is missing.

The Determinant of the Stiffness Matrix

Besides the case when single degrees of freedom have no stiffness, there are two other cases where the system of equations has no unique solution. One of these cases is when the internal structure is constructed in a way which allows motion in the structure (Figure 5.24a). The other is when the boundary conditions are not sufficient to give the structure the external support that is necessary (Figure 5.24b). All three cases mean that we have models of unstable structures and that the determinant of the stiffness matrix is zero

$$\det \mathbf{K} = 0 \tag{5.34}$$

It is not uncommon that computer programs report that a structural model is unstable by announcing that the determinant of the stiffness matrix is zero.

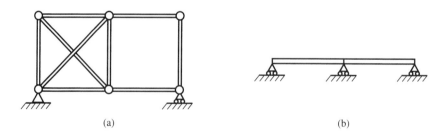

(a)   (b)

Figure 5.24   Unstable structures, $\det \mathbf{K} = 0$

Static Condensation

By static condensation the number of degrees of freedom in a system of equations can be reduced without changing the stiffness described by the system of equations. In Figure 5.25, two examples of this are shown. One application is to compute the *equivalent stiffness* of a structure or a substructure. Using static condensation, a scalar value of the stiffness of the structure can be determined. The stiffness is computed by condensing all degrees of freedom, but the ones where a possible load is applied and those at the supports of the system. Another application is to replace complex structures by simple representations, so-called *substructures*. The use of substructures provides a possibility to construct one's own elements. It may be

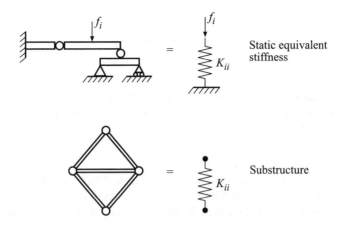

Figure 5.25 Examples of applications for static condensation

elements that replace a repeated part of a structure. Such an element is first built as a separate structure. After that, the internal degrees of freedom are condensed out, so that the remaining structure only has a few degrees of freedom on the external edge.

In Section 1.7, we have described how a system of equations can be partitioned. The purpose there was to handle prescribed variables, that is boundary conditions, when solving the system of equations. To perform a static condensation, we partition the system of equations in the same manner, that is the system of equations

$$\mathbf{Ka} = \mathbf{f} \tag{5.35}$$

is written as

$$\begin{bmatrix} \mathbf{K}_{aa} & \mathbf{K}_{ab} \\ \mathbf{K}_{ba} & \mathbf{K}_{bb} \end{bmatrix} \begin{bmatrix} \mathbf{a}_a \\ \tilde{\mathbf{a}} \end{bmatrix} = \begin{bmatrix} \mathbf{f}_a \\ \mathbf{f}_b \end{bmatrix} \tag{5.36}$$

or

$$\mathbf{K}_{aa}\mathbf{a}_a + \mathbf{K}_{ab}\tilde{\mathbf{a}} = \mathbf{f}_a \tag{5.37}$$

$$\mathbf{K}_{ba}\mathbf{a}_a + \mathbf{K}_{bb}\tilde{\mathbf{a}} = \mathbf{f}_b \tag{5.38}$$

We have divided the displacement vector $\mathbf{a}$ into two parts $\mathbf{a}_a$ and $\tilde{\mathbf{a}}$ and we want to eliminate $\mathbf{a}_a$ without changing the properties of the computational model. We rewrite (5.37) as

$$\mathbf{a}_a = \mathbf{K}_{aa}^{-1}(\mathbf{f}_a - \mathbf{K}_{ab}\tilde{\mathbf{a}}) \tag{5.39}$$

and substitute this relation into (5.38), which gives

$$\mathbf{K}_{ba}\mathbf{K}_{aa}^{-1}(\mathbf{f}_a - \mathbf{K}_{ab}\tilde{\mathbf{a}}) + \mathbf{K}_{bb}\tilde{\mathbf{a}} = \mathbf{f}_b \tag{5.40}$$

or

$$\tilde{\mathbf{K}}\tilde{\mathbf{a}} = \tilde{\mathbf{f}} \tag{5.41}$$

where

$$\tilde{\mathbf{K}} = \mathbf{K}_{bb} - \mathbf{K}_{ba}\mathbf{K}_{aa}^{-1}\mathbf{K}_{ab} \tag{5.42}$$

$$\tilde{\mathbf{f}} = \mathbf{f}_b - \mathbf{K}_{ba}\mathbf{K}_{aa}^{-1}\mathbf{f}_a \tag{5.43}$$

The system of equations (5.41) contains only the degrees of freedom $\tilde{\mathbf{a}}$. Once these are determined, the remaining, that is $\mathbf{a}_a$, can be computed from (5.39) with the same result as would be obtained if the system of equations (5.35) had been solved directly.

Example 5.2 Static condensation – substructure
Consider a bar with varying cross-sectional area which at the midpoint is loaded by a point load P (Figure 1). The bar has the length L, the cross-sectional area A_1 and $A_2 = 2A_1$ and the modulus of elasticity E. A bar element (a substructure) for the entire bar considered is to be formulated, and the element stiffness matrix and element load vector are to be determined by means of static condensation.

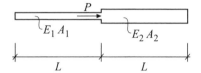

Figure 1   Bar with varying cross-sectional area

Computational model

The bar is first modelled using two bar elements with a total of three degrees of freedom a_1, a_2 and a_3 (Figure 2).

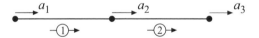

Figure 2   Computational model

System of equations

With the load in question and $D_{EA} = EA$, we get the following system of equations for the respective bar element:

Element 1:

$$\begin{bmatrix} \dfrac{EA_1}{L} & -\dfrac{EA_1}{L} \\ -\dfrac{EA_1}{L} & \dfrac{EA_1}{L} \end{bmatrix} \begin{bmatrix} \bar{u}_1^{(1)} \\ \bar{u}_2^{(1)} \end{bmatrix} = \begin{bmatrix} \bar{P}_1^{(1)} \\ \bar{P}_2^{(1)} \end{bmatrix} \tag{1}$$

Element 2:

$$
\begin{bmatrix}
\dfrac{2EA_1}{L} & -\dfrac{2EA_1}{L} \\
-\dfrac{2EA_1}{L} & \dfrac{2EA_1}{L}
\end{bmatrix}
\begin{bmatrix}
\bar{u}_1^{(2)} \\
\bar{u}_2^{(2)}
\end{bmatrix}
=
\begin{bmatrix}
\bar{P}_1^{(2)} \\
\bar{P}_2^{(2)}
\end{bmatrix}
\tag{2}
$$

Assembling and introducing the point load P gives the system of equations

$$
\begin{bmatrix}
\dfrac{EA_1}{L} & -\dfrac{EA_1}{L} & 0 \\
-\dfrac{EA_1}{L} & \dfrac{3EA_1}{L} & -\dfrac{2EA_1}{L} \\
0 & -\dfrac{2EA_1}{L} & \dfrac{2EA_1}{L}
\end{bmatrix}
\begin{bmatrix}
a_1 \\
a_2 \\
a_3
\end{bmatrix}
=
\begin{bmatrix}
f_1 \\
P \\
f_3
\end{bmatrix}
\tag{3}
$$

Static condensation

A static condensation is based on equation (5.36)

$$
\begin{bmatrix}
\mathbf{K}_{aa} & \mathbf{K}_{ab} \\
\mathbf{K}_{ba} & \mathbf{K}_{bb}
\end{bmatrix}
\begin{bmatrix}
\mathbf{a}_a \\
\tilde{\mathbf{a}}
\end{bmatrix}
=
\begin{bmatrix}
\mathbf{f}_a \\
\mathbf{f}_b
\end{bmatrix}
\tag{4}
$$

where the rows belonging to the degrees of freedom to be condensed are placed first. This can in many cases be achieved already at the initial numbering of the degrees of freedom if one considers giving the degrees of freedom that will condense out the lowest number. If this has not been done a reordering has to be done. The systematics for such a reordering is based on change of the sequence of equations, while the sequence of terms in the equations is changed accordingly. Reordering of (3) and partitioning gives

$$
\left[
\begin{array}{c|cc}
\dfrac{3EA_1}{L} & -\dfrac{EA_1}{L} & -\dfrac{2EA_1}{L} \\
\hline
-\dfrac{EA_1}{L} & \dfrac{EA_1}{L} & 0 \\
-\dfrac{2EA_1}{L} & 0 & \dfrac{2EA_1}{L}
\end{array}
\right]
\begin{bmatrix}
a_2 \\
\hline
a_1 \\
a_3
\end{bmatrix}
=
\begin{bmatrix}
P \\
\hline
f_1 \\
f_3
\end{bmatrix}
\tag{5}
$$

The stiffness matrix for the new bar element is obtained from (5.42)

$$
\tilde{\mathbf{K}} = \mathbf{K}_{bb} - \mathbf{K}_{ba}\mathbf{K}_{aa}^{-1}\mathbf{K}_{ab}
\tag{6}
$$

and a force vector, which takes into account the point load acting in the middle of the bar is obtained from (5.43)

$$
\tilde{\mathbf{f}} = \mathbf{f}_b - \mathbf{K}_{ba}\mathbf{K}_{aa}^{-1}\mathbf{f}_a
\tag{7}
$$

Substituting the components from (5) into (6) and (7), one obtains

$$
\begin{aligned}
\tilde{\mathbf{K}} &=
\begin{bmatrix}
\dfrac{EA_1}{L} & 0 \\
0 & \dfrac{2EA_1}{L}
\end{bmatrix}
-
\begin{bmatrix}
-\dfrac{EA_1}{L} \\
-\dfrac{2EA_1}{L}
\end{bmatrix}
\left[\dfrac{3EA_1}{L}\right]^{-1}
\begin{bmatrix}
-\dfrac{EA_1}{L} & -\dfrac{2EA_1}{L}
\end{bmatrix} \\
&=
\begin{bmatrix}
\dfrac{EA_1}{L} & 0 \\
0 & \dfrac{2EA_1}{L}
\end{bmatrix}
-
\begin{bmatrix}
\dfrac{EA_1}{3L} & \dfrac{2EA_1}{3L} \\
\dfrac{2EA_1}{3L} & \dfrac{4EA_1}{3L}
\end{bmatrix}
=
\begin{bmatrix}
\dfrac{2EA_1}{3L} & -\dfrac{2EA_1}{3L} \\
-\dfrac{2EA_1}{3L} & \dfrac{2EA_1}{3L}
\end{bmatrix}
\end{aligned}
\tag{8}
$$

$$\tilde{\mathbf{f}} = \begin{bmatrix} f_1 \\ f_3 \end{bmatrix} - \begin{bmatrix} -\dfrac{EA_1}{L} \\ -\dfrac{2EA_1}{L} \end{bmatrix} \begin{bmatrix} \dfrac{3EA_1}{L} \end{bmatrix}^{-1} [P] = \begin{bmatrix} f_1 \\ f_3 \end{bmatrix} + \begin{bmatrix} \dfrac{P}{3} \\ \dfrac{2P}{3} \end{bmatrix} \tag{9}$$

Note that the components of the element load vector are different. A stiffer element part carries a larger share of the load. We have earlier in Examples 3.1 and 4.1 computed the equivalent element load vectors, and then the uniformly distributed load was placed at the two nodes with equal shares. This has then been based on the assumption of a constant axial stiffness D_{EA} and a constant bending stiffness D_{EI}.

Example 5.3 Static condensation – equivalent spring stiffness

Consider a console beam that in its free end is loaded by a point load P (Figure 1). The beam has length L, the moment of inertia I and the modulus of elasticity E. To the right of the console beam an elastic spring is shown which represents the vertical stiffness of the beam at the force application point. The value of the spring stiffness k is to be determined by static condensation.

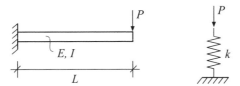

Figure 1   Console beam with point load

Computational model

Figure 2   Computational model

The console beam is modelled using a beam element with four degrees of freedom $\bar{u}_1 - \bar{u}_4$ (Figure 2). We choose to consider the support conditions of the computational model in connection to the static condensation. With this strategy, quite complex two- and three-dimensional support structures may be reduced to a static equivalent spring with the displacement prescribed at one node; see Figures 5.23 and 5.25.

Systems of equations and static condensation

From (4.80), we have the system of equations for a beam element with four degrees of freedom. With $D_{EI} = EI$, we get

$$\frac{EI}{L^3}\begin{bmatrix} 12 & 6L & -12 & 6L \\ 6L & 4L^2 & -6L & 2L^2 \\ -12 & -6L & 12 & -6L \\ 6L & 2L^2 & -6L & 4L^2 \end{bmatrix}\begin{bmatrix} \bar{u}_1 \\ \bar{u}_2 \\ \bar{u}_3 \\ \bar{u}_4 \end{bmatrix} = \begin{bmatrix} \bar{P}_1 \\ \bar{P}_2 \\ \bar{P}_3 \\ \bar{P}_4 \end{bmatrix} \tag{1}$$

For the console beam, the translation a_1 and the rotation a_2 are prescribed to be zero. According to (5.22), the system of equations (1) may thus directly be reduced to

$$\frac{EI}{L^3}\begin{bmatrix} 12 & -6L \\ -6L & 4L^2 \end{bmatrix}\begin{bmatrix} \bar{u}_3 \\ \bar{u}_4 \end{bmatrix} = \begin{bmatrix} \bar{P}_3 \\ \bar{P}_4 \end{bmatrix} \tag{2}$$

Relocation of the lines of the system of equations and partitioning according to (5.36) gives

$$\frac{EI}{L^3}\left[\begin{array}{c|c} 4L^2 & -6L \\ \hline -6L & 12 \end{array}\right]\begin{bmatrix} \bar{u}_4 \\ \bar{u}_3 \end{bmatrix} = \begin{bmatrix} \bar{P}_4 \\ \bar{P}_3 \end{bmatrix} \tag{3}$$

Equation (5.42) finally gives the spring stiffness as

$$\tilde{\mathbf{K}} = \left[\frac{12EI}{L^3}\right] - \left[-\frac{6EI}{L^2}\right]\left[\frac{4EI}{L}\right]^{-1}\left[-\frac{6EI}{L^2}\right]$$

$$= \left[\frac{12EI}{L^3}\right] - \left[\frac{9EI}{L^3}\right] = \left[\frac{3EI}{L^3}\right] \tag{4}$$

that is we have the static equivalent spring stiffness $k = \frac{3EI}{L^3}$.

Example 5.4 Reduction of a degree of freedom for an elementary case

Figure 1   A beam element with four degrees of freedom and a beam element where degree of freedom $\bar{u}_4$ has been condensed out.

Consider a beam element with four degrees of freedom with a uniformly distributed load. We need to eliminate the rotation $\bar{u}_4$ to obtain a beam element with three degrees of freedom according to Figure 1. For this element a reduced element stiffness matrix (3×3) and a reduced nodal load vector (3×1) for a uniformly distributed load shall be determined. From (4.81), we have the element relation

$$\frac{D_{EI}}{L^3}\begin{bmatrix} 12 & 6L & -12 & 6L \\ 6L & 4L^2 & -6L & 2L^2 \\ -12 & -6L & 12 & -6L \\ 6L & 2L^2 & -6L & 4L^2 \end{bmatrix}\begin{bmatrix} \bar{u}_1 \\ \bar{u}_2 \\ \bar{u}_3 \\ \bar{u}_4 \end{bmatrix} = \begin{bmatrix} \bar{P}_1 \\ \bar{P}_2 \\ \bar{P}_3 \\ \bar{P}_4 \end{bmatrix} + q_{\bar{y}}\begin{bmatrix} \frac{L}{2} \\ \frac{L^2}{12} \\ \frac{L}{2} \\ -\frac{L^2}{12} \end{bmatrix} \tag{1}$$

where the components of $\bar{\mathbf{f}}_l^e$ represent a uniformly distributed load according to Example 4.1. Using Equation (4), the displacement $\bar{u}_4$ can be written as

$$\bar{u}_4 = \frac{L}{4D_{EI}}\left(\bar{P}_4 - \frac{q_{\bar{y}}L^2}{12}\right) - \begin{bmatrix} \frac{3}{2L} & \frac{1}{2} & -\frac{3}{2L} \end{bmatrix}\begin{bmatrix} \bar{u}_1 \\ \bar{u}_2 \\ \bar{u}_3 \end{bmatrix} \tag{2}$$

Substituting this into the other three equations then gives

$$\frac{D_{EI}}{L^3}\begin{bmatrix} 3 & 3L & -3 \\ 3L & 3L^2 & -3L \\ -3 & -3L & 3 \end{bmatrix}\begin{bmatrix} \bar{u}_1 \\ \bar{u}_2 \\ \bar{u}_3 \end{bmatrix} = \begin{bmatrix} \bar{P}_1 \\ \bar{P}_2 \\ \bar{P}_3 \end{bmatrix} + \bar{P}_4\begin{bmatrix} -\frac{3}{2L} \\ -\frac{1}{2} \\ \frac{3}{2L} \end{bmatrix} + q_{\bar{y}}\begin{bmatrix} \frac{5L}{8} \\ \frac{L^2}{8} \\ \frac{3L}{8} \end{bmatrix} \tag{3}$$

We have here performed the static condensation without using the systematics of (5.36)–(5.43). Equation (3) is an element relation for a beam element with three degrees of freedom. The element can still rotate around the node to the right where the rotational degree of freedom has been condensed out, but the rotation here cannot be related to rotations of other connected elements.

Canonical Stiffness

From the stiffness matrix $\mathbf{K}$ of a structure, a set of scalar stiffnesses referred to as *canonical stiffnesses* can be computed. Such a computation begins from the assumption that there is a set of displacement modes (deformation patterns) $\mathbf{a}$ that are proportional to a corresponding set of load cases $\mathbf{f}$ acting on the structure (Figure 5.26). For structures modelled with only translation degrees of freedom, such as trusses, this can be described as

$$\mathbf{f} = \lambda\mathbf{a} \tag{5.44}$$

Thus, the eigenvalues λ_i get the dimension [force/length], that is the same dimension as the stiffness of a spring. The system of equations for the structure can now be written as

$$\mathbf{Ka} = \mathbf{f} = \lambda\mathbf{a} \tag{5.45}$$

or

$$(\mathbf{K} - \lambda\mathbf{I})\mathbf{a} = \mathbf{0} \tag{5.46}$$

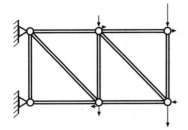 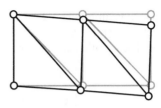

Figure 5.26   Deformation mode and the corresponding proportional load case, $\mathbf{f} = \lambda\mathbf{a}$

where $\mathbf{K}$ is the stiffness matrix of the structure. From (5.46), a set of eigenvalues λ_i and eigenvectors $\mathbf{a}_i$ can be computed. Substitution of these into (5.44), each, by λ_i scaled, deformation mode $\mathbf{a}_i$ can also be interpreted as a load case $\mathbf{f}_i = \lambda_i\mathbf{a}_i$, and the eigenvalue λ_i can be interpreted as the stiffness of the structure against the applied load $\mathbf{f}_i$. As the eigenvalue computation begins from static load cases, the method is referred to as a *static eigenvalue analysis*.

The static eigenvalue analysis leads to three interesting results:

- The eigenvalue λ_i gives a measure of the stiffness of the considered structure, and the eigenmode $\mathbf{a}_i$ gives the corresponding deformation pattern.
- The deformation patterns $\mathbf{a}_i$ are ordered from the most flexible to the stiffest one by the magnitudes of the eigenvalues λ_i.
- Because the deformation patterns $\mathbf{a}_i$ are directly proportional to load vectors $\mathbf{f}_i = \lambda_i\mathbf{a}_i$, the load vectors $\mathbf{f}_i$ can be interpreted as a set of potential load cases applied to the structure. These load cases are sorted so that the load case for which the structure is most flexible corresponds to the load vector for the smallest eigenvalue. With that, the eigenvalue analysis has predicted and ordered a set of load cases $\mathbf{f}_i$ to which the structure is flexible, using only the stiffness $\mathbf{K}$ of the structure.

In the system of equations $\mathbf{Ka} = \mathbf{f}$, the stiffness matrix $\mathbf{K}$, with $n \times n$ components, contains all information about the stiffness of the structure. By a static eigenvalue analysis, the stiffness of the structure has been gathered and ordered as n inherent[5] stiffnesses λ_i. Instead of the abstract concept eigenvalue, we choose in what follows to refer to these stiffnesses λ_i as *canonical stiffnesses*. The above-described form of static eigenvalue analysis is valid for structures with only axial deformations. A relation corresponding to (5.44) can be formulated for structures with rotational degrees of freedom (bending), but this is not discussed further here.

An engineer can use the sorting of load cases from the eigenvalue analysis as a guidance about which load cases the structure should be designed for. This aspect is most interesting when structures of unusual shapes and in unknown situations shall be designed. Where design codes and established practice do not cover the shape and/or possible loadings of the structure, a static eigenvalue analysis can give advice on the decisive load case.

---

[5] Latin *inhærens* with the meaning 'part of the essence of somebody or something'.

Unit Displacement

Consider the frame in Figure 5.27. For a unit displacement in degree of freedom i, while the displacements of all other degrees of freedom are zero, a deformation pattern according to the figure arises. The situation can be formulated in a system of equations

$$\begin{bmatrix} K_{1,1} & \cdot & K_{1,i} & \cdot & K_{1,n} \\ \cdot & \cdot & \cdot & \cdot & \cdot \\ K_{i,1} & \cdot & K_{i,i} & \cdot & K_{i,n} \\ \cdot & \cdot & \cdot & \cdot & \cdot \\ K_{n,1} & \cdot & K_{n,i} & \cdot & K_{n,n} \end{bmatrix} \begin{bmatrix} 0 \\ \cdot \\ 1 \\ \cdot \\ 0 \end{bmatrix} = \begin{bmatrix} f_1 \\ \cdot \\ f_i \\ \cdot \\ f_n \end{bmatrix} \tag{5.47}$$

If we carry out the matrix multiplication on the left-hand side, we obtain

$$\begin{bmatrix} K_{1,i} \\ \cdot \\ K_{i,i} \\ \cdot \\ K_{n,i} \end{bmatrix} = \begin{bmatrix} f_1 \\ \cdot \\ f_i \\ \cdot \\ f_n \end{bmatrix} \tag{5.48}$$

It turns out that column i of the stiffness matrix can be interpreted as the forces necessary to obtain a unit displacement in the direction of degree of freedom i.

In the same manner we can, at the element level, interpret the columns in an element stiffness matrix as the forces required to obtain a unit displacement of the element in the direction of each degree of freedom. For a beam element with pure beam action, we obtain from (4.80) the element stiffness matrix

$$\bar{\mathbf{K}}^e = \frac{D_{EI}}{L^3} \begin{bmatrix} 12 & 6L & -12 & 6L \\ 6L & 4L^2 & -6L & 2L^2 \\ -12 & -6L & 12 & -6L \\ 6L & 2L^2 & -6L & 4L^2 \end{bmatrix} \tag{5.49}$$

If we consider the matrix elements in each column of the matrix in (5.49), we have the forces required to obtain unit displacements in the four degrees of freedom of the element. We can collect these forces into a set of elementary cases, which is shown in Figure 5.28. The elementary cases are useful for manual calculations, which is demonstrated in Section 5.3.

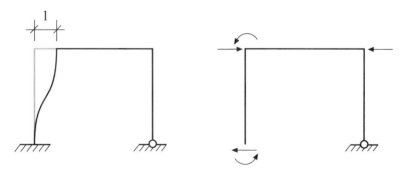

Figure 5.27 A frame deformed by a unit displacement in the direction of degree of freedom i and the forces necessary to obtain this unit displacement

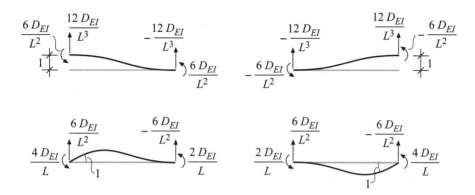

Figure 5.28 Elementary cases for a beam element with four degrees of freedom

Example 5.4 showed among other things how we, with static condensation, can eliminate a rotational degree of freedom to obtain a simplified description of a beam element. If we, in this manner, eliminate first the right and then the left rotational degree of freedom, we obtain six more elementary cases to our collection. Compare the forces shown in the left column of Figure 5.29 with the elements of the stiffness matrix in (3) in Example 5.4.

When a degree of freedom is eliminated using static condensation also the element loads are affected. Element loads for a uniformly distributed load on a beam element with four degrees of freedom are given in (4.80) and are shown in Figure 5.30. For a beam element with three degrees of freedom, equivalent element loads are given from Example 5.4 and are shown in Figure 5.31.

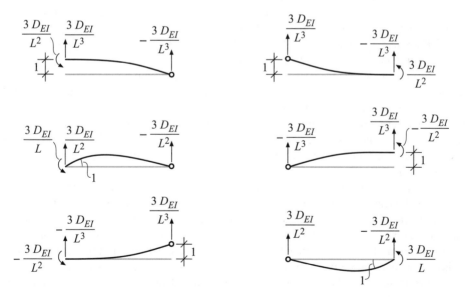

Figure 5.29 Elementary cases for a beam element with three degrees of freedom

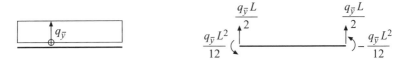

Figure 5.30 Element loads for a beam element with four degrees of freedom

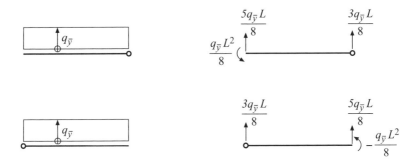

Figure 5.31 Element loads for a beam element with three degrees of freedom

5.3 Structural Design and Simplified Manual Calculations

5.3.1 *Characterising Structures*

General characteristics of mechanical systems are given by the three concepts *mechanism, statically determinate* and *statically indeterminate*.[6] Usually a distinction is made between how these concepts are related to the internal structure and to the external supports. Figure 5.24a is a mechanism with respect to the internal structure and statically determinate with respect to external boundary conditions. In Figure 5.24b, the circumstances are the opposite. In both cases, the structures are referred to as mechanisms.

The term mechanism is used to denote a structure that, through its internal structure or its external supports, does not have a configuration that is sufficient for the structure to be stable. In order to be stable, such a structure has therefore to be given complementary elements or supports. However, even the motion patterns of the unstable mechanisms may be of interest in a design process. They can give a visual guidance to where material must be added to obtain stability. They can also describe a desirable motion pattern of a structure completely or partly intended for motion.

A statically determinate structure has an internal structure and a set of external supports precisely sufficient to obtain stability. A statically determinate construction has the advantageous property that when exposed to an external loading which tries to achieve a change in shape, for example a temperature or a moisture variation, neither the internal structure nor the combination of external supports create any restraints that prevent this change in shape. A disadvantageous property can be that if one single link in the statically determinate system breaks, the entire system becomes a mechanism – and collapses. Modern design thinking has

---

[6] Sometimes the terms hypostatic, isostatic and hyperstatic are used.

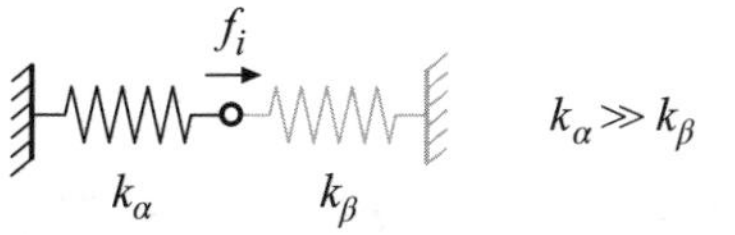

Figure 5.32 A statically indeterminate system. With $k_\alpha \gg k_\beta$, the major part of the load f_i is carried by the left spring

been dominated by statically determinate solutions for a long time. One of the reasons for this is that the structure is able to freely expand and contract, which is of interest especially in changes of temperature. Along with this philosophy comes also that the design process can be divided and the structural members treated one at a time (as a necessary link in the system). These can then with high safety be given dimensions sufficient to obtain required stiffness and strength. However, increased attention has lately been paid in securing the structure with alternative load paths if the main load path collapses.

Old buildings are as a rule static indeterminate with several different possible load paths. Uncritically applying current design philosophy on these structures can be downright devastating. Old brickwork buildings are examples of statically indeterminate constructions where the bearing capacity of the building is based on local flexibility and ability of force redistribution. To formulate statically determinate computational models and reinforce the buildings with guidance of the results from these computations can give most undesired results.

The statically indeterminate structure is characterised by that the internal structure and connections of the structure allow an external load to be carried and distributed along different force paths through the structure. Which load path that becomes active is determined by the relative stiffness of the load paths – the stiffest load path carries the heavy load (Figure 5.32). If a main load path begins to break down at a heavy load, this load path becomes more flexible. This is detected by the structure as a system and it can allow other load paths to successively take over. The weakness of the statically indeterminate structure is that it can be sensitive to, for example, temperature-driven changes in shape. When a member is prevented from expanding by another member undesired stresses can arise. In old constructions, with materials of high deformation capacity, this is rarely a big problem. However, if a new stiff structural component uncritically is added with the aim to strengthen the structure, this component will, due to its stiffness, become the main load path and rearrange the force paths of the entire construction. As a result, unexpected stress concentrations can occur also far away from the component added.

5.3.2 Axial and Bending Stiffness

So far, two types of stiffness at the cross-section level have been derived. For bar action we have the axial stiffness D_{EA} and for beam action the bending stiffness D_{EI}. These stiffnesses can tell how the material and shape of the cross-section contribute to creating stiffness for the respective modes of action. A corresponding stiffness measure at the element level can also be derived. Besides material and cross-sectional shape, the length of the element will then also affect the stiffness. By replacing bar and beam action with equivalent springs, we can obtain scalar expressions for the stiffnesses at the element level (Figure 5.33).

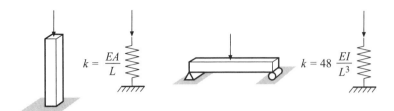

Figure 5.33   Axial and bending stiffness

Steel	$k = \dfrac{EA}{L}$	$k = 48\,\dfrac{EI}{L^3}$
$A = 5 \times 10^{-3}$ $\quad L = 1$	$\sim 10^9$	$\sim 10^7$
$I = 2 \times 10^{-6}$ $\quad L = 10$	$\sim 10^8$	$\sim 10^4$
$A = 5 \times 10^{-3}$ $\quad L = 1$	$\sim 10^9$	$\sim 10^8$
$I = 2 \times 10^{-5}$ $\quad L = 10$	$\sim 10^8$	$\sim 10^5$

Figure 5.34   The influence of the shape, length and mode of action on the stiffness

Figure 5.34 shows how these stiffnesses vary for different cross-sectional shapes and different lengths. Two distinct tendencies can be seen:

- For a particular cross-section, the axial stiffness is essentially larger than the bending stiffness.
- On the bending stiffness, the length has a particularly large effect.

The knowledge that bar action is more effective than beam action is important in the design of structures. In building history, there are numerous examples of how structures are designed to replace beam action (bending) with bar action. Figure 5.35 shows models for two examples of roof trusses where an inner structure with bar action supports the beams on which the external load mainly acts. Spanning over the church of St. Catherine's monastery on the Sinai Peninsula, we have the oldest known wooden roof truss, from ca 500 A.D. Of a considerably more recent date is the type of roof truss which was introduced by the French engineer Camille Polonceau. This roof truss became very popular among buildings with wide spans in its times. An early example is the railway station Gare de Saint Lazare in Paris, completed in 1848.

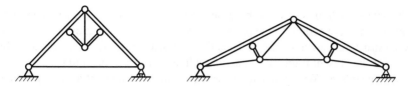

Figure 5.35   The roof truss of the church of St. Catherine's monastery (ca. 500 A.D.) and the Polonceau truss

5.3.3   Reducing the Number of Degrees of Freedom

Using symmetry, constraints and static condensation and prescribing displacements, we have shown different ways to reduce the number of degrees of freedom in a system of equations. Earlier, these methods were used to limit the size of the system of equations in manual calculations. Today, the use of both symmetry and static condensation with introduction of substructures give a possibility to make computer computations more efficient. A reduced computational model which describes a structure in terms of a compact system of equations is easy to get a view of and the elements of the stiffness matrix can be given a clear physical interpretation.

Example 5.5   Reducing systems of equations

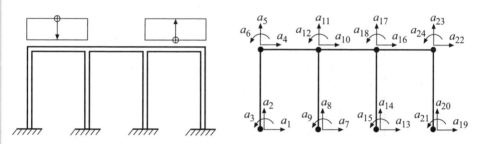

Figure 1   A frame structure modelled with 24 degrees of freedom

Consider a frame with the corresponding computational model (Figure 1). We here show how to reduce the number of degrees of freedom from 24 to 4 with some simple tricks. Using symmetry, the number of degrees of freedom is reduced from 24 to 15; see Figure 2.

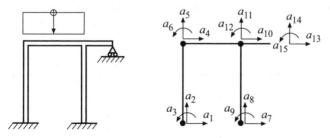

Figure 2   The frame structure according to Figure 1 modelled with consideration taken to the symmetry. The number of degrees of freedom in the computational model has been reduced to 15

The next reduction is possible because the mode of action of the frame in question is dominated by beam action. In the previous section, we pointed out differences in stiffness between structures carrying load by bar action and structures carrying load mainly by beam action. For the present frame, beam action will dominate the magnitudes of deformations and internal stresses. With beam action as the dominating mode of action, the axial deformations can be neglected. This can be formulated as constraints where a_5, a_{10}, a_{11} and a_{13} can be regarded as sub-variables and the other displacements as main variables. Then we obtain the constraints

$$a_5 = a_2; \qquad a_{10} = a_4; \qquad a_{11} = a_8; \qquad a_{13} = a_4 \tag{1}$$

which can be formulated in matrix form as

$$\begin{bmatrix} \mathbf{a}_m \\ \mathbf{a}_s \end{bmatrix} = \begin{bmatrix} \mathbf{I} \\ \mathbf{C} \end{bmatrix} \mathbf{a} \tag{2}$$

If the degrees of freedom are written in numerical order and consequently with main and sub-variables mixed, (2) becomes

$$\begin{bmatrix} a_1 \\ a_2 \\ a_3 \\ a_4 \\ a_5 \\ a_6 \\ a_7 \\ a_8 \\ a_9 \\ a_{10} \\ a_{11} \\ a_{12} \\ a_{13} \\ a_{14} \\ a_{15} \end{bmatrix} = \begin{bmatrix} 1 & 0 & 0 & 0 & 0 & 0 & 0 & 0 & 0 & 0 & 0 \\ 0 & 1 & 0 & 0 & 0 & 0 & 0 & 0 & 0 & 0 & 0 \\ 0 & 0 & 1 & 0 & 0 & 0 & 0 & 0 & 0 & 0 & 0 \\ 0 & 0 & 0 & 1 & 0 & 0 & 0 & 0 & 0 & 0 & 0 \\ 0 & 1 & 0 & 0 & 0 & 0 & 0 & 0 & 0 & 0 & 0 \\ 0 & 0 & 0 & 0 & 1 & 0 & 0 & 0 & 0 & 0 & 0 \\ 0 & 0 & 0 & 0 & 0 & 1 & 0 & 0 & 0 & 0 & 0 \\ 0 & 0 & 0 & 0 & 0 & 0 & 1 & 0 & 0 & 0 & 0 \\ 0 & 0 & 0 & 0 & 0 & 0 & 0 & 1 & 0 & 0 & 0 \\ 0 & 0 & 0 & 1 & 0 & 0 & 0 & 0 & 0 & 0 & 0 \\ 0 & 0 & 0 & 0 & 0 & 0 & 1 & 0 & 0 & 0 & 0 \\ 0 & 0 & 0 & 0 & 0 & 0 & 0 & 0 & 1 & 0 & 0 \\ 0 & 0 & 0 & 1 & 0 & 0 & 0 & 0 & 0 & 0 & 0 \\ 0 & 0 & 0 & 0 & 0 & 0 & 0 & 0 & 0 & 1 & 0 \\ 0 & 0 & 0 & 0 & 0 & 0 & 0 & 0 & 0 & 0 & 1 \end{bmatrix} \begin{bmatrix} a_1 \\ a_2 \\ a_3 \\ a_4 \\ a_5 \\ a_6 \\ a_7 \\ a_8 \\ a_9 \\ a_{10} \\ a_{11} \end{bmatrix} \tag{3}$$

where the old numbering of the degrees of freedom is shown in Figure 2 and the new numbering of the degrees of freedom is shown in Figure 3a. Note that a_i on the left hand side of (3) does not denote the same thing as a_i on the right-hand side.

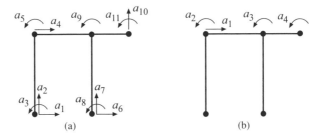

(a)   (b)

Figure 3   A computational model with neglected axial deformations (a) and the substituted boundary conditions (b).

For the reduced frame model, we have the boundary conditions $a_1 = 0$, $a_2 = 0$, $a_3 = 0$, $a_6 = 0$, $a_7 = 0$ and $a_8 = 0$. Moreover, the anti-symmetric load gives the condition $a_{10} = 0$. When solving the system of equations, these rows can be deleted according to (5.22). Then there are four degrees of freedom remaining. These are shown in Figure 3b. With the computational model with four degrees of freedom that we have now, the essential mode of action of the structure can be described.

5.3.4 *Manual Calculation Using Elementary Cases*

To be able to perform simple estimations, it is a good knowledge to know how to reduce the degrees of freedom of a structure and after that, manually, find the elements of the reduced stiffness matrix $\mathbf{K}$ and the corresponding load vector $\mathbf{f}$. We here show this with an example. To our help, we have the elementary cases shown in Figures 5.28 and 5.29.

Example 5.6 Identifying stiffness and element load from elementary cases
Consider the frame in Example 4.2. We have reduced the number of degrees of freedom so that only three degrees of freedom remain (Figure 1). The deformation pattern shown by the structure can then be interpreted as a superposition of the three deformation patterns created by the displacements of these degrees of freedom. Figure 2 shows the deformation patterns that a unit displacement of respective degree of freedom give rise to and how the displacements of the elementary cases and the corresponding forces can be identified.

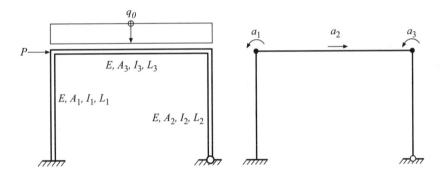

Figure 1   A frame with load and the computational model

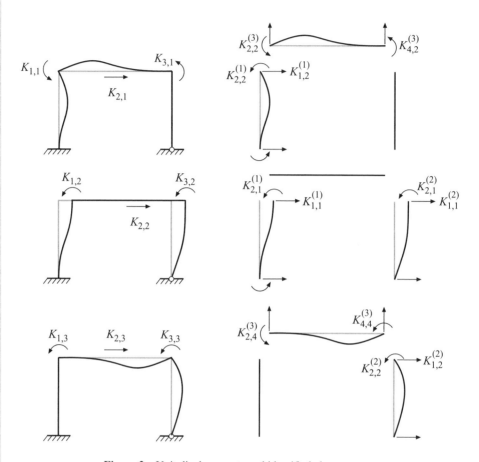

Figure 2   Unit displacements and identified elementary cases

From the elementary cases, we can identify the reduced stiffness matrix **K** as

$$
\mathbf{K} =
\begin{bmatrix}
K_{2,2}^{(1)} + K_{2,2}^{(3)} & K_{2,1}^{(1)} & K_{2,4}^{(3)} \\
K_{1,2}^{(1)} & K_{1,1}^{(1)} + K_{1,1}^{(2)} & K_{1,2}^{(2)} \\
K_{4,2}^{(3)} & K_{2,1}^{(2)} & K_{2,2}^{(2)} + K_{4,4}^{(3)}
\end{bmatrix}
$$

$$
=
\begin{bmatrix}
\dfrac{4EI_1}{L_1} + \dfrac{4EI_3}{L_3} & \dfrac{6EI_1}{L_1^2} & \dfrac{2EI_3}{L_3} \\[2mm]
\dfrac{6EI_1}{L_1^2} & \dfrac{12EI_1}{L_1^3} + \dfrac{3EI_2}{L_2^3} & \dfrac{3EI_2}{L_2^2} \\[2mm]
\dfrac{2EI_3}{L_3} & \dfrac{3EI_2}{L_2^2} & \dfrac{3EI_2}{L_1} + \dfrac{4EI_3}{L_3}
\end{bmatrix}
$$

$$
=
\begin{bmatrix}
10.4 & 1.2 & 3.6 \\
1.2 & 0.75 & 0.6 \\
3.6 & 0.6 & 9.6
\end{bmatrix} 10^6
\tag{1}
$$

In the reduced computational model, nodal loads are introduced in their respective degrees of freedom and distributed loads as element loads. With a horizontal point load $P = 2$ kN in degree of freedom 2 and with element load for a uniformly distributed load $q_0 = 10$ kN/m according to the elementary cases in degree of freedom 1 and 3, we obtain

$$
\mathbf{f} = \begin{bmatrix} \dfrac{q_{\bar{y}}L_3^2}{12} \\ P \\ -\dfrac{q_{\bar{y}}L_3^2}{12} \end{bmatrix} = \begin{bmatrix} -30 \\ 2 \\ 30 \end{bmatrix} 10^3 \tag{2}
$$

Solving the system of equations gives

$$
\mathbf{a} = \begin{bmatrix} -5.3683 \\ 7.5216 \\ 4.6680 \end{bmatrix} 10^{-3} \tag{3}
$$

Compared with the results we obtained in Example 4.2, we can observe that the difference is less than 0.2 %. This means that our simplified model well describes the dominating deformations of the frame.

Exercises

5.1

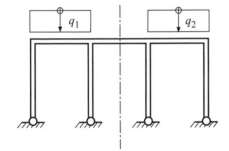 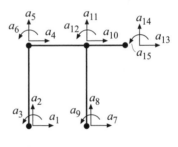

The frame in the figure is symmetric and loaded by two uniformly distributed loads q_1 and q_2. To the right, a computational model for the frame is shown. Note that symmetry has been used.

(a) Assume that the frame is loaded by a symmetric load $q_1 = q_2$. For this case, give the displacements that should be prescribed in the symmetry section.

(b) Assume that the frame is loaded only by the load q_1, that is $q_2 = 0$. Divide the load into a symmetric and an anti-symmetric load case. Show the load cases and give the displacements that should be prescribed in the symmetry section for each load case.

5.2

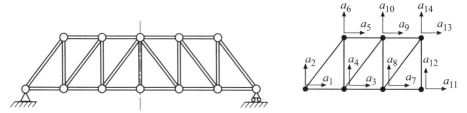

In the figure, a symmetric truss and the corresponding computational model are shown. It is assumed that the truss is loaded by a symmetric load. State the degrees of freedom to be prescribed and describe how to model the bar element at the symmetry section.

5.3

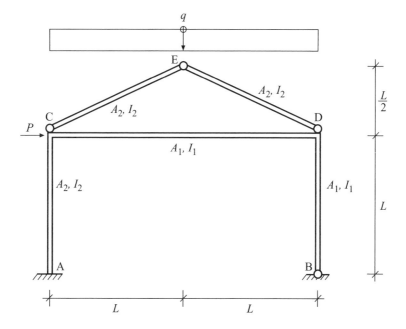

The frame in the figure is constructed with hinges between some of the frame members at C, D and E. It is clamped at A and hinged at B. It is loaded by a point load $P = 20.0\,\text{kN}$ and a uniformly distributed load $q = 40.0\,\text{kN/m}$. Analyse the frame using CALFEM. Give the horizontal and vertical displacement at E. Draw normal force, shear force and moment diagrams. For the frame $L = 3.0$ m, $A_1 = 6.0 \times 10^{-3}$ m^2, $I_1 = 40.0 \times 10^{-6}$ m^4, $A_2 = 10.0 \times 10^{-3}$ m^2, $I_2 = 100.0 \times 10^{-6}$ m^4 and $E = 210$ GPa.

5.4

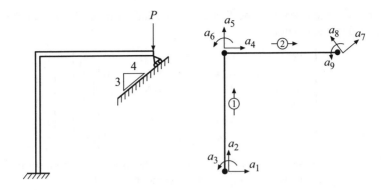

The frame in the figure is fixed to the left and has a roller support which is able to move along a sloping plane to the right. It is loaded with a vertical point load $P = 20\,\text{kN}$ above the roller support. The properties of the beams are $L = 3.0$ m, $A = 0.12\,\text{m}^2$, $I = 1.6 \times 10^{-3}\,\text{m}^4$ and $E = 210$ GPa. To be able to model the roller support on the sloping plane, the computational model has been chosen so that the degrees of freedom at the roller support are directed parallel to and perpendicular to the sloping plane.

(a) Establish a kinematic relation $\mathbf{a}_s = \mathbf{Ca}_m$ for Element 2 between the original degrees of freedom and the new ones.

(b) Express the element stiffness matrix for Element 2 using the new degrees of freedom.

(c) Assemble the element stiffness matrices and compute displacements and support forces.

5.5

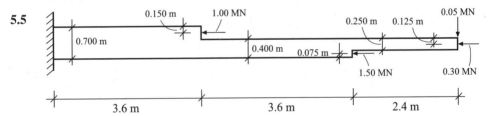

The cantilever beam in the figure consists of three parts with rectangular cross-sections. The elastic modulus is 20 GPa and the beam width is 0.400 m. Determine the maximum deflection and the support forces.

5.6

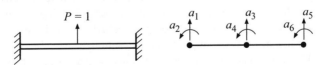

The rigidly fixed beam in the figure is modelled with six degrees of freedom and described by the relation

$$
\begin{bmatrix}
96 & 24 & -96 & 24 & 0 & 0 \\
24 & 8 & -24 & 4 & 0 & 0 \\
-96 & -24 & 192 & 0 & -96 & 24 \\
24 & 4 & 0 & 16 & -24 & 4 \\
0 & 0 & -96 & -24 & 96 & -24 \\
0 & 0 & 24 & 4 & -24 & 8
\end{bmatrix}
\begin{bmatrix}
a_1 \\ a_2 \\ a_3 \\ a_4 \\ a_5 \\ a_6
\end{bmatrix}
=
\begin{bmatrix}
f_1 \\ f_2 \\ f_3 \\ f_4 \\ f_5 \\ f_6
\end{bmatrix}
$$

(a) Reduce the system of equations with consideration taken to prescribed displacements and compute manually the displacements a_3 and a_4.

(b) Return to the original system of equations and let $f_3 = 1$ and $f_4 = 0$. Condense out the degrees of freedom a_3 and a_4 by using static condensation. The result is a beam element with four degrees of freedom where the point load in the middle has been converted to element load. The condensed stiffness matrix agrees with the element stiffness matrix according to (4.80). The example shows a method that is useful for finding element loads for non-continuous load cases.

5.7 Consider the beam system at the top of Figure 5.25 and let the lengths be L, $2L$ and L, respectively. Assume that the external load f_i acts at the midpoint of the beam in the middle and that the beam in the middle has its right support at the middle of the beam to the right. With $L = 4$, $E = 2 \times 10^{11}$ and $I = 4 \times 10^{-5}$, determine a statically equivalent spring stiffness for the system.

5.8

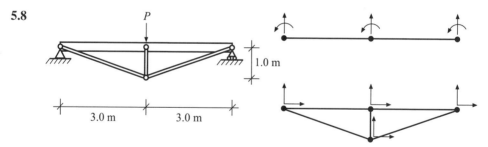

Consider the wooden beam reinforced with a steel construction in the figure to the left. For the wooden beam $E_1 = 15$ GPa, $A_1 = 1 \times 10^{-2}$ m^2, $I_1 = 2 \times 10^{-4}$ m^4 and for the steel construction $E_2 = 210$ GPa and the cross-sectional area is A_2.

To the right, two computational models are shown, one where the construction carries load with pure beam action and one where it carries load with pure bar action.

(a) Assume $A_2 = 1 \times 10^{-4}$ m^2 and compute equivalent stiffnesses for both of the load-carrying structures using static condensation.

(b) Assemble the two systems to a common system and determine the moment diagram for the wooden beam with $P = 10$ kN for $A_2 = 1 \times 10^{-5}$ m^2, $A_2 = 1 \times 10^{-4}$ m^2 and finally for $A_2 = 1 \times 10^{-3}$ m^2.

(c) Compare the results in (b) with the cases infinitely stiff steel construction and infinitely weak steel construction.

5.9 (a) For a spring element with two degrees of freedom, determine two elementary cases in analogy with the elementary cases for a beam element according to Figure 5.28.

(b)

Consider the spring system in Exercise 2.1. For a reduced computational model according to the figure shown here, determine all matrix elements in the stiffness matrix using elementary cases according to (a).

5.10 (a) For a bar element with two degrees of freedom, determine two elementary cases in analogy with the elementary cases for a beam element according to Figure 5.28.

(b) Consider the frame in Exercise 4.8 and determine the matrix elements $K_{4,4}$, $K_{5,4}$, $K_{6,4}$ and $K_{6,6}$ in the stiffness matrix using elementary cases.

5.11

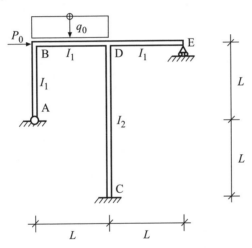

For the frame in the figure, $E = 200$ GPa, $A = 50.0 \times 10^{-4}$ m^2, $I_1 = 3.0 \times 10^{-5}$ m^4, $I_2 = 6.0 \times 10^{-5}$ m^4, $L = 3.0$ m, $q_0 = 0.05$ MN/m and $P_0 = 0.1$ MN.

(a) Define a simplified computational model where axial deformations are neglected.

(b) Establish the stiffness matrix and load vector of the system using elementary cases.

(c) Determine displacements and draw the frame in its deformed state.

6

Flexible Supports

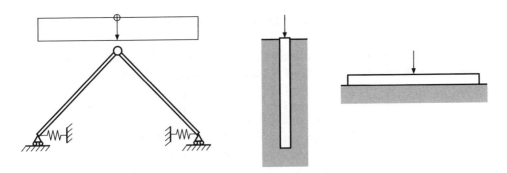

Figure 6.1 Flexible supports

A structure can be supported by its surroundings in different manners. So far, we have modelled supports by prescribing displacements of the nodes. This approach allows only for two different options – completely fixed or completely free to move. We here study the possibility to model deformable supports, which we refer to as flexible supports (Figure 6.1). Flexible supports can exist both at nodes and along elements.

6.1 Flexible Supports at Nodes

The simplest type of flexible supports is obtained by introducing discrete elastic springs at the nodes of the structure. From Chapter 2, we have the element relation for a spring:

$$\mathbf{K}^e \mathbf{a}^e = \mathbf{f}^e \tag{6.1}$$

where

$$\mathbf{K}^e = k \begin{bmatrix} 1 & -1 \\ -1 & 1 \end{bmatrix}; \quad \mathbf{a}^e = \begin{bmatrix} u_1 \\ u_2 \end{bmatrix}; \quad \mathbf{f}^e = \begin{bmatrix} P_1 \\ P_2 \end{bmatrix} \tag{6.2}$$

The element displacements $\mathbf{a}^e$ can be translation as well as rotational degrees of freedom, as shown in Figure 6.2.

Structural Mechanics: Modelling and Analysis of Frames and Trusses, First Edition.
Karl-Gunnar Olsson and Ola Dahlblom.
© 2016 John Wiley & Sons, Ltd. Published 2016 by John Wiley & Sons, Ltd.

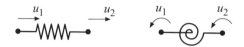

Figure 6.2   Discrete elastic springs

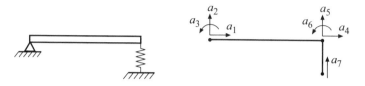

Figure 6.3   A beam on a flexible support

For the beam shown in Figure 6.3, the flexible support can be modelled by assembling a vertical spring between degrees of freedom 5 and 7. Then, we obtain the global system of equations as

$$
\begin{bmatrix}
K_{1,1} & K_{1,2} & K_{1,3} & K_{1,4} & K_{1,5} & K_{1,6} & 0 \\
K_{2,1} & K_{2,2} & K_{2,3} & K_{2,4} & K_{2,5} & K_{2,6} & 0 \\
K_{3,1} & K_{3,2} & K_{3,3} & K_{3,4} & K_{3,5} & K_{3,6} & 0 \\
K_{4,1} & K_{4,2} & K_{4,3} & K_{4,4} & K_{4,5} & K_{4,6} & 0 \\
K_{5,1} & K_{5,2} & K_{5,3} & K_{5,4} & K_{5,5}+k & K_{5,6} & -k \\
K_{6,1} & K_{6,2} & K_{6,3} & K_{6,4} & K_{6,5} & K_{6,6} & 0 \\
0 & 0 & 0 & 0 & -k & 0 & k
\end{bmatrix}
\begin{bmatrix}
a_1 \\ a_2 \\ a_3 \\ a_4 \\ a_5 \\ a_6 \\ a_7
\end{bmatrix}
=
\begin{bmatrix}
f_1 \\ f_2 \\ f_3 \\ f_4 \\ f_5 \\ f_6 \\ f_7
\end{bmatrix}
\tag{6.3}
$$

where $a_7 = 0$ since the bottom end of the spring is assumed to have a non-deformable support. By partitioning (6.3), we can divide the system of equations into two parts:

$$
\begin{bmatrix}
K_{1,1} & K_{1,2} & K_{1,3} & K_{1,4} & K_{1,5} & K_{1,6} \\
K_{2,1} & K_{2,2} & K_{2,3} & K_{2,4} & K_{2,5} & K_{2,6} \\
K_{3,1} & K_{3,2} & K_{3,3} & K_{3,4} & K_{3,5} & K_{3,6} \\
K_{4,1} & K_{4,2} & K_{4,3} & K_{4,4} & K_{4,5} & K_{4,6} \\
K_{5,1} & K_{5,2} & K_{5,3} & K_{5,4} & K_{5,5}+k & K_{5,6} \\
K_{6,1} & K_{6,2} & K_{6,3} & K_{6,4} & K_{6,5} & K_{6,6}
\end{bmatrix}
\begin{bmatrix}
a_1 \\ a_2 \\ a_3 \\ a_4 \\ a_5 \\ a_6
\end{bmatrix}
=
\begin{bmatrix}
f_1 \\ f_2 \\ f_3 \\ f_4 \\ f_5 \\ f_6
\end{bmatrix}
\tag{6.4}
$$

and

$$
-ka_5 = f_7 \tag{6.5}
$$

In the system of equations (6.4), the stiffness k of the spring will only contribute by an addition to the diagonal element $K_{5,5}$. Once the system of equations (6.4) is solved, the support force f_7 in the spring can be computed separately as the product between the spring stiffness k and the node displacement a_5, according to (6.5). This example illustrates a method for modelling discrete springs. This method is general and applicable both on translational and rotational springs.

For the stiffness matrix **K** in (6.4), we have that det **K** = 0. According to the discussion belonging to Equation (5.11), this means that the structure is unstable, which is due to the fact that the stiffness matrix considered is not reduced with respect to the present boundary conditions. The system of equations (6.4) is not solvable until it has been reduced with respect to the prescribed displacements, here $a_1 = 0$ and $a_2 = 0$.

In Chapter 5, we showed how to use static condensation to represent a structure with nothing but discrete springs and how one with constraints for example can move nodes. Here, it means that a flexible support can represent more than just the discrete spring shown and be situated at another position than the support presented in the model. In Section 6.2, an example of the latter is given.

6.2   Foundation on Flexible Support

A common type of support is when a structure is supported by a foundation, which in its turn rests on a flexible support. A reasonable approximation can then be to consider the foundation as a rigid body and consider only the flexibility of the supports of the foundation. We here derive a model for this type of flexible supports (Figure 6.4).

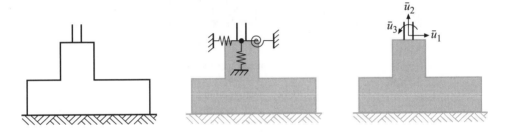

Figure 6.4   Foundation on flexible support

The derivation starts from constitutive relations in the connection point between the foundation and its support. By choosing a reference point on the base surface (contact surface) of the foundation, we can use kinematic and force relations to derive a constitutive relation for the entire base surface. Finally, we can, by using kinematics (constraints) and force relations, move the position of the reference point to the node where the structure meets the foundation. A summary of the quantities and relations of the derivation is shown in Figure 6.5.

6.2.1   The Constitutive Relations of the Connection Point

At an arbitrary point on the contact surface between a foundation and its support we have the constitutive relations

$$p_{x'} = k_{x'} u(x') \tag{6.6}$$

$$p_{y'} = k_{y'} v(x') \tag{6.7}$$

where $p_{x'}$ and $p_{y'}$ are the surface forces acting between the foundation and the support (N/m^2), $k_{x'}$ and $k_{y'}$ are the stiffness of the support (N/m^3), and $u(x')$ and $v(x')$ are the translations of the

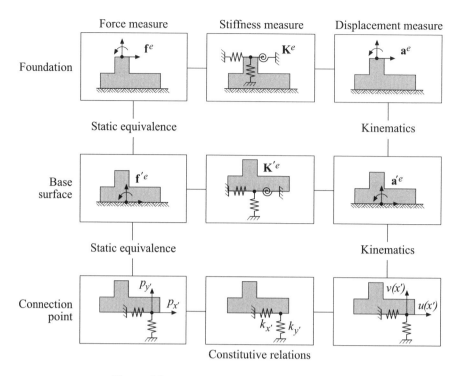

Figure 6.5   From connection point to foundation

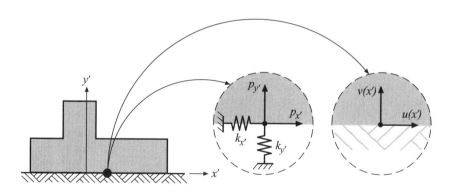

Figure 6.6   The stiffness of the connection point

point [m] horizontally and vertically (Figure 6.6). The relations (6.6) and (6.7) can be written in matrix form as

$$\begin{bmatrix} p_{x'} \\ p_{y'} \end{bmatrix} = \begin{bmatrix} k_{x'} & 0 \\ 0 & k_{y'} \end{bmatrix} \begin{bmatrix} u(x') \\ v(x') \end{bmatrix} \tag{6.8}$$

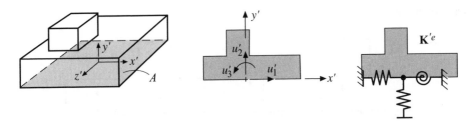

Figure 6.7   The local coordinate system, degrees of freedom and associated stiffnesses of the contact surface

6.2.2   The Constitutive Relation of the Base Surface

To be able to formulate a comprehensive constitutive relation for the entire base surface of the foundation, we have to choose a reference point. This reference point becomes the origin for a local coordinate system (x', y', z') (Figure 6.7). We restrict ourselves to discuss only two-dimensional motion of the foundation and for this we introduce three displacement degrees of freedom u'_1, u'_2 and u'_3 at the reference point. With that we have created the required conditions to establish a stiffness matrix $\mathbf{K}'^e$, which summarises the stiffness of the flexible support of the base surface.

Kinematics

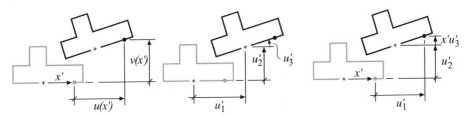

Figure 6.8   The kinematics of the base surface

For a rigid body motion of the foundation, the following kinematic relations between the displacements of the reference point and an arbitrary point on the base surface (Figure 6.8) can be established as

$$u(x') = u'_1 \tag{6.9}$$

$$v(x') = u'_2 + x'u'_3 \tag{6.10}$$

or in matrix form as

$$\begin{bmatrix} u(x') \\ v(x') \end{bmatrix} = \begin{bmatrix} 1 & 0 & 0 \\ 0 & 1 & x' \end{bmatrix} \begin{bmatrix} u'_1 \\ u'_2 \\ u'_3 \end{bmatrix} \tag{6.11}$$

Force Relations

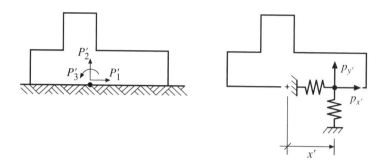

Figure 6.9   Resulting forces

If we consider the foundation in its original state, the resulting forces acting on the support at the reference point of the base surface (Figure 6.9) are obtained as

$$P_1' = \int_A p_{x'} \, dA \tag{6.12}$$

$$P_2' = \int_A p_{y'} \, dA \tag{6.13}$$

$$P_3' = \int_A x' p_{y'} \, dA \tag{6.14}$$

or in matrix form as

$$\begin{bmatrix} P_1' \\ P_2' \\ P_3' \end{bmatrix} = \int_A \begin{bmatrix} 1 & 0 \\ 0 & 1 \\ 0 & x' \end{bmatrix} \begin{bmatrix} p_{x'} \\ p_{y'} \end{bmatrix} dA \tag{6.15}$$

Constitutive Relation

Substituting (6.8) and (6.11) into (6.15) and keeping in mind that the displacements u_1', u_2' and u_3' are quantities that remain unchanged across the cross-section and, therefore, can be moved outside the integral, we obtain

$$\mathbf{K}'^e \mathbf{a}'^e = \mathbf{f}'^e \tag{6.16}$$

where

$$\mathbf{K}'^e = \begin{bmatrix} \int_A k_{x'} \, dA & 0 & 0 \\ 0 & \int_A k_{y'} \, dA & \int_A x' k_{y'} \, dA \\ 0 & \int_A x' k_{y'} \, dA & \int_A x'^2 k_{y'} \, dA \end{bmatrix}; \quad \mathbf{a}'^e = \begin{bmatrix} u_1' \\ u_2' \\ u_3' \end{bmatrix}; \quad \mathbf{f}'^e = \begin{bmatrix} P_1' \\ P_2' \\ P_3' \end{bmatrix} \tag{6.17}$$

By letting the reference point be where $\int_A x' k_{y'} \, dA = 0$ we get

$$\mathbf{K}'^e = \begin{bmatrix} \int_A k_{x'} \, dA & 0 & 0 \\ 0 & \int_A k_{y'} \, dA & 0 \\ 0 & 0 & \int_A x'^2 k_{y'} \, dA \end{bmatrix} \tag{6.18}$$

If the stiffnesses $k_{x'}$ and $k_{y'}$ of the contact surface are constant across the surface, the integral expressions may be simplified and $\mathbf{K}'^e$ written as

$$\mathbf{K}'^e = \begin{bmatrix} k_{x'}A & 0 & 0 \\ 0 & k_{y'}A & 0 \\ 0 & 0 & k_{y'}I_{z'} \end{bmatrix} \tag{6.19}$$

where A is the area of the base surface and $I_{z'}$ is its moment of inertia with respect to rotation about the z'-axis and where the reference point becomes the centroid of the base surface.

6.2.3 Constitutive Relation for the Support Point of the Structure

The goal here is to derive a stiffness representing the foundation and its flexible support and which can be assembled at the degrees of freedom where the global structure meets the foundation (Figure 6.4). This stiffness can be found if we use constraints and static equivalence to move the reference point to where the foundation meets the system line of the structure (Figure 6.10).

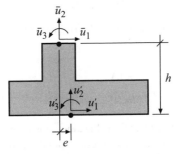

Figure 6.10 Moving the reference point using constraints

Kinematics

If we again consider the foundation as a rigid body, the relation between the reference displacements, u'_1, u'_2 and u'_3, and the displacements, $\bar{u}_1$, $\bar{u}_2$ and $\bar{u}_3$, of the support point of the structure can be formulated as a constraint; see Figure 5.14.

$$\mathbf{a}'^e = \mathbf{C}\bar{\mathbf{a}}^e \tag{6.20}$$

where

$$
\mathbf{a}'^e = \begin{bmatrix} u_1' \\ u_2' \\ u_3' \end{bmatrix}; \quad \mathbf{C} = \begin{bmatrix} 1 & 0 & h \\ 0 & 1 & e \\ 0 & 0 & 1 \end{bmatrix}; \quad \bar{\mathbf{a}}^e = \begin{bmatrix} \bar{u}_1 \\ \bar{u}_2 \\ \bar{u}_3 \end{bmatrix} \tag{6.21}
$$

Force Relation

By using static equivalence, the forces P_1', P_2' and P_3' at the reference point can be expressed in terms of the forces $\bar{P}_1$, $\bar{P}_2$ and $\bar{P}_3$ at the support point. We then obtain the force relation

$$
\bar{\mathbf{f}}^e = \mathbf{C}^T \mathbf{f}'^e \tag{6.22}
$$

where

$$
\bar{\mathbf{f}}^e = \begin{bmatrix} \bar{P}_1 \\ \bar{P}_2 \\ \bar{P}_3 \end{bmatrix}; \quad \mathbf{C}^T = \begin{bmatrix} 1 & 0 & 0 \\ 0 & 1 & 0 \\ h & e & 1 \end{bmatrix}; \quad \mathbf{f}'^e = \begin{bmatrix} P_1' \\ P_2' \\ P_3' \end{bmatrix} \tag{6.23}
$$

Element Relations in Local and Global Coordinate System

Substituting (6.20) and (6.22) into (6.16), we obtain

$$
\bar{\mathbf{K}}^e \bar{\mathbf{a}}^e = \bar{\mathbf{f}}^e \tag{6.24}
$$

where

$$
\bar{\mathbf{K}}^e = \mathbf{C}^T \mathbf{K}'^e \mathbf{C} \tag{6.25}
$$

or if we perform the matrix multiplication at the right-hand side

$$
\bar{\mathbf{K}}^e = \begin{bmatrix} k_{x'}A & 0 & k_{x'}Ah \\ 0 & k_{y'}A & k_{y'}Ae \\ k_{x'}Ah & k_{y'}Ae & k_{y'}I_{z'} + k_{x'}Ah^2 + k_{y'}Ae^2 \end{bmatrix} \tag{6.26}
$$

If the foundation is oriented in a direction that does not coincide with the global directions of the structure, a coordinate transformation corresponding to the one in (4.108) can be performed. Then we obtain for the foundation, in global coordinates, an element relation in the form

$$
\mathbf{K}^e \mathbf{a}^e = \mathbf{f}^e \tag{6.27}
$$

where

$$
\mathbf{K}^e = \mathbf{G}^T \bar{\mathbf{K}}^e \mathbf{G}; \quad \mathbf{f}^e = \mathbf{G}^T \bar{\mathbf{f}}^e; \quad \mathbf{G} = \begin{bmatrix} n_{x\bar{x}} & n_{y\bar{x}} & 0 \\ n_{x\bar{y}} & n_{y\bar{y}} & 0 \\ 0 & 0 & 1 \end{bmatrix} \tag{6.28}
$$

6.3 Bar with Axial Springs

For bar action, axially flexible supports may exist continuously along the longitudinal direction of the bar (Figure 6.11). We here derive a spring stiffness matrix $\mathbf{K}_s^e$, which describes the stiffness of such a continuously flexible support. In the same manner as a discrete spring, k can be modelled by adding its stiffness to a global stiffness matrix, continuous springs can also be modelled by assembling a spring stiffness matrix $\mathbf{K}_s^e$ into the global stiffness matrix.

The presumptions and manner of the approach discussed here are identical to those in Sections 3.1 and 3.2 with two exceptions: the equilibrium equation (3.20) is here affected also by an external force that the deformed spring loads the bar with, and to solve the new differential equation which arises, we choose to introduce an approximate solution.

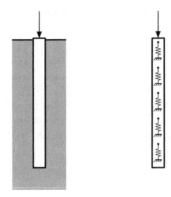

Figure 6.11   A bar with axial springs along its longitudinal direction

6.3.1   The Differential Equation for Bar Action with Axial Springs

We consider a bar surrounded by axial springs according to Figure 6.12. Along the bar, a uniformly distributed load $q_{\bar{x}}(\bar{x})$ acts. Axial springs with the stiffness $k_{\bar{x}}(\bar{x})$ give support in the longitudinal direction of the bar. The derivation starts from the constitutive relation of the bar at the cross-section level (3.23)

$$N(\bar{x}) = D_{EA}(\bar{x})\varepsilon_{\bar{x}}(\bar{x}) \tag{6.29}$$

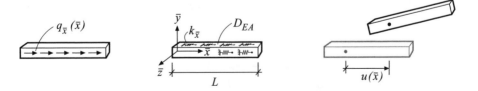

Figure 6.12   The quantities of the differential equation

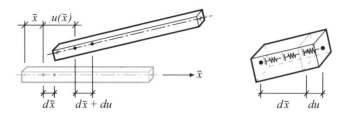

Figure 6.13 Deformed bar element

where $N(\bar{x})$ is the normal force, $\varepsilon_{\bar{x}}(\bar{x})$ is the generalised strain and $D_{EA}(\bar{x})$ is the axial stiffness

$$D_{EA}(\bar{x}) = \int_A E(\bar{x}, \bar{y}, \bar{z})dA \tag{6.30}$$

Kinematics

If we assume that the spring support gives a support symmetric about the system line of the bar, it is reasonable to keep the generalised strain $\varepsilon_{\bar{x}}(\bar{x})$ of the cross-section lamella as the strain measure (Figure 6.13). From (3.19), we then have a kinematic relation between the deformation measure of bar action $u(\bar{x})$ and the deformation measure of the cross-section level $\varepsilon_{\bar{x}}(\bar{x})$

$$\varepsilon_{\bar{x}}(\bar{x}) = \frac{du}{d\bar{x}} \tag{6.31}$$

Equilibrium

Consider a small part $d\bar{x}$ of a bar with deformed axial springs (Figure 6.14). When an axial spring with stiffness $k_{\bar{x}}(\bar{x})$ is displaced by $u(\bar{x})$, a support force

$$p_{\bar{x}}(\bar{x})d\bar{x} = k_{\bar{x}}(\bar{x})u(\bar{x})d\bar{x} \tag{6.32}$$

arises. For the part considered, we have the equilibrium relation

$$-N(\bar{x}) + (N(\bar{x}) + dN) - k_{\bar{x}}(\bar{x})u(\bar{x})d\bar{x} + q_{\bar{x}}(\bar{x})d\bar{x} = 0 \tag{6.33}$$

or

$$\frac{dN}{d\bar{x}} - k_{\bar{x}}(\bar{x})u(\bar{x}) + q_{\bar{x}}(\bar{x}) = 0 \tag{6.34}$$

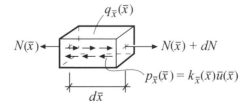

Figure 6.14 Equilibrium for a part $d\bar{x}$ of a bar

The Differential Equation for a Bar with Axial Springs

Substituting the kinematic relation (6.31) into (6.29) gives

$$N(\bar{x}) = D_{EA}(\bar{x})\frac{du}{d\bar{x}} \tag{6.35}$$

Substituting (6.35) into the equilibrium relation (6.34) gives

$$\frac{d}{d\bar{x}}\left(D_{EA}(\bar{x})\frac{du}{d\bar{x}}\right) - k_{\bar{x}}(\bar{x})u(\bar{x}) + q_{\bar{x}}(\bar{x}) = 0 \tag{6.36}$$

With the stiffnesses D_{EA} and $k_{\bar{x}}$ constant along the bar, we obtain

$$D_{EA}\frac{d^2u}{d\bar{x}^2} - k_{\bar{x}}u(\bar{x}) + q_{\bar{x}}(\bar{x}) = 0 \tag{6.37}$$

6.3.2 Bar Element

The bar element with axial springs in Figure 6.15 has two displacement degrees of freedom $\bar{u}_1$ and $\bar{u}_2$ and two nodal forces $\bar{P}_1$ and $\bar{P}_2$.

Figure 6.15 A bar element with axial springs

Solving the Differential Equation

The homogeneous equation associated with (6.37) is

$$D_{EA}\frac{d^2u_h}{d\bar{x}^2} - k_{\bar{x}}u_h(\bar{x}) = 0 \tag{6.38}$$

This equation can be solved exactly[1] and an exact element stiffness matrix can be established, but we choose here an approximate solution that gives us the possibility to express the stiffness of the cross-section lamella and the spring stiffness in two separate stiffness matrices. The force exerted by the springs on the bar according to (6.32) is

$$p_{\bar{x}}(\bar{x}) = k_{\bar{x}}(\bar{x})u(\bar{x}) \tag{6.39}$$

Assuming that the spring stiffness $k_{\bar{x}}$ is constant and that the axial displacement $u_h(\bar{x})$ varies linearly with $\bar{x}$, the force in the springs can be expressed as

$$p_{\bar{x}}(\bar{x}) = k_{\bar{x}}\mathbf{N}\bar{\mathbf{a}}^e \tag{6.40}$$

---

[1] The solution to (6.38) is in the form $u_h(\bar{x}) = C_1\cosh\lambda_{\bar{x}}\bar{x} + C_2\sinh\lambda_{\bar{x}}\bar{x}$ where $\lambda_{\bar{x}} = \sqrt{\dfrac{k_{\bar{x}}}{D_{EA}}}$.

where $\mathbf{N}\bar{\mathbf{a}}^e$ is the displacement of a bar element without springs and without axial load, according to Chapter 3. The differential equation (6.37) can, when we use the approximation (6.40), be written as

$$D_{EA}\frac{d^2u}{d\bar{x}^2} - p_{\bar{x}}(\bar{x}) + q_{\bar{x}}(\bar{x}) = 0 \tag{6.41}$$

and the homogeneous equation associated with this equation is

$$D_{EA}\frac{d^2u_h}{d\bar{x}^2} = 0 \tag{6.42}$$

As we have seen in Chapter 3, the solution to (6.42) is given by (3.41), that is

$$u_h(\bar{x}) = \mathbf{N}\bar{\mathbf{a}}^e \tag{6.43}$$

where according to (3.42) and (3.38)

$$\mathbf{N} = \bar{\mathbf{N}}\mathbf{C}^{-1} = \begin{bmatrix} 1 & \bar{x} \end{bmatrix}\begin{bmatrix} 1 & 0 \\ -\frac{1}{L} & \frac{1}{L} \end{bmatrix} = \begin{bmatrix} 1 - \frac{\bar{x}}{L} & \frac{\bar{x}}{L} \end{bmatrix}; \quad \bar{\mathbf{a}}^e = \begin{bmatrix} \bar{u}_1 \\ \bar{u}_2 \end{bmatrix} \tag{6.44}$$

Differentiation of (6.43) gives

$$\frac{du_h}{d\bar{x}} = \begin{bmatrix} 0 & 1 \end{bmatrix}\begin{bmatrix} 1 & 0 \\ -\frac{1}{L} & \frac{1}{L} \end{bmatrix}\begin{bmatrix} \bar{u}_1 \\ \bar{u}_2 \end{bmatrix} = \begin{bmatrix} -\frac{1}{L} & \frac{1}{L} \end{bmatrix}\begin{bmatrix} \bar{u}_1 \\ \bar{u}_2 \end{bmatrix} \tag{6.45}$$

The general solution to (6.41) can be written as

$$u(\bar{x}) = u_h(\bar{x}) + u_p(\bar{x}) \tag{6.46}$$

The particular solution is obtained by twice integrating (6.41) with (6.40) substituted. For the case with a constant load $q_{\bar{x}}$, we obtain

$$u_p(\bar{x}) = \frac{k_{\bar{x}}}{D_{EA}}\begin{bmatrix} \frac{\bar{x}^2}{2} & \frac{\bar{x}^3}{6} \end{bmatrix}\begin{bmatrix} 1 & 0 \\ -\frac{1}{L} & \frac{1}{L} \end{bmatrix}\begin{bmatrix} \bar{u}_1 \\ \bar{u}_2 \end{bmatrix} + \frac{1}{D_{EA}}\left(-q_{\bar{x}}\frac{\bar{x}^2}{2} + C_1\bar{x} + C_2\right) \tag{6.47}$$

The boundary conditions (3.47) and (3.48) give

$$C_2 = 0 \tag{6.48}$$

$$C_1 = -k_{\bar{x}}\begin{bmatrix} \frac{L}{2} & \frac{L^2}{6} \end{bmatrix}\begin{bmatrix} 1 & 0 \\ -\frac{1}{L} & \frac{1}{L} \end{bmatrix}\begin{bmatrix} \bar{u}_1 \\ \bar{u}_2 \end{bmatrix} + q_{\bar{x}}\frac{L}{2} \tag{6.49}$$

that is

$$u_p(\bar{x}) = \frac{k_{\bar{x}}}{D_{EA}}\begin{bmatrix} \frac{\bar{x}^2-L\bar{x}}{2} & \frac{\bar{x}^3-L^2\bar{x}}{6} \end{bmatrix}\begin{bmatrix} 1 & 0 \\ -\frac{1}{L} & \frac{1}{L} \end{bmatrix}\begin{bmatrix} \bar{u}_1 \\ \bar{u}_2 \end{bmatrix} - \frac{q_{\bar{x}}}{D_{EA}}\left(\frac{\bar{x}^2}{2} - \frac{L\bar{x}}{2}\right) \tag{6.50}$$

Differentiation gives

$$\frac{du_p}{d\bar{x}} = \frac{k_{\bar{x}}}{D_{EA}}\begin{bmatrix} \frac{2\bar{x}-L}{2} & \frac{3\bar{x}^2-L^2}{6} \end{bmatrix}\begin{bmatrix} 1 & 0 \\ -\frac{1}{L} & \frac{1}{L} \end{bmatrix}\begin{bmatrix} \bar{u}_1 \\ \bar{u}_2 \end{bmatrix} - \frac{q_{\bar{x}}}{D_{EA}}\left(\bar{x} - \frac{L}{2}\right) \tag{6.51}$$

or

$$\frac{du_p}{d\bar{x}} = \frac{k_{\bar{x}}}{D_{EA}}\left[-\frac{\bar{x}^2}{2L} + \bar{x} - \frac{L}{3} \quad \frac{\bar{x}^2}{2L} - \frac{L}{6}\right]\begin{bmatrix}\bar{u}_1\\ \bar{u}_2\end{bmatrix} - \frac{q_{\bar{x}}}{D_{EA}}\left(\bar{x} - \frac{L}{2}\right) \tag{6.52}$$

An expression for the normal force $N(\bar{x})$ can be determined by substituting (6.46), (6.45) and (6.52) into (6.35)

$$N(\bar{x}) = D_{EA}\left[-\frac{1}{L} \quad \frac{1}{L}\right]\begin{bmatrix}\bar{u}_1\\ \bar{u}_2\end{bmatrix} + k_{\bar{x}}\left[-\frac{\bar{x}^2}{2L} + \bar{x} - \frac{L}{3} \quad \frac{\bar{x}^2}{2L} - \frac{L}{6}\right]\begin{bmatrix}\bar{u}_1\\ \bar{u}_2\end{bmatrix}$$

$$-q_{\bar{x}}\left(\bar{x} - \frac{L}{2}\right) \tag{6.53}$$

For $\bar{x} = 0$ and $\bar{x} = L$, we have

$$\bar{P}_1 = -N(0) = D_{EA}\left[\frac{1}{L} \quad -\frac{1}{L}\right]\begin{bmatrix}\bar{u}_1\\ \bar{u}_2\end{bmatrix} + k_{\bar{x}}\left[\frac{L}{3} \quad \frac{L}{6}\right]\begin{bmatrix}\bar{u}_1\\ \bar{u}_2\end{bmatrix} - q_{\bar{x}}\frac{L}{2} \tag{6.54}$$

$$\bar{P}_2 = N(L) = D_{EA}\left[-\frac{1}{L} \quad \frac{1}{L}\right]\begin{bmatrix}\bar{u}_1\\ \bar{u}_2\end{bmatrix} + k_{\bar{x}}\left[\frac{L}{6} \quad \frac{L}{3}\right]\begin{bmatrix}\bar{u}_1\\ \bar{u}_2\end{bmatrix} - q_{\bar{x}}\frac{L}{2} \tag{6.55}$$

or

$$\boxed{\bar{\mathbf{K}}^e\,\bar{\mathbf{a}}^e = \bar{\mathbf{f}}^e} \tag{6.56}$$

where

$$\bar{\mathbf{K}}^e = \bar{\mathbf{K}}_0^e + \bar{\mathbf{K}}_s^e; \quad \bar{\mathbf{f}}^e = \bar{\mathbf{f}}_b^e + \bar{\mathbf{f}}_l^e \tag{6.57}$$

$$\bar{\mathbf{K}}_0^e = \frac{D_{EA}}{L}\begin{bmatrix}1 & -1\\ -1 & 1\end{bmatrix}; \quad \bar{\mathbf{K}}_s^e = k_{\bar{x}}L\begin{bmatrix}\frac{1}{3} & \frac{1}{6}\\ \frac{1}{6} & \frac{1}{3}\end{bmatrix} \tag{6.58}$$

$$\bar{\mathbf{a}}^e = \begin{bmatrix}\bar{u}_1\\ \bar{u}_2\end{bmatrix}; \quad \bar{\mathbf{f}}_b^e = \begin{bmatrix}\bar{P}_1\\ \bar{P}_2\end{bmatrix}; \quad \bar{\mathbf{f}}_l^e = \frac{q_{\bar{x}}L}{2}\begin{bmatrix}1\\ 1\end{bmatrix} \tag{6.59}$$

Equation (6.56) formulates the element relations for a bar with axial springs. The element stiffness matrix here consists of two parts: one matrix $\bar{\mathbf{K}}_0^e$ describing the stiffness of a bar without springs and a matrix $\bar{\mathbf{K}}_s^e$ describing the stiffness of the springs.

Example 6.1 Choice of element length for a bar with axial springs

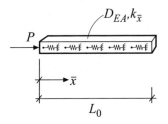

Figure 1 A bar with axial springs

Consider a bar of length L_0 and with stiffness D_{EA} supported by axial elastic springs of stiffness $k_{\bar{x}}$ (Figure 1). At its left end, the bar is loaded by a force P. We choose to study a case where the total stiffness of the springs $k_{\bar{x}}L_0$ is 16 times larger than the stiffness of the bar $\frac{D_{EA}}{L_0}$. The relation between the total spring stiffness $k_{\bar{x}}L_0$, and the stiffness of the bar $\frac{D_{EA}}{L_0}$ then is $\frac{k_{\bar{x}}L_0}{D_{EA}/L_0} = 16$ or $\sqrt{\frac{k_{\bar{x}}}{D_{EA}}}L_0 = 4$. For the bar considered, the displacements $u(\bar{x})$ and the normal forces $N(\bar{x})$ can be determined analytically

$$u(\bar{x}) = \frac{P}{D_{EA}\lambda_{\bar{x}}} \left(\frac{\cosh \lambda_{\bar{x}}L_0}{\sinh \lambda_{\bar{x}}L_0} \cosh \lambda_{\bar{x}}\bar{x} - \sinh \lambda_{\bar{x}}\bar{x} \right) \tag{1}$$

$$N(\bar{x}) = P \left(\frac{\cosh \lambda_{\bar{x}}L_0}{\sinh \lambda_{\bar{x}}L_0} \sinh \lambda_{\bar{x}}\bar{x} - \cosh \lambda_{\bar{x}}\bar{x} \right) \tag{2}$$

where

$$\lambda_{\bar{x}} = \sqrt{\frac{k_{\bar{x}}}{D_{EA}}} \tag{3}$$

The analytically computed displacement and normal force are shown as solid lines in Figures 2 and 3.

In Figures 2 and 3, approximate solutions are shown for one, two and four elements. For four elements, that is with $\sqrt{\frac{k_{\bar{x}}}{D_{EA}}}L = 1$, where L is the length of an element, we have an approximate solution, which is close to the exact one. A rule of thumb is that the error due to the approximate solution (6.46) is small if the stiffness relation $\sqrt{\frac{k_{\bar{x}}}{D_{EA}}}L \leq 1$ is satisfied. Notice that using fewer elements gives a stiffer solution (smaller deformations) and compare this with the discussion on kinematic approximations in Section 5.2.1.

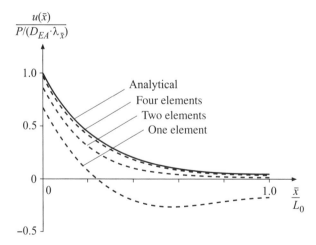

Figure 2   Displacement

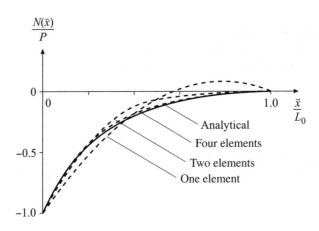

Figure 3 Normal force

6.4 Beam on Elastic Spring Foundation

For beam action, transverse flexible supports may exist continuously along the beam. In a manner corresponding to the one for the bar with axial springs we derive a spring stiffness matrix $\mathbf{K}_s^e$, which describes the stiffness of such a continuously flexible support. The equilibrium equation (4.17) will be affected by the external load exerted by the deformed springs, and similar to that for the bar we introduce an approximate solution to the differential equation.

6.4.1 The Differential Equation for Beam Action with Transverse Springs

We consider a beam resting on a continuous flexible support with the stiffness $k_{\bar{y}}(\bar{x})$ perpendicular to the longitudinal direction of the beam (Figure 6.16). The beam is also loaded by a distributed load $q_{\bar{y}}(\bar{x})$. The derivation starts from the constitutive relation of beam action at the cross-section level (4.12)

$$M(\bar{x}) = D_{EI}(\bar{x})\kappa(\bar{x}) \qquad (6.60)$$

where $M(\bar{x})$ is the bending moment, $\kappa(\bar{x})$ is the curvature and $D_{EI}(\bar{x})$ is the bending stiffness

$$D_{EI}(\bar{x}) = \int_A E(\bar{x}, \bar{y}, \bar{z})\bar{y}^2 \, dA \qquad (6.61)$$

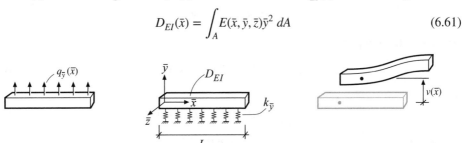

Figure 6.16 The quantities of the differential equation

Kinematics

The flexible support is assumed to give a transverse support along the system line of the beam (Figure 6.17). From (4.16), we have a relation between the deformation measure $v(\bar{x})$ of beam action and the deformation measure $\kappa(\bar{x})$ of the cross-section level

$$\kappa(\bar{x}) = \frac{d^2 v}{d\bar{x}^2} \tag{6.62}$$

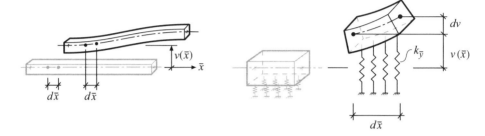

Figure 6.17   Deformed beam element

Equilibrium

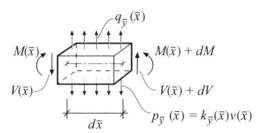

Figure 6.18   Equilibrium for a small part $d\bar{x}$ of a beam

We consider a small part $d\bar{x}$ of a beam with deformed transverse springs (Figure 6.18). When a transverse spring of stiffness $k_{\bar{y}}(\bar{x})$ is displaced by $v(\bar{x})$ a support force

$$p_{\bar{y}}(\bar{x})dx = k_{\bar{y}}(\bar{x})v(\bar{x})dx \tag{6.63}$$

arises. Equilibrium perpendicular to the beam gives

$$-V(\bar{x}) + (V(\bar{x}) + dV) - k_{\bar{y}}(\bar{x})v(\bar{x})d\bar{x} + q_{\bar{y}}(\bar{x})d\bar{x} = 0 \tag{6.64}$$

or

$$\frac{dV}{d\bar{x}} - k_{\bar{y}}(\bar{x})v(\bar{x}) + q_{\bar{y}}(\bar{x}) = 0 \tag{6.65}$$

Moment equilibrium about an axis parallel with the $\bar{z}$-axis at the right end of the beam part in Figure 6.18 gives

$$-M(\bar{x}) + (M(\bar{x}) + dM) + V(\bar{x})d\bar{x} + k_{\bar{y}}(\bar{x})v(\bar{x})d\bar{x}\frac{d\bar{x}}{2} - q_{\bar{y}}(\bar{x})d\bar{x}\frac{d\bar{x}}{2} = 0 \tag{6.66}$$

or

$$\frac{dM}{d\bar{x}} + V(\bar{x}) = 0 \tag{6.67}$$

The two relations (6.65) and (6.67) can be combined to give

$$\frac{d^2M}{d\bar{x}^2} + k_{\bar{y}}(\bar{x})v(\bar{x}) - q_{\bar{y}}(\bar{x}) = 0 \tag{6.68}$$

The Differential Equation for a Beam with Transverse Springs

Substituting the kinematic relation (6.62) into (6.60) gives

$$M(\bar{x}) = D_{EI}(\bar{x})\frac{d^2v}{d\bar{x}^2} \tag{6.69}$$

Substituting (6.69) into the equilibrium relation (6.68) gives

$$\frac{d^2}{d\bar{x}^2}\left(D_{EI}(\bar{x})\frac{d^2v}{d\bar{x}^2}\right) + k_{\bar{y}}(\bar{x})v(\bar{x}) - q_{\bar{y}}(\bar{x}) = 0 \tag{6.70}$$

With the stiffnesses D_{EI} and $k_{\bar{y}}$ constant along the beam, we obtain

$$\boxed{D_{EI}\frac{d^4v}{d\bar{x}^4} + k_{\bar{y}}v(\bar{x}) - q_{\bar{y}}(\bar{x}) = 0} \tag{6.71}$$

6.4.2 Beam Element

The beam element in Figure 6.19 has four displacement degrees of freedom and four nodal forces.

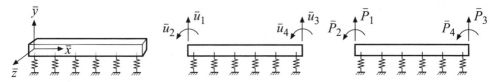

Figure 6.19 A beam element with transverse springs

Solving the Differential Equation

The homogeneous equation associated with (6.71) is

$$D_{EI}\frac{d^4v_h}{d\bar{x}^4} + k_{\bar{y}}(\bar{x})v_h(\bar{x}) = 0 \tag{6.72}$$

Instead of an exact solution[2] we, similar to that for the bar element, introduce an approximate solution. The force exerted by the springs on the beam according to (6.63) is

$$p_{\bar{y}}(\bar{x}) = k_{\bar{y}}(\bar{x})v(\bar{x}) \tag{6.73}$$

---

[2] The solution to (6.72) can be written as $v_h(x) = C_1 \cosh\lambda_{\bar{y}}\bar{x}\cos\lambda_{\bar{y}}\bar{x} + C_2\cosh\lambda_{\bar{y}}\bar{x}\sin\lambda_{\bar{y}}\bar{x} + C_3\sinh\lambda_{\bar{y}}\bar{x}\cos\lambda_{\bar{y}}\bar{x} + C_4\sinh\lambda_{\bar{y}}\bar{x}\sin\lambda_{\bar{y}}\bar{x}$ where $\lambda_{\bar{y}} = \sqrt[4]{\frac{k_{\bar{y}}}{4D_{EI}}}$.

Assuming that the spring stiffness $k_{\bar{y}}$ is constant and that the transverse displacement can be described by a polynomial of degree three, the force in the springs can be expressed as

$$p_{\bar{y}}(\bar{x}) = k_{\bar{y}}\mathbf{N}\bar{\mathbf{a}}^e \qquad (6.74)$$

where $\mathbf{N}\bar{\mathbf{a}}^e$ is the displacement of a beam element without springs and without transverse load, according to Chapter 4. The differential equation (6.71), using the approximation (6.74), can be written as

$$D_{EI}\frac{d^4v}{d\bar{x}^4} + p_{\bar{y}}(\bar{x}) - q_{\bar{y}}(\bar{x}) = 0 \qquad (6.75)$$

and the homogeneous equation associated with this differential equation is

$$D_{EI}\frac{d^4v_h}{d\bar{x}^4} = 0 \qquad (6.76)$$

The solution to (6.76) is given from (4.47) of Chapter 4,

$$v_h(\bar{x}) = \mathbf{N}\bar{\mathbf{a}}^e \qquad (6.77)$$

where

$$\mathbf{N} = \bar{\mathbf{N}}\mathbf{C}^{-1} = \begin{bmatrix} 1 & \bar{x} & \bar{x}^2 & \bar{x}^3 \end{bmatrix} \begin{bmatrix} 1 & 0 & 0 & 0 \\ 0 & 1 & 0 & 0 \\ -\frac{3}{L^2} & -\frac{2}{L} & \frac{3}{L^2} & -\frac{1}{L} \\ \frac{2}{L^3} & \frac{1}{L^2} & -\frac{2}{L^3} & \frac{1}{L^2} \end{bmatrix}; \quad \bar{\mathbf{a}}^e = \begin{bmatrix} \bar{u}_1 \\ \bar{u}_2 \\ \bar{u}_3 \\ \bar{u}_4 \end{bmatrix} \qquad (6.78)$$

Differentiating (6.77) gives

$$\frac{d^2v_h}{d\bar{x}^2} = \mathbf{B}\bar{\mathbf{a}}^e \qquad (6.79)$$

$$\frac{d^3v_h}{d\bar{x}^3} = \frac{d\mathbf{B}}{d\bar{x}}\bar{\mathbf{a}}^e \qquad (6.80)$$

where

$$\mathbf{B} = \frac{d^2\mathbf{N}}{d\bar{x}^2} = \frac{d^2\bar{\mathbf{N}}}{d\bar{x}^2}\mathbf{C}^{-1} = \begin{bmatrix} 0 & 0 & 2 & 6\bar{x} \end{bmatrix} \begin{bmatrix} 1 & 0 & 0 & 0 \\ 0 & 1 & 0 & 0 \\ -\frac{3}{L^2} & -\frac{2}{L} & \frac{3}{L^2} & -\frac{1}{L} \\ \frac{2}{L^3} & \frac{1}{L^2} & -\frac{2}{L^3} & \frac{1}{L^2} \end{bmatrix} \qquad (6.81)$$

$$\frac{d\mathbf{B}}{d\bar{x}} = \frac{d^3\mathbf{N}}{d\bar{x}^3} = \frac{d^3\bar{\mathbf{N}}}{d\bar{x}^3}\mathbf{C}^{-1} = \begin{bmatrix} 0 & 0 & 0 & 6 \end{bmatrix} \begin{bmatrix} 1 & 0 & 0 & 0 \\ 0 & 1 & 0 & 0 \\ -\frac{3}{L^2} & -\frac{2}{L} & \frac{3}{L^2} & -\frac{1}{L} \\ \frac{2}{L^3} & \frac{1}{L^2} & -\frac{2}{L^3} & \frac{1}{L^2} \end{bmatrix} \qquad (6.82)$$

The general solution to (6.75) can be written as

$$v(\bar{x}) = v_h(\bar{x}) + v_p(\bar{x}) \tag{6.83}$$

The particular solution is obtained by integrating (6.75) four times with (6.74) substituted. For the case with a constant load $q_{\bar{y}}$, we obtain

$$v_p(\bar{x}) = -\frac{k_{\bar{y}}}{D_{EI}} \left[\frac{\bar{x}^4}{24} \ \frac{\bar{x}^5}{120} \ \frac{\bar{x}^6}{360} \ \frac{\bar{x}^7}{840} \right] \mathbf{C}^{-1}\bar{\mathbf{a}}^e + \frac{1}{D_{EI}} \left(q_{\bar{y}}\frac{\bar{x}^4}{24} + C_1\frac{\bar{x}^3}{6} + C_2\frac{\bar{x}^2}{2} + C_3\bar{x} + C_4 \right) \tag{6.84}$$

The boundary conditions (4.56)–(4.59) give

$$C_1 = k_{\bar{y}} \left[\frac{L}{2} \ \frac{3L^2}{20} \ \frac{L^3}{15} \ \frac{L^4}{28} \right] \mathbf{C}^{-1}\bar{\mathbf{a}}^e - q_{\bar{y}}\frac{L}{2} \tag{6.85}$$

$$C_2 = -k_{\bar{y}} \left[\frac{L^2}{12} \ \frac{L^3}{30} \ \frac{L^4}{60} \ \frac{L^5}{105} \right] \mathbf{C}^{-1}\bar{\mathbf{a}}^e + q_{\bar{y}}\frac{L^2}{12} \tag{6.86}$$

$$C_3 = 0 \tag{6.87}$$

$$C_4 = 0 \tag{6.88}$$

that is

$$v_p(\bar{x}) = -\frac{k_{\bar{y}}}{D_{EI}} \begin{bmatrix} \frac{\bar{x}^4 - 2L\bar{x}^3 + L^2\bar{x}^2}{24} \\ \frac{\bar{x}^5 - 3L^2\bar{x}^3 + 2L^3\bar{x}^2}{120} \\ \frac{\bar{x}^6 - 4L^3\bar{x}^3 + 3L^4\bar{x}^2}{360} \\ \frac{\bar{x}^7 - 5L^4\bar{x}^3 + 4L^5\bar{x}^2}{840} \end{bmatrix}^T \mathbf{C}^{-1}\bar{\mathbf{a}}^e + \frac{q_{\bar{y}}}{D_{EI}} \left(\frac{\bar{x}^4}{24} - \frac{L\bar{x}^3}{12} + \frac{L^2\bar{x}^2}{24} \right) \tag{6.89}$$

$$\frac{d^2v_p}{d\bar{x}^2} = -\frac{k_{\bar{y}}}{D_{EI}} \begin{bmatrix} \frac{6\bar{x}^2 - 6L\bar{x} + L^2}{12} \\ \frac{10\bar{x}^3 - 9L^2\bar{x} + 2L^3}{60} \\ \frac{5\bar{x}^4 - 4L^3\bar{x} + L^4}{60} \\ \frac{21\bar{x}^5 - 15L^4\bar{x} + 4L^5}{420} \end{bmatrix}^T \mathbf{C}^{-1}\bar{\mathbf{a}}^e + \frac{q_{\bar{y}}}{D_{EI}} \left(\frac{\bar{x}^2}{2} - \frac{L\bar{x}}{2} + \frac{L^2}{12} \right) \tag{6.90}$$

$$\frac{d^3v_p}{d\bar{x}^3} = -\frac{k_{\bar{y}}}{D_{EI}} \begin{bmatrix} \frac{2\bar{x} - L}{2} \\ \frac{10\bar{x}^2 - 3L^2}{20} \\ \frac{5\bar{x}^3 - L^3}{15} \\ \frac{7\bar{x}^4 - L^4}{28} \end{bmatrix}^T \mathbf{C}^{-1}\bar{\mathbf{a}}^e + \frac{q_{\bar{y}}}{D_{EI}} \left(\bar{x} - \frac{L}{2} \right) \tag{6.91}$$

Equation (6.69) together with (6.83) and (6.90) gives an expression for the moment as a function of $\bar{x}$

$$M(\bar{x}) = D_{EI}\mathbf{B}\bar{\mathbf{a}}^e + M_p(\bar{x}) \tag{6.92}$$

where

$$M_p(\bar{x}) = D_{EI}\frac{d^2 v_p}{d\bar{x}^2} \tag{6.93}$$

or

$$M_p(\bar{x}) = -\frac{k_{\bar{y}}}{420L^3}\begin{bmatrix} 42\bar{x}^5 - 105L\bar{x}^4 + 210L^3\bar{x}^2 - 156L^4\bar{x} + 22L^5 \\ 21L\bar{x}^5 - 70L^2\bar{x}^4 + 70L^3\bar{x}^3 - 22L^5\bar{x} + 4L^6 \\ -42\bar{x}^5 + 105L\bar{x}^4 - 54L^4\bar{x} + 13L^5 \\ 21L\bar{x}^5 - 35L^2\bar{x}^4 + 13L^5\bar{x} - 3L^6 \end{bmatrix}^T \begin{bmatrix} \bar{u}_1 \\ \bar{u}_2 \\ \bar{u}_3 \\ \bar{u}_4 \end{bmatrix}$$

$$+ q_{\bar{y}}\left(\frac{\bar{x}^2}{2} - \frac{L\bar{x}}{2} + \frac{L^2}{12}\right) \tag{6.94}$$

Equation (6.69) together with (6.67), (6.83) and (6.91) gives an expression for the shear force as a function of $\bar{x}$

$$V(\bar{x}) = -\frac{dM}{d\bar{x}} = -D_{EI}\frac{d\mathbf{B}}{d\bar{x}}\bar{\mathbf{a}}^e + V_p(\bar{x}) \tag{6.95}$$

where

$$V_p(\bar{x}) = -D_{EI}\frac{d^3 v_p}{d\bar{x}^3} \tag{6.96}$$

or

$$V_p(\bar{x}) = \frac{k_{\bar{y}}}{420L^3}\begin{bmatrix} 210\bar{x}^4 - 420L\bar{x}^3 + 420L^3\bar{x} - 156L^4 \\ 105L\bar{x}^4 - 280L^2\bar{x}^3 + 210L^3\bar{x}^2 - 22L^5 \\ -210\bar{x}^4 + 420L\bar{x}^3 - 54L^4 \\ 105L\bar{x}^4 - 140L^2\bar{x}^3 + 13L^5 \end{bmatrix}^T \begin{bmatrix} \bar{u}_1 \\ \bar{u}_2 \\ \bar{u}_3 \\ \bar{u}_4 \end{bmatrix} - q_{\bar{y}}\left(\bar{x} - \frac{L}{2}\right) \tag{6.97}$$

For $\bar{x} = 0$, we have

$$\bar{P}_1 = -V(0) = D_{EI}\begin{bmatrix} \frac{12}{L^3} \\ \frac{6}{L^2} \\ -\frac{12}{L^3} \\ \frac{6}{L^2} \end{bmatrix}^T \begin{bmatrix} \bar{u}_1 \\ \bar{u}_2 \\ \bar{u}_3 \\ \bar{u}_4 \end{bmatrix} + \frac{k_{\bar{y}}}{420L^3}\begin{bmatrix} 156L^4 \\ 22L^5 \\ 54L^4 \\ -13L^5 \end{bmatrix}^T \begin{bmatrix} \bar{u}_1 \\ \bar{u}_2 \\ \bar{u}_3 \\ \bar{u}_4 \end{bmatrix} - q_{\bar{y}}\frac{L}{2} \tag{6.98}$$

In the same manner, we can determine $\bar{P}_2 = -M(0)$, $\bar{P}_3 = V(L)$ and $\bar{P}_4 = M(L)$ and we get the relation

$$\boxed{\bar{\mathbf{K}}^e\bar{\mathbf{a}}^e = \bar{\mathbf{f}}^e} \tag{6.99}$$

where

$$\bar{\mathbf{K}}^e = \bar{\mathbf{K}}_0^e + \bar{\mathbf{K}}_s^e; \quad \bar{\mathbf{f}}^e = \bar{\mathbf{f}}_b^e + \bar{\mathbf{f}}_l^e \tag{6.100}$$

$$\bar{\mathbf{K}}_0^e = \frac{D_{EI}}{L^3} \begin{bmatrix} 12 & 6L & -12 & 6L \\ 6L & 4L^2 & -6L & 2L^2 \\ -12 & -6L & 12 & -6L \\ 6L & 2L^2 & -6L & 4L^2 \end{bmatrix} \tag{6.101}$$

$$\bar{\mathbf{K}}_s^e = \frac{k_{\bar{y}}L}{420} \begin{bmatrix} 156 & 22L & 54 & -13L \\ 22L & 4L^2 & 13L & -3L^2 \\ 54 & 13L & 156 & -22L \\ -13L & -3L^2 & -22L & 4L^2 \end{bmatrix} \tag{6.102}$$

$$\bar{\mathbf{a}}^e = \begin{bmatrix} \bar{u}_1 \\ \bar{u}_2 \\ \bar{u}_3 \\ \bar{u}_4 \end{bmatrix}; \quad \bar{\mathbf{f}}_b^e = \begin{bmatrix} \bar{P}_1 \\ \bar{P}_2 \\ \bar{P}_3 \\ \bar{P}_4 \end{bmatrix}; \quad \bar{\mathbf{f}}_l^e = q_{\bar{y}} \begin{bmatrix} \frac{L}{2} \\ \frac{L^2}{12} \\ \frac{L}{2} \\ -\frac{L^2}{12} \end{bmatrix} \tag{6.103}$$

In the element relations for a beam on an elastic spring foundation (6.99), the element stiffness matrix consists of two parts: the element stiffness matrix $\bar{\mathbf{K}}_0^e$ for a beam without springs and the matrix $\bar{\mathbf{K}}_s^e$ that describes the stiffness of the springs.

A beam element with six degrees of freedom can be obtained by combining the expressions for the bar element and the beam element. The element relations for the beam with six degrees of freedom can then also be written in the form (6.99), with $\bar{\mathbf{K}}_0^e$ according to (4.87) and

$$\bar{\mathbf{K}}_s^e = \frac{L}{420} \begin{bmatrix} 140k_{\bar{x}} & 0 & 0 & 70k_{\bar{x}} & 0 & 0 \\ 0 & 156k_{\bar{y}} & 22k_{\bar{y}}L & 0 & 54k_{\bar{y}} & -13k_{\bar{y}}L \\ 0 & 22k_{\bar{y}}L & 4k_{\bar{y}}L^2 & 0 & 13k_{\bar{y}}L & -3k_{\bar{y}}L^2 \\ 70k_{\bar{x}} & 0 & 0 & 140k_{\bar{x}} & 0 & 0 \\ 0 & 54k_{\bar{y}} & 13k_{\bar{y}}L & 0 & 156k_{\bar{y}} & -22k_{\bar{y}}L \\ 0 & -13k_{\bar{y}}L & -3k_{\bar{y}}L^2 & 0 & -22k_{\bar{y}}L & 4k_{\bar{y}}L^2 \end{bmatrix} \tag{6.104}$$

Example 6.2 Choice of element length for a beam with transverse springs

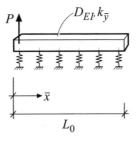

Figure 1 A beam with a transverse flexible support

Consider a beam of length L_0 and with stiffness D_{EI} supported by transverse elastic springs of stiffness $k_{\bar{y}}$ (Figure 1). At its left end, the beam is loaded by a force P. We choose to study a case where the total stiffness of the springs $k_{\bar{y}}L_0$ is 65 536 times larger than the stiffness of the beam $\dfrac{D_{EI}}{L_0^3}$. The relation between the total spring stiffness $k_{\bar{y}}L_0$ and the stiffness of the beam $\dfrac{D_{EI}}{L_0^3}$ then is $\dfrac{k_{\bar{y}}L_0}{D_{EI}/L_0^3} = 65536$ or $\sqrt[4]{\dfrac{k_{\bar{y}}}{D_{EI}}}L_0 = 16$. For the beam considered, displacements $v(\bar{x})$, moment $M(\bar{x})$ and shear forces $V(\bar{x})$ can be determined Analytically

$$v(\bar{x}) = \frac{P}{2D_{EI}\lambda_{\bar{y}}^3}\left(\frac{\cos\lambda_{\bar{y}}L_0\sin\lambda_{\bar{y}}L_0 - \cosh\lambda_{\bar{y}}L_0\sinh\lambda_{\bar{y}}L_0}{\sin^2\lambda_{\bar{y}}L_0 - \sinh^2\lambda_{\bar{y}}L_0}\cosh\lambda_{\bar{y}}\bar{x}\cos\lambda_{\bar{y}}\bar{x} \right.$$

$$+\frac{\sin^2\lambda_{\bar{y}}L_0}{\sin^2\lambda_{\bar{y}}L_0 - \sinh^2\lambda_{\bar{y}}L_0}\cosh\lambda_{\bar{y}}\bar{x}\sin\lambda_{\bar{y}}\bar{x}$$

$$\left. +\frac{\sinh^2\lambda_{\bar{y}}L_0}{\sin^2\lambda_{\bar{y}}L_0 - \sinh^2\lambda_{\bar{y}}L_0}\sinh\lambda_{\bar{y}}\bar{x}\cos\lambda_{\bar{y}}\bar{x} \right) \tag{1}$$

$$M(\bar{x}) = \frac{P}{\lambda_{\bar{y}}}\left(\frac{-\sinh^2\lambda_{\bar{y}}L_0}{\sin^2\lambda_{\bar{y}}L_0 - \sinh^2\lambda_{\bar{y}}L_0}\cosh\lambda_{\bar{y}}\bar{x}\sin\lambda_{\bar{y}}\bar{x} \right.$$

$$+\frac{\sin^2\lambda_{\bar{y}}L_0}{\sin^2\lambda_{\bar{y}}L_0 - \sinh^2\lambda_{\bar{y}}L_0}\sinh\lambda_{\bar{y}}\bar{x}\cos\lambda_{\bar{y}}\bar{x}$$

$$\left. -\frac{\cos\lambda_{\bar{y}}L_0\sin\lambda_{\bar{y}}L_0 - \cosh\lambda_{\bar{y}}L_0\sinh\lambda_{\bar{y}}L_0}{\sin^2\lambda_{\bar{y}}L_0 - \sinh^2\lambda_{\bar{y}}L_0}\sinh\lambda_{\bar{y}}\bar{x}\sin\lambda_{\bar{y}}\bar{x} \right) \tag{2}$$

$$V(\bar{x}) = -P\left(\cosh\lambda_{\bar{y}}\bar{x}\cos\lambda_{\bar{y}}\bar{x} \right.$$

$$-\frac{\cos\lambda_{\bar{y}}L_0\sin\lambda_{\bar{y}}L_0 - \cosh\lambda_{\bar{y}}L_0\sinh\lambda_{\bar{y}}L_0}{\sin^2\lambda_{\bar{y}}L_0 - \sinh^2\lambda_{\bar{y}}L_0}\cosh\lambda_{\bar{y}}\bar{x}\sin\lambda_{\bar{y}}\bar{x}$$

$$-\frac{\cos\lambda_{\bar{y}}L_0\sin\lambda_{\bar{y}}L_0 - \cosh\lambda_{\bar{y}}L_0\sinh\lambda_{\bar{y}}L_0}{\sin^2\lambda_{\bar{y}}L_0 - \sinh^2\lambda_{\bar{y}}L_0}\sinh\lambda_{\bar{y}}\bar{x}\cos\lambda_{\bar{y}}\bar{x}$$

$$\left. -\frac{\sinh^2\lambda_{\bar{y}}L_0 + \sin^2\lambda_{\bar{y}}L_0}{\sin^2\lambda_{\bar{y}}L_0 - \sinh^2\lambda_{\bar{y}}L_0}\sinh\lambda_{\bar{y}}\bar{x}\sin\lambda_{\bar{y}}\bar{x} \right) \tag{3}$$

where

$$\lambda_{\bar{y}} = \sqrt[4]{\frac{k_{\bar{y}}}{4D_{EI}}} \tag{4}$$

Analytically computed displacement, moment and shear force are shown with solid lines in Figures 2–4.

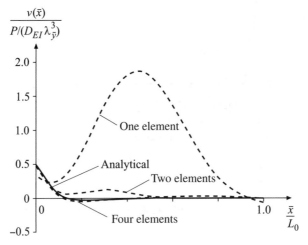

Figure 2   Displacement

In Figures 2–4, approximate solutions are shown for one, two and four elements. For four elements, that is with $\sqrt[4]{k_{\bar{y}}/D_{EI}}\ L = 4$, where L is the length of an element, we have an approximate solution which is close to the exact one. A rule of thumb is that the error due to the approximate solution (6.83) is small if the stiffness relation $\sqrt[4]{k_{\bar{y}}/D_{EI}}L \le 4$ is satisfied. Notice that using fewer elements gives a stiffer solution.

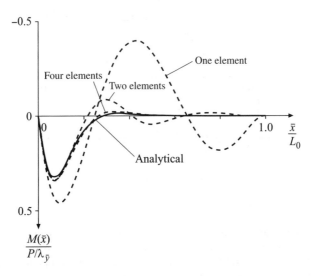

Figure 3   Moment

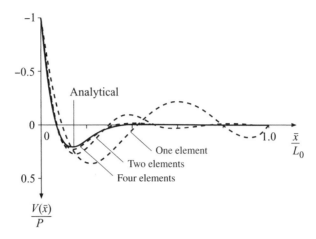

Figure 4   Shear force

Exercises

6.1 Consider the beam in Exercise 4.11 loaded with a uniformly distributed downwards directed load $q = 40.0$ kN/m along the left part of the beam. Replace the mid-support with a spring with stiffness $k = 4.0$ MN/m. Determine the vertical displacement of the flexible mid-support and the support force (the spring force). Also, draw the moment diagram of the beam.

6.2

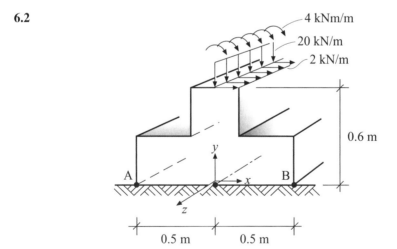

A long base plate is supported by a flexible support with the spring stiffnesses $k_x = 10$ MN/m^3 and $k_y = 10$ MN/m^3. The base plate is loaded by distributed loads $q_x = 2$ kN/m, $q_y = -20$ kN/m and $q_\omega = -4$ kNm/m according to the figure. Determine the displacements horizontally and vertically for points A and B at the base of the plate.

6.3

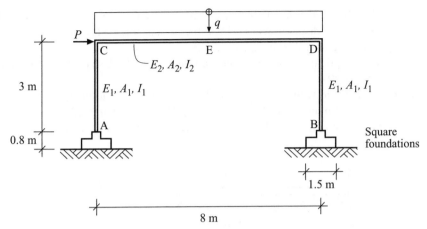

A frame with dimensions according to the figure and with the properties $E_1 = 210$ GPa, $A_1 = 3.0 \times 10^{-3}$ m^2, $I_1 = 1.0 \times 10^{-5}$ m^4, $E_2 = 210$ GPa, $A_2 = 5.0 \times 10^{-3}$ m^2 and $I_2 = 2.0 \times 10^{-5}$ m^4 is fixed to two base plates and loaded by a point load $P = 10$ kN and a uniformly distributed load $q = 10$ kN/m. Compute the horizontal displacement at C and draw a moment diagram for the frame assuming

(a) the support at the foundation to be rigid.
(b) the support at the foundation to be flexible and have spring stiffnesses $k_{x'} = 20$ MN/m^3 and $k_{y'} = 10$ MN/m^3.

6.4

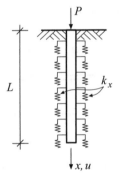

The pile in the figure has axial stiffness D_{EA} and is supported by the surrounding soil. The support of the soil is modelled as distributed springs with stiffness k_x along the pile. The pile is loaded by a force P at its upper end. The bottom end is free. Using the analytical expressions for displacement and normal force that are given in Example 6.1, draw the spring force distribution along the pile for different values of $\lambda_x L = \sqrt{\dfrac{k_x}{D_{EA}}} L$

(a) $\lambda_x L = 1$, which means that the spring stiffness is small compared with the axial stiffness of the pile.
(b) $\lambda_x L = 5$, which means that the spring stiffness is large compared with the axial stiffness of the pile.

For $E = 1000$ MPa, $A = 0.04$ m^2, $L = 20$ m, $k_x = 6.0$ MN/m^2 and $P = 0.1$ MN:

(c) Determine the displacement $u(x)$ and the normal force $N(x)$ along the pile. Draw by hand diagrams showing $u(x)$ and $N(x)$.

(d) Model the pile using one element according to (6.56). Use CALFEM to compute $u(0)$ and $u(20)$.

(e) model the pile with two and four elements and compute $u(0)$, $u(10)$ and $u(20)$. Compare with the results from (c) and (d). Comment on the differences.

(f) compute, using the model with four elements, the normal force at a depth of 10 m, $N(10)$.

6.5

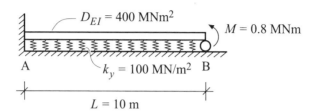

A 10.0 m long beam with the stiffness $D_{EI} = 400.0$ MN m^2 has a flexible support with the spring stiffness $k_y = 100.0$ MN/m^2. At its left end, the beam is fixed. At its right end, the beam has a roller support and is loaded by a moment $M = 0.8$ MN m.

(a) Model the beam with one element and determine the rotation at B.

(b) Determine using the condition $L \le 4\sqrt[4]{\dfrac{D_{EI}}{k_{\bar{y}}}}$ an appropriate element division and determine using CALFEM the rotation at B.

6.6

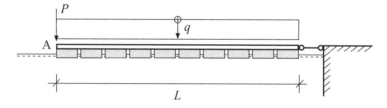

The pontoon system in the figure consists of a beam with the length $L = 50$ m resting on pontoons of width $b = 4.0$ m perpendicular to the plane shown. The bending stiffness of the beam is $D_{EI} = 4000$ kNm^2 and the pontoons are handled as a flexible support.

(a) Use 10.0 kN/m^3 as the specific weight of water and show by using Archimedes's principle that the spring stiffnesses of the pontoons are $k_y = 40$ kN/m^2.

(b) For a downwards directed point load $P = 60$ kN at A and a downwards directed uniformly distributed load $q = 3$ kN/m along the entire beam, compute the distribution of displacement and moment. Compare the results from element divisions of two and four elements with each other.

7

Three-Dimensional Structures

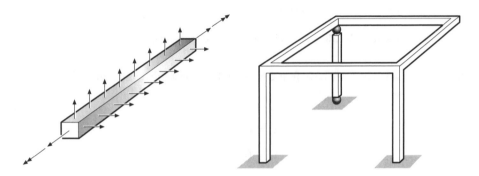

Figure 7.1   Three-dimensional beam and three-dimensional frame

A three-dimensional structure (Figure 7.1) carries load in four different manners:

- bar action
- beam action in the local $\bar{x}\bar{y}$-plane
- beam action in the local $\bar{x}\bar{z}$-plane
- torsional action.

These can, with appropriate choices of reference axes, be formulated as four independent differential equations (Figure 7.2). With constant stiffnesses D_{EA}, $D_{EI_{\bar{z}}}$, $D_{EI_{\bar{y}}}$ and D_{GK}, we have the four differential equations

$$D_{EA}\, \frac{d^2 u}{d\bar{x}^2} + q_{\bar{x}}(\bar{x}) = 0 \tag{7.1}$$

$$D_{EI_{\bar{z}}}\, \frac{d^4 v}{d\bar{x}^4} - q_{\bar{y}}(\bar{x}) = 0 \tag{7.2}$$

$$D_{EI_{\bar{y}}}\, \frac{d^4 w}{d\bar{x}^4} - q_{\bar{z}}(\bar{x}) = 0 \tag{7.3}$$

$$D_{GK}\, \frac{d^2 \varphi}{d\bar{x}^2} + q_{\omega}(\bar{x}) = 0 \tag{7.4}$$

Structural Mechanics: Modelling and Analysis of Frames and Trusses, First Edition.
Karl-Gunnar Olsson and Ola Dahlblom.
© 2016 John Wiley & Sons, Ltd. Published 2016 by John Wiley & Sons, Ltd.

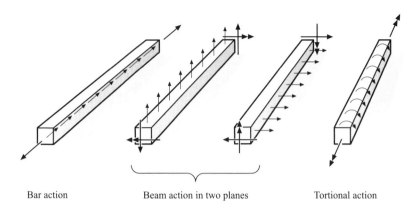

| Bar action | Beam action in two planes | Tortional action |

Figure 7.2 Modes of action for a three-dimensional beam element

At the cross-section level, we have four corresponding constitutive relations

$$N(\bar{x}) = D_{EA}(\bar{x})\,\varepsilon(\bar{x}) \tag{7.5}$$

$$M_{\bar{z}}(\bar{x}) = D_{EI_{\bar{z}}}(\bar{x})\,\kappa_{\bar{z}}(\bar{x}) \tag{7.6}$$

$$M_{\bar{y}}(\bar{x}) = D_{EI_{\bar{y}}}(\bar{x})\,\kappa_{\bar{y}}(\bar{x}) \tag{7.7}$$

$$T(\bar{x}) = D_{GK}(\bar{x})\,\theta(\bar{x}) \tag{7.8}$$

From each differential equation, we can derive an element relation in local directions. If we combine these relations, we obtain a three-dimensional beam element with 12 degrees of freedom (Figure 7.3). Here, we have introduced double arrows to represent rotations and moments. Figure 7.4 shows how a double arrow describes rotation or torque about an arbitrary axis. The three-dimensional beam element can then be transformed from the local coordinate system to a global one and be assembled into a general three-dimensional structure (Figure 7.5).

An important special case among three-dimensional structures is the three-dimensional truss. By allowing only bar action and excluding all other modes of action, we obtain a

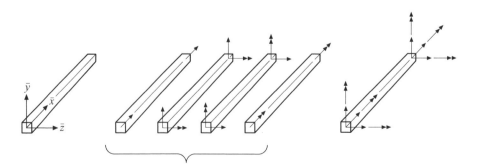

Figure 7.3 A three-dimensional beam element

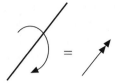

Figure 7.4   Different representations of positive rotation and positive torque

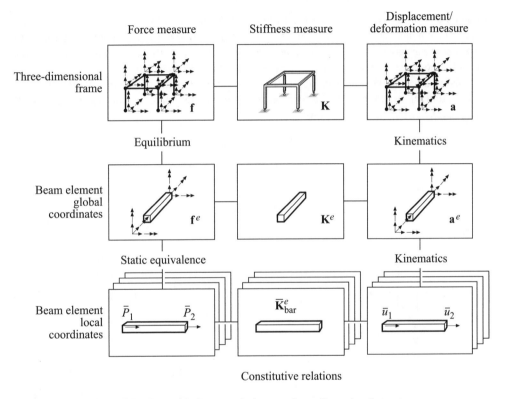

Figure 7.5   From the element relation to a three-dimensional structure.

structure with elements that carry load the most efficiently. For a truss, as a computational model, the external loads can only be applied as axial loads along the bars and as concentrated loads at the joints. Moreover, the joints are assumed to be frictionless hinges. By the consistent use of frictionless hinges throughout the entire structure, the elements in the truss will not be exposed to bending or torsion. Section 7.1 shows how the local bar element from Chapter 3 can be transformed into a global three-dimensional coordinate system, and Section 7.2 shows how the element is assembled into a three-dimensional truss using the established method from the previous chapters.

From Chapters 3 and 4, we have bar action (7.5) and beam action in the $\bar{x}\bar{y}$-plane (7.6) and by changing the local coordinates we have beam action in the $\bar{x}\bar{z}$-plane as well (7.7). To fully establish the three-dimensional beam, we must add torsional action. In Section 7.3, a constitutive relation at the cross-section level (7.8) and the differential equation for torsional action (7.4) are derived. In Section 7.4, the element relations for the three-dimensional beam element are formulated. Finally, Section 7.5 shows how a computational model for a three-dimensional frame is established.

7.1 Three-Dimensional Bar Element

To be able to place the bar element (3.60) in a three-dimensional truss, we have to use forces and displacements in the global coordinate system (x, y, z) of the truss. Here, the displacements of the bar are described by the displacement components u_1, u_2, u_3, u_4, u_5 and u_6 and its nodal forces by the force components P_1, P_2, P_3, P_4, P_5 and P_6 (Figure 7.6).

By using direction cosines, analogous to (3.73) and (3.74), the displacements $\bar{u}_1$ and $\bar{u}_2$ in the longitudinal direction of the bar can be expressed in the global displacement components u_1, u_2, u_3, u_4 u_5 and u_6 according to

$$\bar{u}_1 = n_{x\bar{x}}u_1 + n_{y\bar{x}}u_2 + n_{z\bar{x}}u_3 \tag{7.9}$$

$$\bar{u}_2 = n_{x\bar{x}}u_4 + n_{y\bar{x}}u_5 + n_{z\bar{x}}u_6 \tag{7.10}$$

or in matrix form

$$\bar{\mathbf{a}}^e = \mathbf{G}\mathbf{a}^e \tag{7.11}$$

where

$$\bar{\mathbf{a}}^e = \begin{bmatrix} \bar{u}_1 \\ \bar{u}_2 \end{bmatrix} ; \quad \mathbf{G} = \begin{bmatrix} n_{x\bar{x}} & n_{y\bar{x}} & n_{z\bar{x}} & 0 & 0 & 0 \\ 0 & 0 & 0 & n_{x\bar{x}} & n_{y\bar{x}} & n_{z\bar{x}} \end{bmatrix} ; \quad \mathbf{a}^e = \begin{bmatrix} u_1 \\ u_2 \\ u_3 \\ u_4 \\ u_5 \\ u_6 \end{bmatrix} \tag{7.12}$$

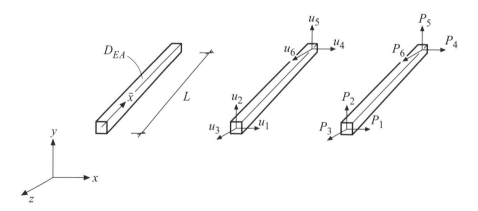

Figure 7.6 Three-dimensional bar element with local and global degrees of freedom

The components P_1, P_2, P_3, P_4, P_5 and P_6 of the nodal forces $\bar{P}_1$ and $\bar{P}_2$, analogous to (3.77)–(3.80), can be written as

$$P_1 = n_{x\bar{x}}\bar{P}_1 \tag{7.13}$$

$$P_2 = n_{y\bar{x}}\bar{P}_1 \tag{7.14}$$

$$P_3 = n_{z\bar{x}}\bar{P}_1 \tag{7.15}$$

$$P_4 = n_{x\bar{x}}\bar{P}_2 \tag{7.16}$$

$$P_5 = n_{y\bar{x}}\bar{P}_2 \tag{7.17}$$

$$P_6 = n_{z\bar{x}}\bar{P}_2 \tag{7.18}$$

or in matrix form

$$\mathbf{f}_b^e = \mathbf{G}^T \bar{\mathbf{f}}_b^e \tag{7.19}$$

where

$$\mathbf{f}_b^e = \begin{bmatrix} P_1 \\ P_2 \\ P_3 \\ P_4 \\ P_5 \\ P_6 \end{bmatrix}; \quad \mathbf{G}^T = \begin{bmatrix} n_{x\bar{x}} & 0 \\ n_{y\bar{x}} & 0 \\ n_{z\bar{x}} & 0 \\ 0 & n_{x\bar{x}} \\ 0 & n_{y\bar{x}} \\ 0 & n_{z\bar{x}} \end{bmatrix}; \quad \bar{\mathbf{f}}_b^e = \begin{bmatrix} \bar{P}_1 \\ \bar{P}_2 \end{bmatrix} \tag{7.20}$$

The relation between equivalent nodal loads $\mathbf{f}_l^e$ in a global system and equivalent nodal loads $\bar{\mathbf{f}}_l^e$ in a local system can in the corresponding manner be written as

$$\mathbf{f}_l^e = \mathbf{G}^T \bar{\mathbf{f}}_l^e \tag{7.21}$$

An element relation with quantities expressed in the directions of the global coordinate system is obtained if the transformations (7.19), (7.11) and (7.21) are substituted into the element relation (3.58)

$$\boxed{\mathbf{f}_b^e = \mathbf{K}^e \mathbf{a}^e - \mathbf{f}_l^e} \tag{7.22}$$

where

$$\mathbf{K}^e = \mathbf{G}^T \bar{\mathbf{K}}^e \mathbf{G} \tag{7.23}$$

When the matrix multiplication in Equation (7.23) is performed, the components of the element stiffness matrix $\mathbf{K}^e$ are obtained for a bar element in the global three-dimensional system

$$\mathbf{K}^e = \frac{D_{EA}}{L}\begin{bmatrix} \mathbf{C} & -\mathbf{C} \\ -\mathbf{C} & \mathbf{C} \end{bmatrix}; \quad \mathbf{C} = \begin{bmatrix} n_{x\bar{x}}n_{x\bar{x}} & n_{x\bar{x}}n_{y\bar{x}} & n_{x\bar{x}}n_{z\bar{x}} \\ n_{y\bar{x}}n_{x\bar{x}} & n_{y\bar{x}}n_{y\bar{x}} & n_{y\bar{x}}n_{z\bar{x}} \\ n_{z\bar{x}}n_{x\bar{x}} & n_{z\bar{x}}n_{y\bar{x}} & n_{z\bar{x}}n_{z\bar{x}} \end{bmatrix} \tag{7.24}$$

7.2   Three-Dimensional Trusses

To define a three-dimensional truss model, we define a set of global displacement degrees of freedom and gather them in a global displacement vector $\mathbf{a}$. From the element relations for the separate bar elements, we have element stiffness matrices $\mathbf{K}^e$ and element load vectors $\mathbf{f}^e$ expressed by six local displacement degrees of freedom in the element displacement vector $\mathbf{a}^e$. Based on compatibility conditions and equilibrium conditions, a stiffness matrix $\mathbf{K}$ and a load vector $\mathbf{f}_l$ are established in the previously described manner. When considering the present boundary conditions, the displacements and support forces can be computed and after that the normal force distribution can be determined.

Example 7.1 Truss

The truss in Figure 1 consists of four bars with the elastic modulus $E = 200.0$ GPa and with the cross-sectional areas $A_1 = 6.0 \times 10^{-4}$ m^2, $A_2 = 3.0 \times 10^{-4}$ m^2, $A_3 = 4.0 \times 10^{-4}$ m^2 and $A_4 = 10.0 \times 10^{-4}$ m^2. The truss is fixed at four joints and loaded with a force $P = 80$ kN directed downwards in the fifth.

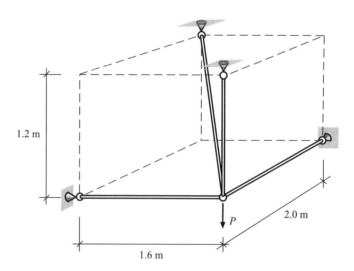

Figure 1   A three-dimensional truss consisting of four bars

Computational model

The truss model consists of four bar elements, denoted as 1, 2, 3 and 4 (Figure 2). The model has the displacement degrees of freedom a_1–a_{15}. The downwards directed force acting at degree of freedom 5 implies that $f_5 = -80$ kN. In degrees of freedom $a_1, a_2, a_3, a_7, a_8, a_9$, $a_{10}, a_{11}, a_{12}, a_{13}, a_{14}$ and a_{15}, the displacement is prescribed to be zero.

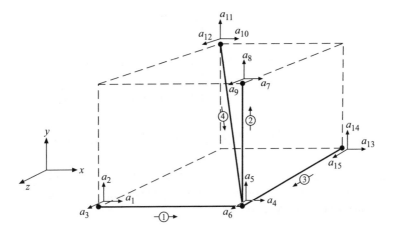

Figure 2   Computational model

Element matrices

For each bar element, an element relation $\mathbf{K}^e \mathbf{a}^e = \mathbf{f}^e_b$ can be established. The element stiffness matrices $\mathbf{K}^e$ for the four elements are given by (7.24).

Element 1:

$$\frac{EA_1}{L_1} = \frac{200.0 \times 10^9 \cdot 6.0 \times 10^{-4}}{1.6} = 75.0 \times 10^6 \qquad (1)$$

The local $\bar{x}$-axis coincides with the global x-axis. The direction cosines for the angles between these are, therefore, $n_{x\bar{x}} = \cos(x, \bar{x}) = 1$, $n_{y\bar{x}} = \cos(y, \bar{x}) = 0$ and $n_{z\bar{x}} = \cos(z, \bar{x}) = 0$, respectively, which give us the element stiffness matrix

$$\mathbf{K}^1 = 75.0 \times 10^6 \begin{bmatrix} 1 & 0 & 0 & -1 & 0 & 0 \\ 0 & 0 & 0 & 0 & 0 & 0 \\ 0 & 0 & 0 & 0 & 0 & 0 \\ -1 & 0 & 0 & 1 & 0 & 0 \\ 0 & 0 & 0 & 0 & 0 & 0 \\ 0 & 0 & 0 & 0 & 0 & 0 \end{bmatrix} \qquad (2)$$

Element 2:

$$\frac{EA_2}{L_2} = \frac{200.0 \times 10^9 \cdot 3.0 \times 10^{-4}}{1.2} = 50.0 \times 10^6 \qquad (3)$$

The local $\bar{x}$-axis coincides with the global y-axis. The direction cosines are, therefore, $n_{x\bar{x}} = \cos(x, \bar{x}) = 0$, $n_{y\bar{x}} = \cos(y, \bar{x}) = 1$ and $n_{z\bar{x}} = \cos(y, \bar{x}) = 0$, respectively, which give us the element stiffness matrix

$$\mathbf{K}^2 = 50.0 \times 10^6 \begin{bmatrix} 0 & 0 & 0 & 0 & 0 & 0 \\ 0 & 1 & 0 & 0 & -1 & 0 \\ 0 & 0 & 0 & 0 & 0 & 0 \\ 0 & 0 & 0 & 0 & 0 & 0 \\ 0 & -1 & 0 & 0 & 1 & 0 \\ 0 & 0 & 0 & 0 & 0 & 0 \end{bmatrix} \tag{4}$$

Element 3:

$$\frac{EA_3}{L_3} = \frac{200.0 \times 10^9 \cdot 4.0 \times 10^{-4}}{2.0} = 40.0 \times 10^6 \tag{5}$$

The local $\bar{x}$-axis coincides with the global z-axis. The direction cosines are, therefore, $n_{x\bar{x}} = \cos(x, \bar{x}) = 0$, $n_{y\bar{x}} = \cos(y, \bar{x}) = 0$ and $n_{z\bar{x}} = \cos(z, \bar{x}) = 1$, respectively, which give us the element stiffness matrix

$$\mathbf{K}^3 = 40.0 \times 10^6 \begin{bmatrix} 0 & 0 & 0 & 0 & 0 & 0 \\ 0 & 0 & 0 & 0 & 0 & 0 \\ 0 & 0 & 1 & 0 & 0 & -1 \\ 0 & 0 & 0 & 0 & 0 & 0 \\ 0 & 0 & 0 & 0 & 0 & 0 \\ 0 & 0 & 1 & 0 & 0 & -1 \end{bmatrix} \tag{6}$$

Element 4:

$$\frac{EA_4}{L_4} = \frac{200.0 \times 10^9 \cdot 10.0 \times 10^{-4}}{2.828} = 70.71 \times 10^6 \tag{7}$$

The element length is $L_4 = \sqrt{1.6^2 + 1.2^2 + 2.0^2} = 2.828$. The direction cosines are $n_{x\bar{x}} = \cos(x, \bar{x}) = 1.6/2.828 = 0.566$, $n_{y\bar{x}} = \cos(y, \bar{x}) = -1.2/2.828 = -0.424$ and $n_{z\bar{x}} = \cos(z, \bar{x}) = 2.0/2.828 = 0.707$, respectively. These give us the element stiffness matrix

$$\mathbf{K}^4 = 70.71 \times 10^6 \begin{bmatrix} 0.32 & -0.24 & 0.40 & -0.32 & 0.24 & -0.40 \\ -0.24 & 0.18 & -0.30 & 0.24 & -0.18 & 0.30 \\ 0.40 & -0.30 & 0.50 & -0.40 & 0.30 & -0.50 \\ -0.32 & 0.24 & -0.40 & 0.32 & -0.24 & 0.40 \\ 0.24 & -0.18 & 0.30 & -0.24 & 0.18 & -0.30 \\ -0.40 & 0.30 & -0.50 & 0.40 & -0.30 & 0.50 \end{bmatrix} \tag{8}$$

Compatibility conditions

The topology matrix shows how the local degrees of freedom for Elements 1–4 correspond to the global ones,

$$\text{topology} = \begin{bmatrix} 1 & 1 & 2 & 3 & 4 & 5 & 6 \\ 2 & 4 & 5 & 6 & 7 & 8 & 9 \\ 3 & 13 & 14 & 15 & 4 & 5 & 6 \\ 4 & 10 & 11 & 12 & 4 & 5 & 6 \end{bmatrix} \tag{9}$$

Assembling

Assembling the element stiffness matrices according to the topology matrix results in

$$
\mathbf{K} = \begin{bmatrix}
75.00 & 0 & 0 & -75.00 & 0 & 0 & 0 & 0 & 0 & 0 & 0 & 0 & 0 & 0 & 0 \\
0 & 0 & 0 & 0 & 0 & 0 & 0 & 0 & 0 & 0 & 0 & 0 & 0 & 0 & 0 \\
0 & 0 & 0 & 0 & 0 & 0 & 0 & 0 & 0 & 0 & 0 & 0 & 0 & 0 & 0 \\
-75.00 & 0 & 0 & 97.63 & -16.97 & 28.28 & 0 & 0 & 0 & -22.63 & 16.97 & -28.28 & 0 & 0 & 0 \\
0 & 0 & 0 & -16.97 & 62.73 & -21.21 & 0 & -50.00 & 0 & 16.97 & -12.73 & 21.21 & 0 & 0 & 0 \\
0 & 0 & 0 & 28.28 & -21.21 & 75.36 & 0 & 0 & 0 & -28.28 & 21.21 & -35.36 & 0 & 0 & -40.00 \\
0 & 0 & 0 & 0 & 0 & 0 & 0 & 0 & 0 & 0 & 0 & 0 & 0 & 0 & 0 \\
0 & 0 & 0 & 0 & -50.00 & 0 & 0 & 50.00 & 0 & 0 & 0 & 0 & 0 & 0 & 0 \\
0 & 0 & 0 & 0 & 0 & 0 & 0 & 0 & 0 & 0 & 0 & 0 & 0 & 0 & 0 \\
0 & 0 & 0 & -22.63 & 16.97 & -28.28 & 0 & 0 & 0 & 22.63 & -16.97 & 28.28 & 0 & 0 & 0 \\
0 & 0 & 0 & 16.97 & -12.73 & 21.21 & 0 & 0 & 0 & -16.97 & 12.73 & -21.21 & 0 & 0 & 0 \\
0 & 0 & 0 & -28.28 & 21.21 & -35.36 & 0 & 0 & 0 & 28.28 & -21.21 & 35.35 & 0 & 0 & 0 \\
0 & 0 & 0 & 0 & 0 & 0 & 0 & 0 & 0 & 0 & 0 & 0 & 0 & 0 & 0 \\
0 & 0 & 0 & 0 & 0 & 0 & 0 & 0 & 0 & 0 & 0 & 0 & 0 & 0 & 0 \\
0 & 0 & 0 & 0 & 0 & -40.00 & 0 & 0 & 0 & 0 & 0 & 0 & 0 & 0 & 40.00
\end{bmatrix} 10^6
\tag{10}
$$

Boundary conditions and nodal loads

With the present load and boundary conditions (bc) we have

$$
\mathbf{f}_l = \begin{bmatrix} 0 \\ 0 \\ 0 \\ 0 \\ -80 \\ 0 \\ 0 \\ 0 \\ 0 \\ 0 \\ 0 \\ 0 \\ 0 \\ 0 \\ 0 \end{bmatrix} 10^3; \quad
\mathrm{bc} = \begin{bmatrix} 1 & 0 \\ 2 & 0 \\ 3 & 0 \\ 7 & 0 \\ 8 & 0 \\ 9 & 0 \\ 10 & 0 \\ 11 & 0 \\ 12 & 0 \\ 13 & 0 \\ 14 & 0 \\ 15 & 0 \end{bmatrix}; \quad
\mathbf{a} = \begin{bmatrix} 0 \\ 0 \\ 0 \\ a_4 \\ a_5 \\ a_6 \\ 0 \\ 0 \\ 0 \\ 0 \\ 0 \\ 0 \\ 0 \\ 0 \\ 0 \end{bmatrix}; \quad
\mathbf{f}_b = \begin{bmatrix} f_{b,1} \\ f_{b,2} \\ f_{b,3} \\ 0 \\ 0 \\ 0 \\ f_{b,7} \\ f_{b,8} \\ f_{b,9} \\ f_{b,10} \\ f_{b,11} \\ f_{b,12} \\ f_{b,13} \\ f_{b,14} \\ f_{b,15} \end{bmatrix}
\tag{11}
$$

where the load vector $\mathbf{f}_l$ contains known external loads, the boundary condition matrix specifies prescribed displacements, the displacement vector $\mathbf{a}$ contains both known and unknown displacements and the boundary force vector $\mathbf{f}_b$ contains unknown support (boundary) forces.

Solving the system of equations

Solving the system of equations gives

$$\begin{bmatrix} a_4 \\ a_5 \\ a_6 \end{bmatrix} = \begin{bmatrix} -0.1484 \\ -1.4331 \\ -0.3477 \end{bmatrix} 10^{-3}; \qquad \begin{bmatrix} f_{b,1} \\ f_{b,2} \\ f_{b,3} \\ f_{b,7} \\ f_{b,8} \\ f_{b,9} \\ f_{b,10} \\ f_{b,11} \\ f_{b,12} \\ f_{b,13} \\ f_{b,14} \\ f_{b,15} \end{bmatrix} = \begin{bmatrix} 11.13 \\ 0 \\ 0 \\ 0 \\ 71.65 \\ 0 \\ -11.13 \\ 8.35 \\ -13.91 \\ 0 \\ 0 \\ 13.91 \end{bmatrix} 10^3 \qquad (12)$$

This implies that the node is displaced 0.15 mm in the negative x-direction, 1.43 mm in the negative y-direction and 0.35 mm in the negative z-direction. The external load and the computed support forces are illustrated in Figure 3. We can conclude that the sum of

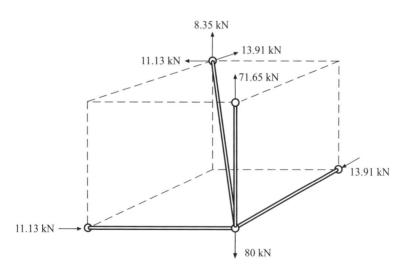

8.35 kN

13.91 kN

11.13 kN

71.65 kN

13.91 kN

11.13 kN

80 kN

Figure 3 External load and computed support forces

the external forces in the x-, y- and z-direction, respectively, is equal to zero. Thus, the equilibrium is satisfied.

Internal forces

Using (7.11), the displacements can be expressed in the local coordinate system of the element. For Element 1, we obtain:

$$\bar{\mathbf{a}}^1 = \mathbf{G}\mathbf{a}^1 = \begin{bmatrix} 1 & 0 & 0 & 0 & 0 & 0 \\ 0 & 0 & 0 & 1 & 0 & 0 \end{bmatrix} \begin{bmatrix} 0 \\ 0 \\ 0 \\ -0.1484 \\ -1.4331 \\ -0.3477 \end{bmatrix} 10^{-3} = \begin{bmatrix} 0 \\ -0.1484 \end{bmatrix} 10^{-3} \quad (13)$$

With the local displacements known, the normal force in Element 1 can be computed using (3.52)

$$N^{(1)} = EA_1\mathbf{B}\bar{\mathbf{a}}^1$$

$$= 200.0 \times 10^9 \times 6.0 \times 10^{-4} \frac{1}{1.6} \begin{bmatrix} -1 & 1 \end{bmatrix} \begin{bmatrix} 0 \\ -0.1484 \end{bmatrix} 10^{-3}$$

$$= -11.13 \times 10^3 \quad (14)$$

For Elements 2–4 we have, with the corresponding computations

$$N^{(2)} = 71.65 \times 10^3 \quad (15)$$

$$N^{(3)} = -13.91 \times 10^3 \quad (16)$$

$$N^{(4)} = 19.67 \times 10^3 \quad (17)$$

These results imply that the normal forces in the four elements are $-11.13, 71.65, -13.91$ and 19.67 kN, respectively; see Figure 4.

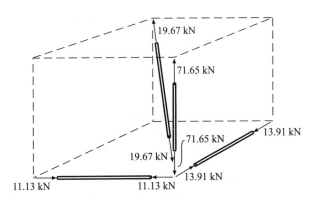

Figure 4   Normal forces in the bars

7.3   The Differential Equation for Torsional Action

Torsional action is expressed as a relation between torsional loading and the cross-section rotations that arise in a twisted beam (Figure 7.7). In the same manner as for bar action and beam action, the basis is a constitutive relation that relates strain to stress, but instead of normal strain and normal stress, shear strain and shear stress arise. Via a kinematic condition and force equivalence, we can derive the constitutive relation of the cross-section. Via another constitutive relation and with equilibrium, we finally reach the differential equation for torsional action.

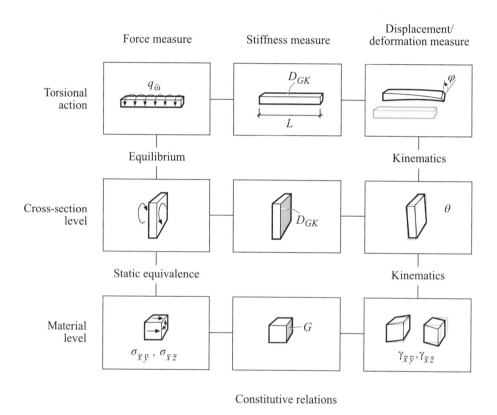

Figure 7.7   From material to torsional action

7.3.1   Definitions

The twisted beam has its main extension along the $\bar{x}$-axis of a local coordinate system $(\bar{x}, \bar{y}, \bar{z})$. The quantities of torsional action are illustrated in Figure 7.8. Shear stresses $\sigma_{\bar{x}\bar{y}}(\bar{x}, \bar{y}, \bar{z})$ and $\sigma_{\bar{x}\bar{z}}(\bar{x}, \bar{y}, \bar{z})$ act on the material and give rise to shear strains $\gamma_{\bar{x}\bar{y}}(\bar{x}, \bar{y}, \bar{z})$ and $\gamma_{\bar{x}\bar{z}}(\bar{x}, \bar{y}, \bar{z})$. The material has a shear stiffness $G(\bar{x}, \bar{y}, \bar{z})$. At the cross-section level, the quantities of the material are summarised to the generalised measures *torque* $T(\bar{x})$ (i.e. torsional moment or twisting moment) and *rate of twist* $\theta(\bar{x})$. Along the system line of the beam, the local $\bar{x}$-axis, a distributed torsional moment $q_{\bar{\omega}}(\bar{x})$ acts and gives rise to a twist angle $\varphi(\bar{x})$.

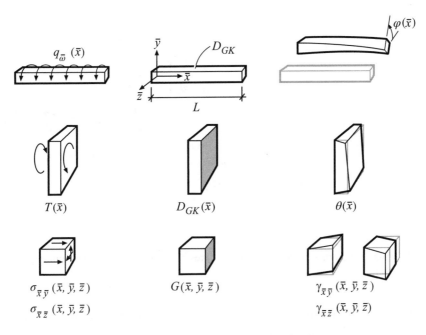

Figure 7.8   The quantities of torsional action

The following derivation of the differential equation for torsional action assumes a circular cross-section, but the differential equation can with some modifications be applied also to other cross-sectional shapes.

7.3.2   The Material Level

Strain

A material point can be understood as three fibres in space, perpendicular to each other; cf. Section 3.1.2. The deformation can be divided into two parts: the relative axial deformation of the fibres and the relative angular deformation of the fibres. So far, only the deformation measure for the relative length change of an axial fibre has been of interest, but for torsion it is instead the relative angular deformation that arises. We consider the two lines AB and AC in a plane parallel to the $\bar{x}\bar{y}$-plane. The lines are in the undeformed state perpendicular to each other. In a deformed state, the plane has been translated and deformed so that the lines have been transformed into $A'B'$ and $A'C'$. We seek the angle $\gamma_{\bar{x}\bar{y}}$, which is the change of the angle between the lines. The angle $\gamma_{\bar{x}\bar{y}}$ is called *shear angle* or just *shear* and is the sum of the two angles β_1 and β_2,

$$\gamma_{\bar{x}\bar{y}} = \beta_1 + \beta_2 \tag{7.25}$$

With the assumption that the angles are small $(\sin \beta \approx \beta)$ this can, using Figure 7.9, be written as

$$\gamma_{\bar{x}\bar{y}} = \frac{du}{|A'C'|} + \frac{dv}{|A'B'|} \tag{7.26}$$

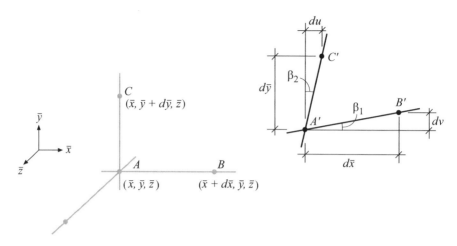

Figure 7.9 Angular deformation for two material fibres in the cross-section plane, initially perpendicular to each other

Under the assumption that the deformations are small, the lengths of the lines are given by

$$|A'B'| = d\bar{x} \tag{7.27}$$

$$|A'C'| = d\bar{y} \tag{7.28}$$

The local changes du and dv of the displacements can by use of the chain rule be written as

$$du = \frac{\partial u}{\partial \bar{x}}d\bar{x} + \frac{\partial u}{\partial \bar{y}}d\bar{y} + \frac{\partial u}{\partial \bar{z}}d\bar{z} \tag{7.29}$$

$$dv = \frac{\partial v}{\partial \bar{x}}d\bar{x} + \frac{\partial v}{\partial \bar{y}}d\bar{y} + \frac{\partial v}{\partial \bar{z}}d\bar{z} \tag{7.30}$$

For the fibre AC, we study du along the $\bar{y}$-axis, that is $d\bar{x} = 0$ and $d\bar{z} = 0$. For the fibre AB we study dv along the $\bar{x}$-axis, that is $d\bar{y} = 0$ and $d\bar{z} = 0$. The changes du and dv then become

$$du = \frac{\partial u}{\partial \bar{y}}d\bar{y} \tag{7.31}$$

$$dv = \frac{\partial v}{\partial \bar{x}}d\bar{x} \tag{7.32}$$

Substituting (7.27), (7.28), (7.31) and (7.32) into (7.26), the shear in the $\bar{x}\bar{y}$-plane can be written as

$$\gamma_{\bar{x}\bar{y}} = \frac{\partial u}{\partial \bar{y}} + \frac{\partial v}{\partial \bar{x}} \tag{7.33}$$

In the corresponding manner, we can establish the relation

$$\gamma_{\bar{x}\bar{z}} = \frac{\partial u}{\partial \bar{z}} + \frac{\partial w}{\partial \bar{x}} \tag{7.34}$$

for shear in the $\bar{x}\bar{z}$-plane.

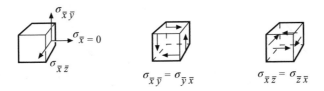

Figure 7.10   Stress components related to torsion

Stress

In Section 3.1, we have defined the three stress components that act on a sectional surface with the normal along the $\bar{x}$-axis. For bar action, the stress component present was the normal stress $\sigma_{\bar{x}}$, while the two shear stress components were equal to zero. For torsional action, we have instead that $\sigma_{\bar{x}\bar{y}}$ and $\sigma_{\bar{x}\bar{z}}$, defined in (3.8), in general, are different from zero, while the normal stress is equal to zero.

In Figure 7.10, the material point is represented by an infinitesimally small cuboid with six sectional surfaces. For the cuboid to be in equilibrium, two stresses equal in magnitude but oppositely directed must act on opposite sides of the cuboid. For the cuboid to be in moment equilibrium, two stress components must act on the sectional surfaces with a normal along the $\bar{y}$- and $\bar{z}$-direction, that is $\sigma_{\bar{y}\bar{x}} = \sigma_{\bar{x}\bar{y}}$ and $\sigma_{\bar{z}\bar{x}} = \sigma_{\bar{x}\bar{z}}$.

The Constitutive Relations of the Material

The material is assumed to be linear elastic, which means that the stress $\sigma_{\bar{x}\bar{y}}$ is proportional to the strain $\gamma_{\bar{x}\bar{y}}$ and the stress $\sigma_{\bar{x}\bar{z}}$ is proportional to the strain $\gamma_{\bar{x}\bar{z}}$

$$\sigma_{\bar{x}\bar{y}}(\bar{x}, \bar{y}, \bar{z}) = G(\bar{x})\, \gamma_{\bar{x}\bar{y}}(\bar{x}, \bar{y}, \bar{z}) \tag{7.35}$$

$$\sigma_{\bar{x}\bar{z}}(\bar{x}, \bar{y}, \bar{z}) = G(\bar{x})\, \gamma_{\bar{x}\bar{z}}(\bar{x}, \bar{y}, \bar{z}) \tag{7.36}$$

where G is the shear modulus of the material (Figure 7.11). Here, we assume that the shear modulus is constant over the cross-section, that is $G = G(\bar{x})$. The material may be isotropic or transversely isotropic. For transversely isotropic materials, G denotes the shear modulus for the $\bar{x}\bar{y}$- and $\bar{x}\bar{z}$-planes.

7.3.3   The Cross-Section Level

When a cross-section is loaded by a torque $T(\bar{x})$, the torque will be partitioned into two parts, each of them carried by a separate stress pattern

$$T(\bar{x}) = T_{sv}(\bar{x}) + T_w(\bar{x}) \tag{7.37}$$

These are referred to as St. Venant torsion and *Vlasov torsion*, respectively. St. Venant torsion is associated with shear stresses which build up a torque $T_{sv}(\bar{x})$ by making closed stress

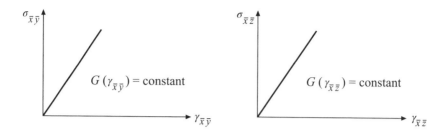

Figure 7.11 Linear elastic material relations in shear

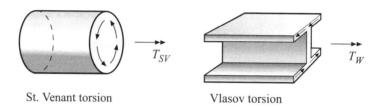

St. Venant torsion Vlasov torsion

Figure 7.12 St. Venant torsion and Vlasov torsion

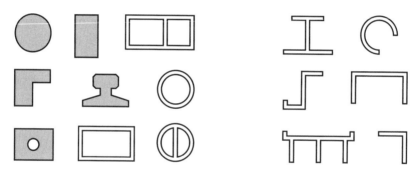

Figure 7.13 Solid and closed thin-walled cross-sections (St. Venant torsion) and open thin-walled cross-sections (Vlasov torsion)

trajectories in the cross-section, whereas in Vlasov torsion, the torque $T_w(\bar{x})$ induces shear stresses which result in open trajectories; cf. Figure 7.12.

For circular cross-sections, there is only St. Venant torsion. For solid or closed thin-walled cross-sections, St. Venant torsion is dominating, whereas Vlasov torsion is dominating for open thin-walled cross-sections (Figure 7.13). In what follows, only St. Venant torsion is discussed, that is we consider only cross-sectional shapes where $T(\bar{x}) \approx T_{sv}(\bar{x})$.

Kinematics

For bar and beam action, we assumed that the shape of the cross-section remains unchanged. We make this assumption also for torsional action. Furthermore, we previously assumed that

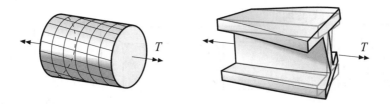

Figure 7.14 Circular cross-section – plane cross-sectional surfaces remain plane. I-shaped cross-section – the cross-sectional surface is deformed in the $\bar{x}$-direction (warping)

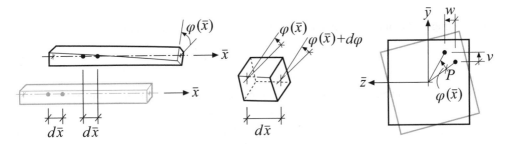

Figure 7.15 The rotation of the cross-section about the reference axis (twist angle) $\varphi(\bar{x})$ and change of twist angle $d\varphi$

plane cross-sections remain plane during deformation. For torsion, this assumption is strictly fulfilled only if the cross-section is circular. All other cross-sectional shapes cause deformations perpendicular to the cross-sectional surface (Figure 7.14). Despite this, here, we choose to formulate the kinematics of the cross-section with the assumption that plane cross-sections remain plane, which for non-circular cross-sections yields a slightly overestimated stiffness. The overestimated stiffness can be handled by multiplying the stiffness of the cross-section with a correction factor, whose magnitude depends on the shape of the cross-section.

The description of the kinematics of torsional action (Figure 7.15) starts with a reference axis, the local $\bar{x}$-axis. At torsional loading, each cross-section rotates an angle $\varphi(\bar{x})$. This angle is called *twist angle*. Between two adjacent points with distance $d\bar{x}$ between them, the twist angle $\varphi(\bar{x})$ is changed by $d\varphi$. This change is referred to as the *rate of twist* and is denoted $\theta(\bar{x})$

$$\theta(\bar{x}) = \frac{d\varphi}{d\bar{x}} \tag{7.38}$$

We now consider a point P with coordinates $(\bar{y}, \bar{z})$ on the cross-sectional surface. In the deformed state, the cross-section is rotated an angle $\varphi(\bar{x})$, thus P is displaced by

$$v = -\varphi(\bar{x})\bar{z} \tag{7.39}$$

$$w = \varphi(\bar{x})\bar{y} \tag{7.40}$$

Since the cross-section is assumed to remain plane

$$u = 0 \tag{7.41}$$

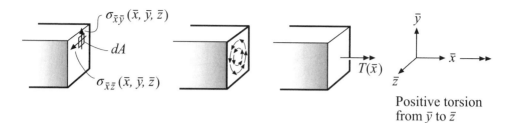

Figure 7.16   Shear stresses and torque

Substituting (7.39) and (7.41) into (7.33) and by using (7.38), we obtain

$$\gamma_{\bar{x}\bar{y}}(\bar{x}, \bar{y}, \bar{z}) = -\theta(\bar{x})\bar{z} \tag{7.42}$$

Substituting (7.40) and (7.41) into (7.34), we obtain in the corresponding manner

$$\gamma_{\bar{x}\bar{z}}(\bar{x}, \bar{y}, \bar{z}) = \theta(\bar{x})\bar{y} \tag{7.43}$$

Equations (7.42) and (7.43) are the kinematic relations that relate the generalised strain of the cross-section level, the rate of twist $\theta(\bar{x})$, to the shear strains $\gamma_{\bar{x}\bar{y}}$ and $\gamma_{\bar{x}\bar{z}}$ of the material level.

Force Relations

The forces acting on a small part dA of the cross-sectional surface is $\sigma_{\bar{x}\bar{y}}\, dA$ and $\sigma_{\bar{x}\bar{z}}\, dA$. The torque of these forces with respect to the reference axis then becomes $\sigma_{\bar{x}\bar{z}}\bar{y}\, dA - \sigma_{\bar{x}\bar{y}}\bar{z}\, dA$ and the resulting torque T (Figure 7.16) on the whole cross-section becomes

$$T(\bar{x}) = \int_A (\sigma_{\bar{x}\bar{z}}(\bar{x}, \bar{y}, \bar{z})\bar{y} - \sigma_{\bar{x}\bar{y}}(\bar{x}, \bar{y}, \bar{z})\bar{z})dA \tag{7.44}$$

The torque is defined as positive when it turns from $\bar{y}$ towards $\bar{z}$.

The Constitutive Relations of the Cross-Section

Substituting the kinematic relations (7.42) and (7.43) as well as the material relations (7.35) and (7.36) in (7.44), a constitutive relation for the cross-section is obtained

$$T(\bar{x}) = \int_A G(\bar{x})(\theta(\bar{x})\bar{y}^2 + \theta(\bar{x})\bar{z}^2)dA \tag{7.45}$$

The rate of twist $\theta(\bar{x})$ is independent of $\bar{y}$ and $\bar{z}$ and can therefore be put outside the integral, which gives

$$T(\bar{x}) = D_{GK}(\bar{x})\theta(\bar{x}) \tag{7.46}$$

where

$$D_{GK}(\bar{x}) = \int_A G(\bar{x})(\bar{y}^2 + \bar{z}^2)dA \qquad (7.47)$$

D_{GK} is the St. Venant torsion stiffness of the cross-section and is built up from the material stiffness G and the cross-sectional shape. If the shear modulus is constant across the cross-section, that is independent of $\bar{y}$ and $\bar{z}$, we have

$$D_{GK} = GK_v \qquad (7.48)$$

where K_v is a measure of the portion of the cross-sectional stiffness that is due to the cross-sectional shape. For a circular cross-section,

$$K_v = I_p \qquad (7.49)$$

$$
\left.
\begin{aligned}
T(\bar{x}) &= \int_A \left(\sigma_{\bar{x}\bar{z}}(\bar{x}, \bar{y}, \bar{z})\, \bar{y} - \sigma_{\bar{x}\bar{y}}(\bar{x}, \bar{y}, \bar{z})\, \bar{z} \right) dA \quad (7.44)\\
\sigma_{\bar{x}\bar{y}}(\bar{x}, \bar{y}, \bar{z}) &= G(\bar{x})\, \gamma_{\bar{x}\bar{y}}(\bar{x}, \bar{y}, \bar{z}) \quad (7.35)\\
\sigma_{\bar{x}\bar{z}}(\bar{x}, \bar{y}, \bar{z}) &= G(\bar{x})\, \gamma_{\bar{x}\bar{z}}(\bar{x}, \bar{y}, \bar{z}) \quad (7.36)\\
\gamma_{\bar{x}\bar{y}}(\bar{x}, \bar{y}, \bar{z}) &= -\theta(\bar{x})\bar{z} \quad (7.42)\\
\gamma_{\bar{x}\bar{z}}(\bar{x}, \bar{y}, \bar{z}) &= \theta(\bar{x})\bar{y} \quad (7.43)
\end{aligned}
\right\}
\Rightarrow
\begin{aligned}
&T(\bar{x}) = D_{GK}(\bar{x})\theta(\bar{x}) \quad (7.46)\\
&\text{where}\\
&D_{GK}(\bar{x}) = \int_A G(\bar{x})(\bar{y}^2 + \bar{z}^2)\, dA
\end{aligned}
$$

Figure 7.17 From the material level to the cross-section level

where $I_p = \int_A r^2 dA$ is the polar moment of inertia. Here, r is the distance from the centre of the cross-section, $r = \sqrt{\bar{y}^2 + \bar{z}^2}$. The polar moment of inertia I_p can be expressed in terms of the moments of inertia $I_{\bar{y}}$ and $I_{\bar{z}}$ as

$$I_p = I_{\bar{y}} + I_{\bar{z}} \qquad (7.50)$$

where $I_{\bar{y}} = \int_A \bar{z}^2 dA$ and $I_{\bar{z}} = \int_A \bar{y}^2 dA$. Figure 7.17 shows how kinematic relations, material relations and a resultant relation together form a relation for the cross-section.

The above-mentioned derivation presumed that plane cross-sections remain plane under deformation and is therefore strictly valid only for circular cross-sections. The corresponding derivation for other cross-sectional shapes such as rectangular cross-sections is more complicated to perform. At the corners of a rectangular cross-section, the shear stresses must be zero and this causes that plane cross-sections do not remain plane but are slightly warped. For rectangular cross-sections, Table 7.1 gives a correction factor for I_p which compensates for the overestimated stiffness,

$$K_v = \alpha I_p \qquad (7.51)$$

where the coefficient α depends on the height–width relation h/b.

Table 7.1 The coefficient α for rectangular massive cross-sections with different height–width relations h/b

h/b	1.0	1.5	2.0	2.5	3.0	4.0	6.0	10.0
α	0.846	0.724	0.550	0.412	0.316	0.198	0.097	0.037

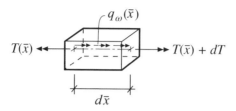

Figure 7.18   Equilibrium for a slice $d\bar{x}$ of a twisted beam

7.3.4   Torsional Action

Kinematics

The deformation of a twisted beam is described by the rate of twist $\theta(\bar{x})$, which arises along the system line of the beam, the $\bar{x}$-axis. From (7.38), we have a relation between the twist angle $\varphi(\bar{x})$ of the cross-section and the rate of twist $\theta(\bar{x})$

$$\theta(\bar{x}) = \frac{d\varphi}{d\bar{x}} \tag{7.52}$$

that is a relation between the deformation measure $\varphi(\bar{x})$ of torsional action and the deformation measure $\theta(\bar{x})$ of the cross-section level.

Equilibrium

Consider a thin slice $d\bar{x}$ of an undeformed beam loaded with an external torsional load $q_\omega(\bar{x})$, according to Figure 7.18. For the part considered, we have the equilibrium relation

$$- T(\bar{x}) + (T(\bar{x}) + dT) + q_\omega(\bar{x})d\bar{x} = 0 \tag{7.53}$$

where $T(\bar{x})$ is the torque at $\bar{x}$ and $T(\bar{x}) + dT$ the torque at $\bar{x} + d\bar{x}$. The relation can be simplified to

$$dT + q_\omega(\bar{x})d\bar{x} = 0 \tag{7.54}$$

or

$$\frac{dT}{d\bar{x}} + q_\omega(\bar{x}) = 0 \tag{7.55}$$

which is the equilibrium relation relating the loading $T(\bar{x})$ on the cross-section to the loading $q_\omega(\bar{x})$ on the beam.

The Differential Equation for Torsional Action

Substituting the kinematic relation (7.52) into Equation (7.46) gives

$$T(\bar{x}) = D_{GK}(\bar{x})\frac{d\varphi}{d\bar{x}} \tag{7.56}$$

Substitution into the equilibrium relation (7.48) then gives

$$\frac{d}{d\bar{x}}\left(D_{GK}(\bar{x})\frac{d\varphi}{d\bar{x}}\right) + q_\omega(\bar{x}) = 0 \tag{7.57}$$

This differential equation describes the relation between torsional load q_ω and the rate of twist φ for torsional action. If the stiffness D_{GK} is constant along the beam, the expression can be written as

$$\boxed{D_{GK}\frac{d^2\varphi}{d\bar{x}^2} + q_\omega(\bar{x}) = 0} \tag{7.58}$$

When the shear modulus is constant across the cross-section, that is independent of $\bar{y}$ and $\bar{z}$, we obtain according to (7.48)

$$GK_v\frac{d^2\varphi}{d\bar{x}^2} + q_\omega(\bar{x}) = 0 \tag{7.59}$$

where the torsional stiffness of the beam is described by the product of the shear modulus G and the sectional torsion constant K_v. Figure 7.19 shows how a kinematic relation, a constitutive relation and equilibrium relations are combined to a relation for the beam.

$$\left.\begin{array}{ll}\dfrac{dT}{d\bar{x}} + q_\omega(\bar{x}) = 0 & (7.55) \\[2mm] T(\bar{x}) = D_{GK}(\bar{x})\quad\theta(\bar{x}) & (7.46) \\[2mm] \theta(\bar{x}) = \dfrac{d\varphi}{d\bar{x}} & (7.52)\end{array}\right\} \Rightarrow \begin{array}{l} D_{GK}\dfrac{d^2\varphi}{d\bar{x}^2} + q_\omega(\bar{x}) = 0 \quad (7.58) \\[2mm] \text{for constant } D_{GK} \end{array}$$

Figure 7.19   From the cross-section level to twist action

The boundary conditions necessary to solve the differential equation can be prescribed twist angle (rotation) φ or prescribed torque T at the endpoints of the beam.

7.4   Three-Dimensional Beam Element

Based on the differential equations for bar action (3.25), beam action (4.27) and torsional action (7.52), the relations between forces and displacements for a three-dimensional beam element are derived. First, a relation for an element with two degrees of freedom is established for torsional action. This is combined with the elements describing bar action according to Chapter 3 and the elements describing beam action in bending with respect to two axes, the local $\bar{y}$-axis and the local $\bar{z}$-axis, according to Chapter 4. A beam element with 12 degrees of freedom is formed from this combination (Figure 7.20). To enable this beam element to be placed with arbitrary orientation in a three-dimensional frame, a transformation from the local coordinate system to a global one is performed. Thereafter, the element is placed in the global structure by use of compatibility and equilibrium (Figure 7.5).

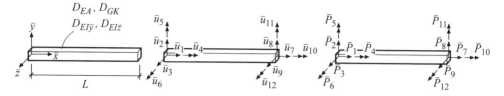

Figure 7.20 Three-dimensional beam element

Figure 7.21 Element for pure torsional action

7.4.1 Element for Torsional Action

We start by formulating an element for pure torsional action. Such an element has two displacement degrees of freedom, $\bar{u}_1$ and $\bar{u}_2$ (Figure 7.21). The degrees of freedom describe the rotation about the $\bar{x}$-axis for $\bar{x} = 0$ and for $\bar{x} = L$. The torques acting at $\bar{x} = 0$ and $\bar{x} = L$ are denoted $\bar{P}_1$ and $\bar{P}_2$, respectively, and are positive when directed as the rotations $\bar{u}_1$ and $\bar{u}_2$.

The differential equation for torsional action (7.52) is built in the same manner as the differential equation for bar action (3.25), but with different variables and constants. Thus, the solution is obtained in the same manner. Therefore, we can by using the solution for bar action establish the solution for torsional action in matrix form as

$$\bar{\mathbf{K}}^e \bar{\mathbf{a}}^e = \bar{\mathbf{f}}^e \tag{7.60}$$

where

$$\bar{\mathbf{K}}^e = \frac{D_{GK}}{L} \begin{bmatrix} 1 & -1 \\ -1 & 1 \end{bmatrix} \tag{7.61}$$

$$\bar{\mathbf{a}}^e = \begin{bmatrix} \bar{u}_1 \\ \bar{u}_2 \end{bmatrix} \tag{7.62}$$

$$\bar{\mathbf{f}}^e = \bar{\mathbf{f}}_b^e + \bar{\mathbf{f}}_l^e \tag{7.63}$$

$$\bar{\mathbf{f}}_b^e = \begin{bmatrix} \bar{P}_1 \\ \bar{P}_2 \end{bmatrix} = \begin{bmatrix} -T(0) \\ T(L) \end{bmatrix} \tag{7.64}$$

$$\bar{\mathbf{f}}_l^e = \begin{bmatrix} T_p(0) \\ -T_p(L) \end{bmatrix} \tag{7.65}$$

7.4.2   Beam Element with 12 Degrees of Freedom

In Section 3.2.2, we have derived the element equations for a bar element with two degrees of freedom, $\bar{u}_{1,\text{bar}}$ and $\bar{u}_{2,\text{bar}}$ (bar action). In the corresponding manner, we have in Section 4.2.2 derived the element equations for a beam element with four degrees of freedom. This can be used to model bending about the $\bar{z}$-axis as well as about the $\bar{y}$-axis. We have $\bar{u}_{1,\text{beam},\bar{z}} - \bar{u}_{4,\text{beam},\bar{z}}$ for bending about the $\bar{z}$-axis (beam action) and $\bar{u}_{1,\text{beam},\bar{y}} - \bar{u}_{4,\text{beam},\bar{y}}$ for bending about the $\bar{y}$-axis (beam action). For the beam element with six degrees of freedom, we concluded that if the location for the local $\bar{x}$-axis (the system line) is chosen in an appropriate way, then the two modes of action are independent of each other. In the same manner, the four modes of action for the three-dimensional beam element with 12 degrees of freedom are independent of each other if $\int_A E \, \bar{y} \, dA = 0$ and $\int_A E \, \bar{z} \, dA = 0$ and if the direction of the local $\bar{y}$-axis is chosen such that $\int_A E \, \bar{y}\bar{z} \, dA = 0$. From Section 7.4.1, we have an element for torsional action with two degrees of freedom $\bar{u}_{1,\text{torsion}}$ and $\bar{u}_{2,\text{torsion}}$. This collected enables introduction of a new beam element with 12 degrees of freedom, $\bar{u}_1 - \bar{u}_{12}$, which includes bar action, beam action with respect to two axes and torsional action. Figure 7.3 shows how the elements for bar action, beam action and torsional action can be combined to a beam element with 12 degrees of freedom. This combination can be expressed as a kinematic condition (compatibility) and a force relation (static equivalence),

$$
\bar{\mathbf{a}}^e =
\begin{bmatrix}
\bar{u}_1 \\
\bar{u}_2 \\
\bar{u}_3 \\
\bar{u}_4 \\
\bar{u}_5 \\
\bar{u}_6 \\
\bar{u}_7 \\
\bar{u}_8 \\
\bar{u}_9 \\
\bar{u}_{10} \\
\bar{u}_{11} \\
\bar{u}_{12}
\end{bmatrix}
=
\begin{bmatrix}
\bar{u}_{1,\text{bar}} \\
\bar{u}_{1,\text{beam},\bar{z}} \\
\bar{u}_{1,\text{beam},\bar{y}} \\
\bar{u}_{1,\text{torsion}} \\
-\bar{u}_{2,\text{beam},\bar{y}} \\
\bar{u}_{2,\text{beam},\bar{z}} \\
\bar{u}_{2,\text{bar}} \\
\bar{u}_{3,\text{beam},\bar{z}} \\
\bar{u}_{3,\text{beam},\bar{y}} \\
\bar{u}_{2,\text{torsion}} \\
-\bar{u}_{4,\text{beam},\bar{y}} \\
\bar{u}_{4,\text{beam},\bar{z}}
\end{bmatrix}
; \quad
\bar{\mathbf{f}}^e =
\begin{bmatrix}
\bar{P}_1 \\
\bar{P}_2 \\
\bar{P}_3 \\
\bar{P}_4 \\
\bar{P}_5 \\
\bar{P}_6 \\
\bar{P}_7 \\
\bar{P}_8 \\
\bar{P}_9 \\
\bar{P}_{10} \\
\bar{P}_{11} \\
\bar{P}_{12}
\end{bmatrix}
=
\begin{bmatrix}
\bar{P}_{1,\text{bar}} \\
\bar{P}_{1,\text{beam},\bar{z}} \\
\bar{P}_{1,\text{beam},\bar{y}} \\
\bar{P}_{1,\text{torsion}} \\
-\bar{P}_{2,\text{beam},\bar{y}} \\
\bar{P}_{2,\text{beam},\bar{z}} \\
\bar{P}_{2,\text{bar}} \\
\bar{P}_{3,\text{beam},\bar{z}} \\
\bar{P}_{3,\text{balk},\bar{y}} \\
\bar{P}_{2,\text{torsion}} \\
-\bar{P}_{4,\text{beam},\bar{y}} \\
\bar{P}_{4,\text{beam},\bar{z}}
\end{bmatrix}
\tag{7.66}
$$

Combination of the relations (3.60), (4.82), (7.60) and (7.66) gives element

$$
\boxed{\bar{\mathbf{K}}^e \bar{\mathbf{a}}^e = \bar{\mathbf{f}}^e}
\tag{7.67}
$$

where

$$
\bar{\mathbf{K}}^e =
\begin{bmatrix}
\frac{D_{EA}}{L} & 0 & 0 & 0 & 0 & 0 & -\frac{D_{EA}}{L} & 0 & 0 & 0 & 0 & 0 \\
0 & \frac{12D_{EI_{\bar{z}}}}{L^3} & 0 & 0 & 0 & \frac{6D_{EI_{\bar{z}}}}{L^2} & 0 & -\frac{12D_{EI_{\bar{z}}}}{L^3} & 0 & 0 & 0 & \frac{6D_{EI_{\bar{z}}}}{L^2} \\
0 & 0 & \frac{12D_{EI_{\bar{y}}}}{L^3} & 0 & -\frac{6D_{EI_{\bar{y}}}}{L^2} & 0 & 0 & 0 & -\frac{12D_{EI_{\bar{y}}}}{L^3} & 0 & -\frac{6D_{EI_{\bar{y}}}}{L^2} & 0 \\
0 & 0 & 0 & \frac{D_{GK}}{L} & 0 & 0 & 0 & 0 & 0 & -\frac{D_{GK}}{L} & 0 & 0 \\
0 & 0 & -\frac{6D_{EI_{\bar{y}}}}{L^2} & 0 & \frac{4D_{EI_{\bar{y}}}}{L} & 0 & 0 & 0 & \frac{6D_{EI_{\bar{y}}}}{L^2} & 0 & \frac{2D_{EI_{\bar{y}}}}{L} & 0 \\
0 & \frac{6D_{EI_{\bar{z}}}}{L^2} & 0 & 0 & 0 & \frac{4D_{EI_{\bar{z}}}}{L} & 0 & -\frac{6D_{EI_{\bar{z}}}}{L^2} & 0 & 0 & 0 & \frac{2D_{EI_{\bar{z}}}}{L} \\
-\frac{D_{EA}}{L} & 0 & 0 & 0 & 0 & 0 & \frac{D_{EA}}{L} & 0 & 0 & 0 & 0 & 0 \\
0 & -\frac{12D_{EI_{\bar{z}}}}{L^3} & 0 & 0 & 0 & -\frac{6D_{EI_{\bar{z}}}}{L^2} & 0 & \frac{12D_{EI_{\bar{z}}}}{L^3} & 0 & 0 & 0 & -\frac{6D_{EI_{\bar{z}}}}{L^2} \\
0 & 0 & -\frac{12D_{EI_{\bar{y}}}}{L^3} & 0 & \frac{6D_{EI_{\bar{y}}}}{L^2} & 0 & 0 & 0 & \frac{12D_{EI_{\bar{y}}}}{L^3} & 0 & \frac{6D_{EI_{\bar{y}}}}{L^2} & 0 \\
0 & 0 & 0 & -\frac{D_{GK}}{L} & 0 & 0 & 0 & 0 & 0 & \frac{D_{GK}}{L} & 0 & 0 \\
0 & 0 & -\frac{6D_{EI_{\bar{y}}}}{L^2} & 0 & \frac{2D_{EI_{\bar{y}}}}{L} & 0 & 0 & 0 & \frac{6D_{EI_{\bar{y}}}}{L^2} & 0 & \frac{4D_{EI_{\bar{y}}}}{L} & 0 \\
0 & \frac{6D_{EI_{\bar{z}}}}{L^2} & 0 & 0 & 0 & \frac{2D_{EI_{\bar{z}}}}{L} & 0 & -\frac{6D_{EI_{\bar{z}}}}{L^2} & 0 & 0 & 0 & \frac{4D_{EI_{\bar{z}}}}{L}
\end{bmatrix}
\tag{7.68}
$$

and where

$$
\bar{\mathbf{f}}^e = \bar{\mathbf{f}}^e_b + \bar{\mathbf{f}}^e_l
\tag{7.69}
$$

with

$$
\bar{\mathbf{f}}^e_b =
\begin{bmatrix}
\bar{P}_1 \\
\bar{P}_2 \\
\bar{P}_3 \\
\bar{P}_4 \\
\bar{P}_5 \\
\bar{P}_6 \\
\bar{P}_7 \\
\bar{P}_8 \\
\bar{P}_9 \\
\bar{P}_{10} \\
\bar{P}_{11} \\
\bar{P}_{12}
\end{bmatrix}
; \quad
\bar{\mathbf{f}}^e_l =
\begin{bmatrix}
N_p(0) \\
V_{\bar{z}p}(0) \\
V_{\bar{y}p}(0) \\
T_p(0) \\
-M_{\bar{y}p}(0) \\
M_{\bar{z}p}(0) \\
-N_p(L) \\
-V_{\bar{z}p}(L) \\
-V_{\bar{y}p}(L) \\
T_p(L) \\
M_{\bar{y}p}(L) \\
-M_{\bar{z}p}(L)
\end{bmatrix}
\tag{7.70}
$$

7.4.3 From Local to Global Directions

In the element relation of the three-dimensional beam (7.67), the nodal force vector $\bar{\mathbf{f}}^e_b$, the element displacement vector $\bar{\mathbf{a}}^e$ and the element load vector $\bar{\mathbf{f}}^e_l$ are expressed in the local coordinate system of the beam $(\bar{x}, \bar{y}, \bar{z})$. To enable the beam element to be placed in a three-dimensional frame, we have to establish an element relation where forces and displacements are expressed in the global coordinate system (x, y, z) of the frame. For the two-dimensional beam, we have a transformation of the element relation from <u>three</u> degrees of freedom in each node, local coordinate system, to <u>three</u> degrees of freedom with new directions for the beam element in the global coordinate system. Here, we go from <u>six</u> degrees of freedom in each node in a local coordinate system to *six* degrees of freedom with new directions for the beam element in the global coordinate system.

The transformation of displacements between local and global coordinate system is performed separately for each degree of freedom. From (7.9), we have that the translation $\bar{u}_1$ in the direction of the local $\bar{x}$-axis can be written as

$$\bar{u}_1 = n_{x\bar{x}}u_1 + n_{y\bar{x}}u_2 + n_{z\bar{x}}u_3 \tag{7.71}$$

In the corresponding manner, the translations $\bar{u}_2$ in the direction of the local $\bar{y}$-axis and $\bar{u}_3$ in the direction of the local $\bar{z}$-axis can be written as

$$\bar{u}_2 = n_{x\bar{y}}u_1 + n_{y\bar{y}}u_2 + n_{z\bar{y}}u_3 \tag{7.72}$$

$$\bar{u}_3 = n_{x\bar{z}}u_1 + n_{y\bar{z}}u_2 + n_{z\bar{z}}u_3 \tag{7.73}$$

The rotations $\bar{u}_4$, $\bar{u}_5$ and $\bar{u}_6$ are also transformed in the same manner

$$\bar{u}_4 = n_{x\bar{x}}u_4 + n_{y\bar{x}}u_5 + n_{z\bar{x}}u_6 \tag{7.74}$$

$$\bar{u}_5 = n_{x\bar{y}}u_4 + n_{y\bar{y}}u_5 + n_{z\bar{y}}u_6 \tag{7.75}$$

$$\bar{u}_6 = n_{x\bar{z}}u_4 + n_{y\bar{z}}u_5 + n_{z\bar{z}}u_6 \tag{7.76}$$

For the node in the other end of the bar, the corresponding relations can be established. In matrix form, this can be expressed as

$$\boxed{\bar{\mathbf{a}}^e = \mathbf{G}\mathbf{a}^e} \tag{7.77}$$

where

$$\bar{\mathbf{a}}^e = \begin{bmatrix} \bar{u}_1 \\ \bar{u}_2 \\ \bar{u}_3 \\ \bar{u}_4 \\ \bar{u}_5 \\ \bar{u}_6 \\ \bar{u}_7 \\ \bar{u}_8 \\ \bar{u}_9 \\ \bar{u}_{10} \\ \bar{u}_{11} \\ \bar{u}_{12} \end{bmatrix} ; \quad \mathbf{G} = \begin{bmatrix} \mathbf{C} & 0 & 0 & 0 \\ 0 & \mathbf{C} & 0 & 0 \\ 0 & 0 & \mathbf{C} & 0 \\ 0 & 0 & 0 & \mathbf{C} \end{bmatrix} ; \quad \mathbf{a}^e = \begin{bmatrix} u_1 \\ u_2 \\ u_3 \\ u_4 \\ u_5 \\ u_6 \\ u_7 \\ u_8 \\ u_9 \\ u_{10} \\ u_{11} \\ u_{12} \end{bmatrix} \tag{7.78}$$

and

$$\mathbf{C} = \begin{bmatrix} n_{x\bar{x}} & n_{y\bar{x}} & n_{z\bar{x}} \\ n_{x\bar{y}} & n_{y\bar{y}} & n_{z\bar{y}} \\ n_{x\bar{z}} & n_{y\bar{z}} & n_{z\bar{z}} \end{bmatrix} \tag{7.79}$$

The components $\bar{P}_1, \bar{P}_2$ and $\bar{P}_3$ of the nodal forces P_1, P_2 and P_3 can, using the relations (3.69), be expressed as

$$P_1 = n_{x\bar{x}}\bar{P}_1 + n_{x\bar{y}}\bar{P}_2 + n_{x\bar{z}}\bar{P}_3 \tag{7.80}$$

$$P_2 = n_{y\bar{x}}\bar{P}_1 + n_{y\bar{y}}\bar{P}_2 + n_{y\bar{z}}\bar{P}_3 \tag{7.81}$$

$$P_3 = n_{z\bar{x}}\bar{P}_1 + n_{z\bar{y}}\bar{P}_2 + n_{z\bar{z}}\bar{P}_3 \tag{7.82}$$

and for the moments we have in the same manner

$$P_4 = n_{x\bar{x}}\bar{P}_4 + n_{x\bar{y}}\bar{P}_5 + n_{x\bar{z}}\bar{P}_6 \tag{7.83}$$

$$P_5 = n_{y\bar{x}}\bar{P}_4 + n_{y\bar{y}}\bar{P}_5 + n_{y\bar{z}}\bar{P}_6 \tag{7.84}$$

$$P_6 = n_{z\bar{x}}\bar{P}_4 + n_{z\bar{y}}\bar{P}_5 + n_{z\bar{z}}\bar{P}_6 \tag{7.85}$$

The corresponding relation is valid also in the other end of the beam. In matrix form, these relations can be written as

$$\mathbf{f}_b^e = \mathbf{G}^T \bar{\mathbf{f}}_b^e \tag{7.86}$$

where

$$\mathbf{f}_b^e = \begin{bmatrix} P_1 \\ P_2 \\ P_3 \\ P_4 \\ P_5 \\ P_6 \\ P_7 \\ P_8 \\ P_9 \\ P_{10} \\ P_{11} \\ P_{12} \end{bmatrix}; \quad \mathbf{G}^T = \begin{bmatrix} \mathbf{C}^T & 0 & 0 & 0 \\ 0 & \mathbf{C}^T & 0 & 0 \\ 0 & 0 & \mathbf{C}^T & 0 \\ 0 & 0 & 0 & \mathbf{C}^T \end{bmatrix}; \quad \bar{\mathbf{f}}_b^e = \begin{bmatrix} \bar{P}_1 \\ \bar{P}_2 \\ \bar{P}_3 \\ \bar{P}_4 \\ \bar{P}_5 \\ \bar{P}_6 \\ \bar{P}_7 \\ \bar{P}_8 \\ \bar{P}_9 \\ \bar{P}_{10} \\ \bar{P}_{11} \\ \bar{P}_{12} \end{bmatrix} \tag{7.87}$$

and $\mathbf{G}^T$ and $\mathbf{C}^T$ are the transposes of $\mathbf{G}$ and $\mathbf{C}$, respectively, which were defined previously.

The relation between element loads $\mathbf{f}_l^e$ in a global system and element nodal loads $\bar{\mathbf{f}}_l^e$ in a local system can in the corresponding manner be written as

$$\mathbf{f}_l^e = \mathbf{G}^T \bar{\mathbf{f}}_l^e \tag{7.88}$$

Substituting the transformations (7.86), (7.77) and (7.88) into the element relation (7.67) gives an element relation with quantities expressed in the directions of the global coordinate system,

$$\mathbf{K}^e \mathbf{a}^e = \mathbf{f}^e \tag{7.89}$$

where

$$\mathbf{K}^e = \mathbf{G}^T \bar{\mathbf{K}}^e \mathbf{G}; \quad \mathbf{f}^e = \mathbf{f}_b^e + \mathbf{f}_l^e \tag{7.90}$$

How transformations of displacements and forces between different coordinate systems lead to a relation for the beam element in global coordinates is shown in Figure 7.22.

$$\left.\begin{aligned}
\mathbf{f}_b^e &= \mathbf{G}^T\bar{\mathbf{f}}_b^e &(7.86)\\
\mathbf{f}_l^e &= \mathbf{G}^T\bar{\mathbf{f}}_l^e &(7.88)\\
\bar{\mathbf{f}}^e &= \bar{\mathbf{K}}^e\bar{\mathbf{a}}^e &(7.67)\\
\bar{\mathbf{f}}^e &= \bar{\mathbf{f}}_b^e + \bar{\mathbf{f}}_l^e &(7.69)\\
\bar{\mathbf{a}}^e &= \mathbf{G}\mathbf{a}^e &(7.77)
\end{aligned}\right\} \Rightarrow \begin{aligned} &\mathbf{K}^e\mathbf{a}^e = \mathbf{f}^e \quad (7.89)\\[4pt] &\text{where}\\[4pt] &\mathbf{K}^e = \mathbf{G}^T\bar{\mathbf{K}}^e\mathbf{G}; \quad \mathbf{f}^e = \mathbf{f}_b^e + \mathbf{f}_l^e \end{aligned}$$

Figure 7.22   From local coordinates to global coordinates

7.5   Three-Dimensional Frames

We introduce a global numbering of all the displacement degrees of freedom and gather them in a global displacement vector $\mathbf{a}$. From the element relations for the separate beam elements, we have a local numbering of the displacements $u_1 - u_{12}$. Based on compatibility and equilibrium conditions, the stiffness matrix $\mathbf{K}$ and the load vector $\mathbf{f}_l$ are established in the same manner as in previous chapters. By considering the present boundary conditions, the displacements and the support forces can be computed and thereafter, the section force distributions determined.

Example 7.2 Frame

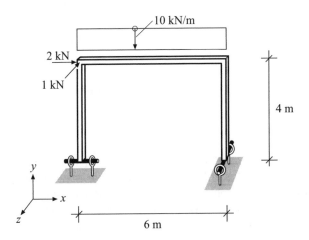

Figure 1   Three-dimensional frame with three beams

The frame in Figure 1 is constructed of three beams with the cross-sectional areas $A_1 = 2.0 \times 10^{-3}$ m², $A_2 = 2.0 \times 10^{-3}$ m² and $A_3 = 6.0 \times 10^{-3}$ m², the moments of inertia $I_{\bar{z},1} = 1.6 \times 10^{-5}$ m⁴, $I_{\bar{z},2} = 1.6 \times 10^{-5}$ m⁴, $I_{\bar{z},3} = 5.4 \times 10^{-5}$ m⁴ $I_{\bar{y},1} = 3.2 \times 10^{-5}$ m⁴, $I_{\bar{y},2} = 3.2 \times 10^{-5}$ m⁴ and $I_{\bar{y},3} = 5.4 \times 10^{-5}$ m⁴, the sectional torsion constants $K_{v,1} = 4.0 \times 10^{-5}$ m⁴, $K_{v,2} = 4.0 \times 10^{-5}$ m⁴, $K_{v,3} = 12.0 \times 10^{-5}$ m⁴ and has the modulus of elasticity $E = 200.0$ GPa and the shear modulus $G = 80.0$ GPa. All beams are oriented such that the local $\bar{z}$-direction coincides with the global z-direction. The lengths of the beams are $L_1 = 4.0$ m, $L_2 = 4.0$ m and $L_3 = 6.0$ m. Along the horizontal beam, the frame is loaded by a uniformly distributed load $q = 10$ kN/m and in the upper left corner a point load $P_1 = 2$ kN acts in the direction of the x-axis and a point load $P_2 = 1$ kN acts in the negative z-direction. At the lower left end, the structure has a support that is hinged with respect to rotation about the x-axis but prevents rotation about the y- and z-axis. At the lower right end, the structure has a support that is hinged with respect to rotation about the z-axis but prevents rotation about the x- and y-axis.

Computational model

The frame model is built up by three beam elements, denoted 1, 2 and 3, respectively (Figure 2). The model has the displacement degrees of freedom a_1–a_{24}. The forces that act at the corner give $f_7 = 2$ kN and $f_9 = -1$ kN. In the degrees of freedom a_1, a_2, a_3, a_5, a_6, a_{19}, a_{20}, a_{21}, a_{22} and a_{23}, the displacements are prescribed to be zero.

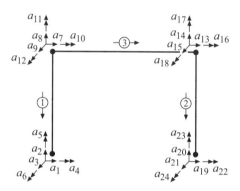

Figure 2 Computational model

Element matrices

For each beam element, an element relation $\mathbf{K}^e \mathbf{a}^e = \mathbf{f}_b^e$ can be established. The element stiffness matrices in local coordinates for the separate elements are given in (7.67) with quantities for the element properties inserted. The element matrices are transformed to the global coordinate system with (7.89) where the direction cosines for the elements are given according to the following:

Element 1:

$$n_{x\bar{x}} = 0; \qquad n_{y\bar{x}} = -1; \qquad n_{z\bar{x}} = 0$$
$$n_{x\bar{y}} = 1; \qquad n_{y\bar{y}} = 0; \qquad n_{z\bar{y}} = 0 \qquad (1)$$
$$n_{x\bar{z}} = 0; \qquad n_{y\bar{z}} = 0; \qquad n_{z\bar{z}} = 1$$

Element 2:

$$n_{x\bar{x}} = 0; \qquad n_{y\bar{x}} = -1; \qquad n_{z\bar{x}} = 0$$
$$n_{x\bar{y}} = 1; \qquad n_{y\bar{y}} = 0; \qquad n_{z\bar{y}} = 0 \qquad (2)$$
$$n_{x\bar{z}} = 0; \qquad n_{y\bar{z}} = 0; \qquad n_{z\bar{z}} = 1$$

Element 3:

$$n_{x\bar{x}} = 1; \qquad n_{y\bar{x}} = 0; \qquad n_{z\bar{x}} = 0$$
$$n_{x\bar{y}} = 0; \qquad n_{y\bar{y}} = 1; \qquad n_{z\bar{y}} = 0 \qquad (3)$$
$$n_{x\bar{z}} = 0; \qquad n_{y\bar{z}} = 0; \qquad n_{z\bar{z}} = 1$$

Compatibility conditions and assembling

The topology matrix expresses how the local degrees of freedom for Elements 1–3 are related to the global degrees of freedom,

$$\text{topology} = \begin{bmatrix} 1 & 7 & 8 & 9 & 10 & 11 & 12 & 1 & 2 & 3 & 4 & 5 & 6 \\ 2 & 13 & 14 & 15 & 16 & 17 & 18 & 19 & 20 & 21 & 22 & 23 & 24 \\ 3 & 7 & 8 & 9 & 10 & 11 & 12 & 13 & 14 & 15 & 16 & 17 & 18 \end{bmatrix} \qquad (4)$$

The topology matrix is used to assemble the element stiffness matrices into a global stiffness matrix.

Boundary conditions and nodal loads

With the present loads and boundary conditions (bc), we have

$$\mathbf{f}_l = \begin{bmatrix} 0 \\ 0 \\ 0 \\ 0 \\ 0 \\ 0 \\ 2.0000 \\ -30.0000 \\ -3.4641 \\ 0 \\ 0 \\ -30.0000 \\ 0 \\ -30.0000 \\ 0 \\ 0 \\ 0 \\ 30.0000 \\ 0 \\ 0 \\ 0 \\ 0 \\ 0 \\ 0 \end{bmatrix} 10^3; \quad \text{bc} = \begin{bmatrix} 1 & 0 \\ 2 & 0 \\ 3 & 0 \\ 5 & 0 \\ 6 & 0 \\ 19 & 0 \\ 20 & 0 \\ 21 & 0 \\ 22 & 0 \\ 23 & 0 \end{bmatrix}; \quad \mathbf{a} = \begin{bmatrix} 0 \\ 0 \\ 0 \\ a_4 \\ 0 \\ 0 \\ a_7 \\ a_8 \\ a_9 \\ a_{10} \\ a_{11} \\ a_{12} \\ a_{13} \\ a_{14} \\ a_{15} \\ a_{16} \\ a_{17} \\ a_{18} \\ 0 \\ 0 \\ 0 \\ 0 \\ 0 \\ a_{24} \end{bmatrix}; \quad \mathbf{f}_b = \begin{bmatrix} f_{b,1} \\ f_{b,2} \\ f_{b,3} \\ 0 \\ f_{b,5} \\ f_{b,6} \\ 0 \\ 0 \\ 0 \\ 0 \\ 0 \\ 0 \\ 0 \\ 0 \\ 0 \\ 0 \\ 0 \\ 0 \\ f_{b,19} \\ f_{b,20} \\ f_{b,21} \\ f_{b,22} \\ f_{b,23} \\ 0 \end{bmatrix} \qquad (5)$$

where the load vector $\mathbf{f}_l$ contains known external loads and element loads, the boundary condition matrix specifies prescribed degrees of freedom, the displacement vector $\mathbf{a}$ contains both known and unknown displacements and the boundary force vector $\mathbf{f}_b$ contains unknown support (boundary) forces.

Solving the system of equations

Solving the system of equations yields

$$
\begin{bmatrix} a_4 \\ a_7 \\ a_8 \\ a_9 \\ a_{10} \\ a_{11} \\ a_{12} \\ a_{13} \\ a_{14} \\ a_{15} \\ a_{16} \\ a_{17} \\ a_{18} \\ a_{24} \end{bmatrix}
=
\begin{bmatrix} -3.9583 \\ 7.5357 \\ -0.2874 \\ -14.9305 \\ -3.2812 \\ -1.7188 \\ -5.3735 \\ 7.5161 \\ -0.3126 \\ -4.2361 \\ -1.9271 \\ -1.7188 \\ 4.6656 \\ -5.1513 \end{bmatrix} 10^{-3};
\qquad
\begin{bmatrix} f_{b,1} \\ f_{b,2} \\ f_{b,3} \\ f_{b,5} \\ f_{b,6} \\ f_{b,19} \\ f_{b,20} \\ f_{b,21} \\ f_{b,22} \\ f_{b,23} \end{bmatrix}
=
\begin{bmatrix} 1.9268 \\ 28.7409 \\ 0.5417 \\ 1.3750 \\ 0.4453 \\ -3.9268 \\ 31.2591 \\ 0.4583 \\ 4.0000 \\ 1.3750 \end{bmatrix} 10^3
\qquad (6)
$$

This implies that in the xy-plane, the frame is deformed in the same manner as in Example 4.2. Moreover, the node loaded in the direction of the z-axis is displaced 14.9 mm in the direction of the load. Figure 3 shows the external loads and computed support forces. We can ascertain that the sum of the forces is zero in the x-, y- and z-direction, respectively. That is, we have force equilibrium in all three directions and moment equilibria about all three axes are satisfied.

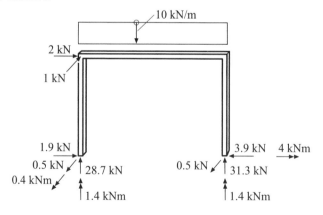

Figure 3   External load and computed support forces

Internal forces

Using (7.77), the displacements can be expressed in the local coordinate system of each element, respectively. Computed section forces are shown in Figure 4.

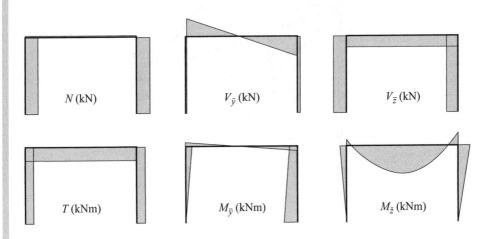

Figure 4   Section forces in the elements

Exercises

7.1

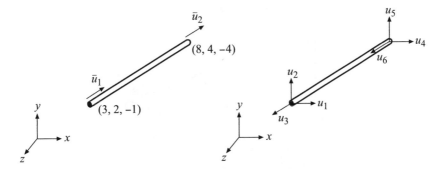

The bar element in the figure has the degrees of freedom $\bar{u}_1$ and $\bar{u}_2$ along its local $\bar{x}$-axis. Express $\bar{u}_1$ and $\bar{u}_2$ as functions of u_1, u_2, u_3, u_4, u_5 and u_6 in a global xyz-system, that is determine the coefficients in the matrix $\mathbf{G}$ in expression (7.11).

7.2

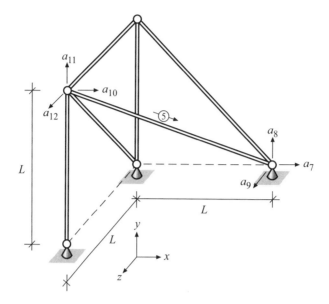

For the truss in the figure, $L = 3.0$ m, $A = 1.0 \times 10^{-3}$ m^2 and $E = 210$ GPa. The displacements a_{10}, a_{11} and a_{12} have been computed to $a_{10} = 11.464$ mm, $a_{11} = -2.857$ mm and $a_{12} = 6.898$ mm, respectively

(a) Determine the displacements $\bar{u}_1^{(5)}$ and $\bar{u}_2^{(5)}$ for Element 5.
(b) Determine the axial deformation of the bar and also the normal force and stress in it.

7.3

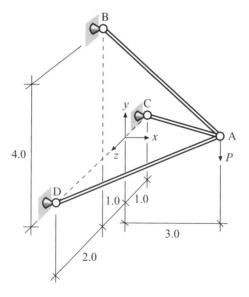

For the truss in the figure, $A = 4.0 \times 10^{-4}$ m^2, $E = 200$ GPa and $P = 50$ kN. Compute the displacement at point A in the x-direction, y-direction and z-direction and also the normal force in each of the three bars.

7.4

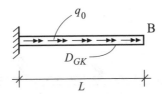

Consider a beam rigidly fixed at its left end and unconstrained at its right end. The beam is loaded by a constant load $q_\omega(\bar{x}) = q_0$.

(a) Determine the element loads $\bar{\mathbf{f}}_l^e$.

(b) Determine the rotation of point B.

7.5

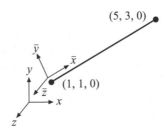

The figure shows a three-dimensional beam element with end point coordinates given in a global xyz-system. The beam is in a global xy-plane and its local $\bar{z}$-axis coincides with the global z-axis. Determine the coefficients in the matrix $\mathbf{G}$ in expression (7.77).

7.6

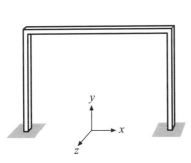

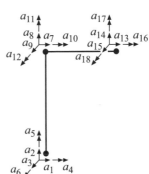

The frame in the figure to the left is symmetric. The right figure illustrates a computational model where the symmetry of the frame is considered.

(a) Assume that the frame is loaded by a symmetric load and specify which displacements should be prescribed at the symmetry plane.

(b) Assume that the frame is loaded by an anti-symmetric load and specify which displacements should be prescribed at the symmetry plane.

7.7

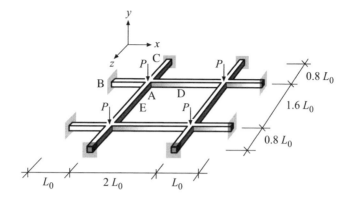

The beam grid in the figure is composed of a number of beams with $A = 1.5 \times 10^{-3}$ m^2, $I_{\bar{y}} = I_{\bar{z}} = 2.0 \times 10^{-6}$ m^4, $K_v = 3.0 \times 10^{-6}$ m^4, $E = 210$ GPa and $G = 80$ GPa. The length $L_0 = 1.0$ m. The grid is loaded by four downwards directed point loads, each of magnitude $P = 20$ kN. The structure is symmetric about both the y- and z-axis and is rigidly fixed at all ends. Determine the largest vertical displacement of the grid and the section forces at the supports.

8

Flows in Networks

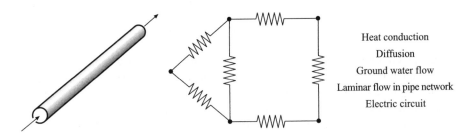

Heat conduction
Diffusion
Ground water flow
Laminar flow in pipe network
Electric circuit

Figure 8.1 A network of different types of flows

Structural mechanics is part of a greater area called *applied mechanics*. In applied mechanics, different causal physical phenomena are studied. Many of them take departure from similar basic mechanisms and principles, mainly the constitutive relationship and the continuity and balance conditions. We now take advantage of this and use the systematics introduced in this textbook to expand the possibilities of creating computational models and perform analyses to an extended set of fields in applied mechanics (Figure 8.1). Here, we have chosen fields that have a common mode of action; they describe flows of energy or substance in different states. We refer to them as *flow problems*. The similarity is based on the fact that we use constitutive relations at the material level and establish a system description using continuity and balance conditions similar to the compatibility and equilibrium conditions used in structural mechanics.

In structural mechanics, we have used the constitutive relation Hooke's law (3.9) which, with substitution of the kinematic relation $\varepsilon = \frac{du}{d\bar{x}}$, can be written as

$$\sigma = E\frac{du}{d\bar{x}} \tag{8.1}$$

The relation implies that a displacement gradient, that is a difference in displacement between two adjacent points, generates a stress and that the magnitude of this stress is proportional to

Structural Mechanics: Modelling and Analysis of Frames and Trusses, First Edition.
Karl-Gunnar Olsson and Ola Dahlblom.
© 2016 John Wiley & Sons, Ltd. Published 2016 by John Wiley & Sons, Ltd.

the elastic modulus of the material. The corresponding constitutive relation for *heat conduction* is *Fourier's law*

$$q = -k\frac{dT}{d\bar{x}} \tag{8.2}$$

Fourier's law expresses that a *temperature gradient*, that is a difference in temperature between two adjacent points, $\frac{dT}{d\bar{x}}$, generates a heat flow q and that the flow is proportional to the *thermal conductivity* k of the material. The minus sign indicates that heat flows from a higher temperature to a lower one. Two further examples of flow problems are *diffusion*, which is described by *Fick's law*

$$q = -D\frac{dc}{d\bar{x}} \tag{8.3}$$

and *groundwater flow*, which is described by *Darcy's law*

$$q = -k\frac{dh}{d\bar{x}} \tag{8.4}$$

All these equations give, at the material level, the constitutive relations and they express that a local difference of potential drives the system. It can be the difference between two adjacent displacements or between two adjacent temperatures.

To the material level, a balance condition is related as well. The concept of stress is based on the fact that the stresses acting at a rectangular cuboid keep it in equilibrium. For the flow through a corresponding rectangular cuboid, we have the corresponding balance: inflow = outflow (Figure 8.2). While the stress in a material point is illustrated by oppositely directed stress arrows, the heat flow in the material point has only one direction and can therefore be represented by one single flow arrow. This flow is called *flux* and for heat conduction it has the dimension (W/m^2).

If we progress from the material level via the element level to the system level, we have, within the area of structural mechanics, used kinematics and force relations to link the different levels together. The corresponding connections between lower and higher levels for flow problems will be *continuity* and *flow balance*.

The flow problems are usually also termed *field problems* and the concept field is then associated with continuous flows in two and three dimensions. We here restrict ourselves to networks of bodies with extension in one dimension and to systems where the flow is one-dimensional (Figure 8.3). With heat transfer as an example, we demonstrate the systematics to build computational models for these networks. In Section 8.1, the basic relations for heat conduction, convection and radiation are presented. By an appropriate definition of the element nodal flows, a matrix relation between heat flows and nodal temperatures is established in Section 8.2, analogous to the element relation for a bar element. After that, a global computational model is built by assembling the local elements into a global network, Section 8.3. The assembling

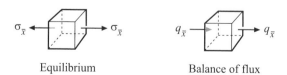

Equilibrium   Balance of flux

Figure 8.2   Local equilibrium and local flow balance

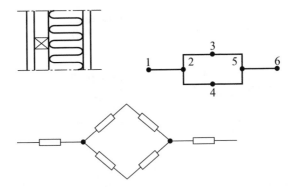

Figure 8.3   One-dimensional heat conduction and electrical circuits are examples of field problems that can be described as networks of one-dimensional flows

is based on the continuity of temperature and flow balances at the nodes of the network. Temperature and flow are scalar quantities and thereby independent of direction. The network can therefore be built without introducing a global coordinate system. Figure 8.4 shows how computation methods for transport of heat energy can be formulated from the material level to a network. Finally, in Section 8.4, we give examples of more flow problems – moisture diffusion, electrical networks, groundwater flow and laminar pipe flow – that can be treated using the same systematics.

8.1   Heat Transport

Heat energy can be transported in three different ways: by *conduction*, *convection* and *radiation*. We start with conduction, which can be handled in direct analogy with bar action. In Section 8.1.5, we thereafter briefly discuss the two other mechanisms for heat transport: convection and radiation.

For heat transport by heat conduction, we seek a relation between the *heat* supplied to a body and the resulting *temperatures* along the body (Figure 8.5). The derivation consists of two steps: from the material level to the cross-section level and from the cross-section level to a body with one-dimensional heat flow.

8.1.1   Definitions

A body with the length L and the cross-sectional area A is conducting heat along a local $\bar{x}$-axis. The quantities of heat conduction are illustrated in Figure 8.6. At the material level, all quantities are free to vary in three-dimensional space. For one-dimensional heat conduction, the quantities at the material point level are flux $q_{\bar{x}}(\bar{x}, \bar{y}, \bar{z})$, temperature gradient $\frac{dT}{d\bar{x}}(\bar{x}, \bar{y}, \bar{z})$ and material conductivity $k(\bar{x}, \bar{y}, \bar{z})$. By presuming that the temperature gradient is constant over the cross-section, $\frac{dT}{d\bar{x}}(\bar{x}, \bar{y}, \bar{z}) = \frac{dT}{d\bar{x}}(\bar{x})$ and by introducing a generalised flow measure $H(\bar{x})$, a generalised conductivity measure $k(\bar{x})$ for the cross-section level of heat conduction can be

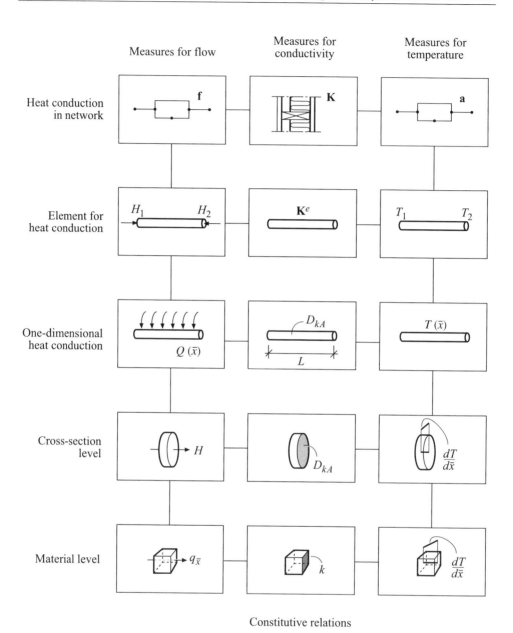

Figure 8.4   The quantities and relations of heat conduction

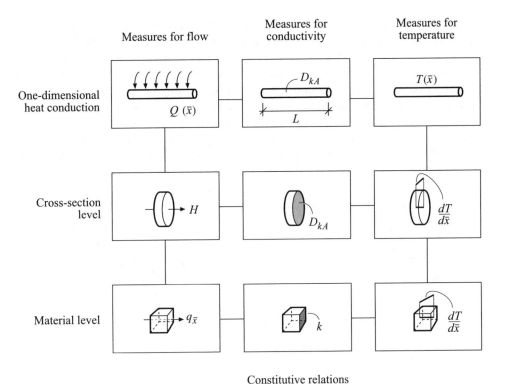

Figure 8.5   From the material level to one-dimensional heat conduction

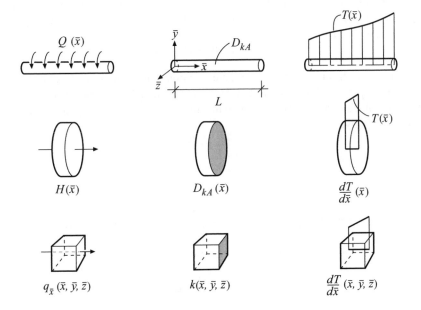

Figure 8.6   The quantities of heat conduction

derived. After that, the one-dimensional equation of heat conduction can be formulated as a relation between the flow $Q(\bar{x})$ supplied along the body and the temperature $T(\bar{x})$ of the body.

An appropriate choice for the location of the local $\bar{x}$-axis on the cross-sectional surface is the point which fulfils the conditions $\int_A k\bar{y}dA = 0$ and $\int_A k\bar{z}dA = 0$. For a temperature gradient that is constant across the cross-section, $\frac{dT}{d\bar{x}} = \text{constant}$, the $\bar{x}$-axis (the system line) will coincide with the centre of gravity of the flux across the cross-section. Also, if the conductivity is constant across the cross-section, that is $k(\bar{y}, \bar{z}) = \text{constant}$, the system line will coincide with the centroid of the cross-section. As opposed to the system lines we have considered so far, we allow the system line of heat conduction to be curved. The length is then considered to be the length of the curved system line (Figure 8.7).

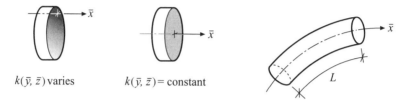

$k(\bar{y}, \bar{z})$ varies $k(\bar{y}, \bar{z}) = \text{constant}$ L

Figure 8.7   The system line of heat conduction

8.1.2   The Material Level

Temperature Gradient

One-dimensional heat conduction is driven by a difference ΔT in temperature between two adjacent material points A and B along a straight or a curved local $\bar{x}$-axis, where point A has the coordinate $\bar{x}$ and point B the coordinate $(\bar{x} + d\bar{x})$. In the limit $d\bar{x} \to 0$, the length $|AB|$ of line AB is (Figure 8.8)

$$|AB| = d\bar{x} \qquad (8.5)$$

To formulate a constitutive relation for a material point that has no spatial extension, we have to be able to express a difference in temperature as a length-independent measure. Such a measure is obtained if we divide the change in temperature ΔT by the distance $|AB|$, which in the limit $d\bar{x} \to 0$ gives

$$\frac{\Delta T}{|AB|} = \frac{dT}{d\bar{x}} \qquad (8.6)$$

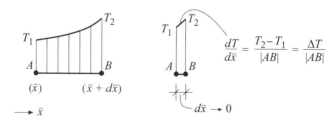

Figure 8.8   Temperature gradient

where $\frac{dT}{d\bar{x}}$ expresses the change in temperature and is referred to as *temperature gradient*. The temperature gradient has the dimension (K/m).

Heat Flux

The amount of heat energy that flows through a material point per unit of area and per unit of time is referred to as *heat flux* (Figure 8.9). The heat flux is denoted $q_{\bar{x}}$ and has the dimension (W/m^2) or (J/m^{2}s). Normally, the heat flux is defined as positive when heat flows in the direction of the local $\bar{x}$-axis.

Figure 8.9   Heat flux

Material Relation

For many materials and within limited temperature intervals, one can with a good accuracy assume a linear relation between the temperature gradient and the heat flux, that is

$$q_{\bar{x}}(\bar{x}, \bar{y}, \bar{z}) = -k(\bar{x}, \bar{y}, \bar{z})\frac{dT}{d\bar{x}}(\bar{x}, \bar{y}, \bar{z}) \tag{8.7}$$

where k is the material conductivity with the dimension (W/mK). The minus sign indicates that a positive temperature gradient gives a negative heat flux, that is heat flows from warmer to colder, Figure 8.10. The material can be isotropic or orthotropic. For orthotropic materials, k denotes the conductivity in the longitudinal direction of the considered body. Equation (8.7) is the constitutive relation of heat conduction, which is also often referred to as Fourier's law after the French mathematician Joseph Fourier (1768–1830).[1]

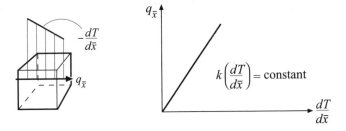

Figure 8.10   Heat flows from warmer to colder, linear material relation

---

[1] Fourier formulated the heat equation and its solution in Théorie de la chaleur (1822) and for this, the mathematical method Fourier analysis is named after him. Among other things, Fourier is also ascribed the discovery of the greenhouse effect.

8.1.3 The Cross-Section Level

Temperature and Temperature Gradient

For one-dimensional heat conduction, at the cross-section level we need to express both a *temperature* and a temperature gradient as functions of $\bar{x}$. This is achieved by letting the temperature $T(\bar{x})$ be the *temperature mean value* of the cross-section

$$T(\bar{x}) = \int_A T(\bar{x}, \bar{y}, \bar{z}) dA / A \tag{8.8}$$

and by assuming that the temperature gradient is constant across the cross-section

$$\boxed{\frac{dT}{d\bar{x}}(\bar{x}, \bar{y}, \bar{z}) = \frac{dT}{d\bar{x}}(\bar{x})} \tag{8.9}$$

where $\frac{dT}{d\bar{x}}(\bar{x})$ is referred to as the *generalised temperature gradient*.

Heat Flow

The total amount of heat energy that flows through a cross-section per unit of time is referred to as *heat flow*. The heat flow is denoted H and has the dimension (W) or (J/s). The heat flow is obtained as the integral (sum) of the heat flux $q_{\bar{x}}$ across the cross-sectional area (Figure 8.11)

$$\boxed{H(\bar{x}) = \int_A q_{\bar{x}}(\bar{x}, \bar{y}, \bar{z}) dA} \tag{8.10}$$

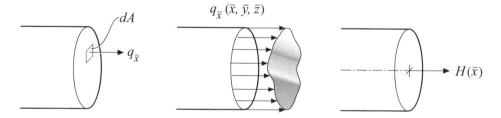

Figure 8.11   Heat flows

The Constitutive Relation of the Cross-Section Level

The resultant expression (8.10), the constitutive relation of the material level (8.7) and the expression for the temperature gradient (8.9) can be combined to

$$H(\bar{x}) = -\int_A k(\bar{x}, \bar{y}, \bar{z}) \frac{dT}{d\bar{x}}(\bar{x}) dA \tag{8.11}$$

The generalised temperature gradient $\frac{dT}{d\bar{x}}(\bar{x})$ is independent of $\bar{y}$ and $\bar{z}$, thus it can be moved outside the integral, which gives

$$H(\bar{x}) = -D_{kA}(\bar{x})\frac{dT}{d\bar{x}}(\bar{x}) \tag{8.12}$$

where

$$D_{kA}(\bar{x}) = \int_A k(\bar{x}, \bar{y}, \bar{z})dA \tag{8.13}$$

is the *conductivity of the cross-section level*. If k is constant across the cross-section, that is independent of $\bar{y}$ and $\bar{z}$, then

$$D_{kA}(\bar{x}) = k(\bar{x})A(\bar{x}) \tag{8.14}$$

Figure 8.12 shows a summary of the relations of the cross-section level.

$$H(\bar{x}) = \int_A q_{\bar{x}}(\bar{x}, \bar{y}, \bar{z})dA \tag{8.10}$$

$$q_{\bar{x}}(\bar{x}, \bar{y}, \bar{z}) = -k(\bar{x}, \bar{y}, \bar{z})\frac{dT}{d\bar{x}}(\bar{x}, \bar{y}, \bar{z}) \tag{8.7}$$

$$\frac{dT}{d\bar{x}}(\bar{x}, \bar{y}, \bar{z}) = \frac{dT}{d\bar{x}}(\bar{x}) \tag{8.9}$$

$$\Rightarrow H(\bar{x}) = -D_{kA}(\bar{x})\frac{dT}{d\bar{x}}(\bar{x}) \tag{8.12}$$

where

$$D_{kA}(\bar{x}) = \int_A k(\bar{x}, \bar{y}, \bar{z})dA$$

Figure 8.12 From the material level to the cross-section level

8.1.4 The Equation for Heat Conduction

Temperature

The temperature $T(\bar{x})$ is taken as the mean value of the temperature across the cross-sections along the local $\bar{x}$-axis of the body; see (8.8).

Heat Balance (Energy Balance)

Consider a small part $d\bar{x}$ of a one-dimensionally heat-conducting body, which is supplied with an external stationary (time-independent) heat flow $Q(\bar{x})$ (W/m), Figure 8.13. The external heat flow is called a *heat source*. For stationary conditions, the following *energy balance* can be established

$$H(\bar{x}) + Q(\bar{x})d\bar{x} - (H(\bar{x}) + dH) = 0 \tag{8.15}$$

where $H(\bar{x})$ is the heat flow at $\bar{x}$ and $(H(\bar{x}) + dH)$ is the heat flow at $(\bar{x} + d\bar{x})$, and where both flows are defined to be positive in direction of the $\bar{x}$-axis. The balance equation expresses that the sum of heat energy supplied to the body per unit of time (heat flow) is equal to zero. The

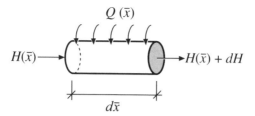

Figure 8.13 Heat balance (energy balance)

relation can be simplified to

$$- dH + Q(\bar{x})d\bar{x} = 0 \tag{8.16}$$

or

$$\boxed{\frac{dH}{d\bar{x}} - Q(\bar{x}) = 0} \tag{8.17}$$

The Differential Equation for Heat Conduction

Substituting (8.12) into (8.17) gives

$$\frac{d}{d\bar{x}}\left(D_{kA}(\bar{x})\frac{dT}{d\bar{x}}\right) + Q(\bar{x}) = 0 \tag{8.18}$$

which is the differential equation for one-dimensional stationary heat conduction. If the conductivity D_{kA} is constant along the body, the expression can be written as

$$\boxed{D_{kA}\frac{d^2T}{d\bar{x}^2} + Q(\bar{x}) = 0} \tag{8.19}$$

Note the similarity between the equation for heat conduction (8.19) and the differential equation for bar action (3.25)

$$D_{EA}\frac{d^2u}{d\bar{x}^2} + q_{\bar{x}}(\bar{x}) = 0 \tag{8.20}$$

If the material conductivity k is constant across the cross-section, we obtain according to (8.14)

$$kA\frac{d^2T}{d\bar{x}^2} + Q(\bar{x}) = 0 \tag{8.21}$$

where the conductivity of the body is the product between the material conductivity k and the cross-sectional area A. How the constitutive relation of the cross-section level together with a flow balance give the differential equation for one-dimensional stationary heat conduction is shown in Figure 8.14.

For a body that is not supplied with or drained of heat along its extension, we have that $Q(\bar{x}) = 0$ and (8.19) then becomes the homogeneous equation

$$D_{kA}(\bar{x})\frac{d^2T}{d\bar{x}^2} = 0 \tag{8.22}$$

$$\left.\begin{array}{l} \dfrac{dH}{d\bar{x}} - Q(\bar{x}) = 0 \qquad (8.17) \\[2mm] H(\bar{x}) = -D_{kA}(\bar{x})\dfrac{dT}{d\bar{x}}(\bar{x}) \quad (8.12) \end{array}\right\} \Rightarrow D_{kA}\dfrac{d^2T}{d\bar{x}^2} + Q(\bar{x}) = 0 \;\;(8.19)$$

$$\text{at constant } D_{kA}$$

Figure 8.14   From the cross-section level to one-dimensional heat conduction

The boundary conditions of the differential equation can be prescribed temperature T or prescribed flow H at the end points of the body.

8.1.5   Convection and Radiation

Convection

The concept of convection stands in general meaning for movement of molecules in liquids and gases (fluids). The molecules can move in a liquid or a gas by random, so called *Brownian motion* but also by *advection* where heat energy related to local movements of molecules is transported by the fact that molecules join the major movement of the fluid. For heat transport by the movements of liquids and gases, most often the term convection is used as a unified term for both advection and Brown's movements of molecules. An example of convective heat transport is the heat transport that locally takes place between a wall surface and its adjacent air. In the interface between wall and air, local air flows take place (Figure 8.15). When molecules that follow a local air flow touch the wall surface, an exchange of heat energy takes place. The amount of heat transport is not only due to the temperature difference ΔT between the wall surface and the adjacent air, but also due to the characteristics of the wall surface and the local air flow. The latter can be summarised as conductivity over the interface and be expressed as a *convective heat transfer coefficient* h_c (c for convection) with the unit (W/m^{2}K). We can thereby formulate a local constitutive relation for heat transport by convection as

$$q_{\bar{x}}(\bar{y},\bar{z}) = -h_c(\bar{y},\bar{z})\,\Delta T(\bar{y},\bar{z}) \tag{8.23}$$

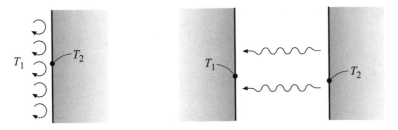

Figure 8.15   Convection and radiation

where the temperature difference ΔT is positive for increasing temperature in the direction of the local $\bar{x}$-axis.

For a cross-section with the area A, we have the conductive heat flow H given by the integral of the heat flux $q_{\bar{x}}$ according to (8.10). With the cross-section temperature T regarded as the average temperature according to (8.8) and with $h_c(\bar{y}, \bar{z}) = $ constant, we have

$$q_{\bar{x}} = -h_c \Delta T \tag{8.24}$$

Substituting (8.24) into (8.10) the constitutive relation at the cross-section level can be written as

$$H = -h_c \, A \, \Delta T \tag{8.25}$$

Radiation

The third phenomenon for transport of heat energy is radiation. All bodies with a temperature above absolute zero send out heat radiation. Heat radiation is an electromagnetic radiation with different spectra of frequencies depending on the temperature of the body. The higher the temperature is, the higher is the average frequency. It has been shown that the radiated flux is proportional to the absolute temperature to the power of 4. The local constitutive relation for heat radiation is formulated by Stefan–Boltzmann's law as

$$q_{\bar{x}} = \varepsilon \sigma \left(T^4 - T_0^4 \right) \tag{8.26}$$

where $q_{\bar{x}}$ is the total radiated flux. T is the temperature of the radiating surface and T_0 is the temperature of the environment or an opposite surface given in Kelvin (K) (Figure 8.15). The factor ε is a constant in the interval $[0, 1]$ which describes the magnitude of the radiation. For most bodies, the value $\varepsilon = 1$ is approximately used, which corresponds to an ideal black body. The factor σ is called the Stefan–Boltzmann constant and has the value

$$\sigma = \frac{2\pi^5 k_B^4}{15c^2 h^3} = 5.67 \times 10^{-8} \qquad \text{(W/m}^2\text{K}^4) \tag{8.27}$$

where k_B is the Boltzmann constant, c is the speed of light and h is the Planck constant.

By reformulation of the Stefan–Boltzmann law (8.26), we can formulate a constitutive relation for radiation analogous to Fourier's law (8.7) and the constitutive relation for convective heat transport (8.23). We have

$$\left(T^4 - T_0^4 \right) = \left(T^2 + T_0^2 \right) \left(T^2 - T_0^2 \right) = \left(T^2 + T_0^2 \right) \left(T + T_0 \right) \left(T - T_0 \right) \tag{8.28}$$

which means that (8.26) can be written as

$$q_{\bar{x}} = \varepsilon \sigma \left(T^2 + T_0^2 \right) \left(T + T_0 \right) \left(T - T_0 \right) \tag{8.29}$$

or

$$q_{\bar{x}} = -h_r \Delta T \tag{8.30}$$

where

$$h_r = \varepsilon \sigma \left(T^2 + T_0^2 \right) \left(T + T_0 \right) \tag{8.31}$$

and where the temperature difference ΔT is positive for increasing temperature in the direction of the local $\bar{x}$-axis.

The factor h_r includes the radiation properties of the body as well as the present temperature levels of the body and the environment. This means that a computation that includes a detailed modelling of radiation has to be performed using an iterative process where the value of h_r is modified as the computation yields current temperatures. For temperatures in a limited interval, such as for buildings at service temperature, one may disregard variations of h_r.

The heat flow of the radiation H we have as the integral of the heat flux $q_{\bar{x}}$ across the cross-sectional area, according to (8.10). Substituting (8.30) in (8.10) and with the cross-section temperature T regarded as the average temperature of a cross-section according to (8.9), the constitutive relation of the cross-section can be written as

$$H = -h_r \, A \, \Delta T \tag{8.32}$$

8.2   Element for Heat Transport

We here establish element equations, which can be used for heat conduction as well as for heat transport by convection and radiation. First, element equations for a heat-conducting element in analogy with the bar element are formulated. Thereafter, it is shown how element equations can be established for the parts of a system where convection and/or radiation drives the heat transport (Figure 8.16).

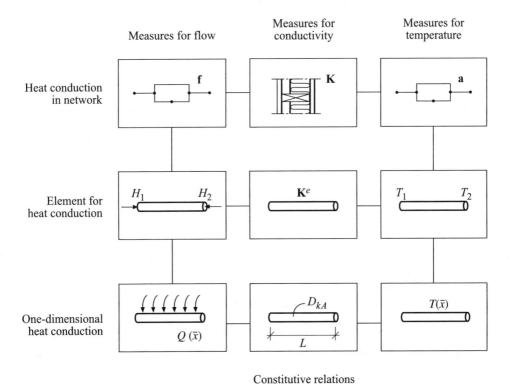

Constitutive relations

Figure 8.16   From one-dimensional heat conduction to a heat-conducting element

8.2.1 Definitions

The element in Figure 8.17 has the *nodal temperatures* $T_1 = T(0)$ and $T_2 = T(L)$. To be able to the use the systematics of Chapter 3, we introduce an *element flow* $H^e(\bar{x}) = -H(\bar{x})$, which is positive when it is directed as the negative direction of the local $\bar{x}$-axis (Figure 8.17). The constitutive relations of the cross-section level (8.12) can analogous to (3.23) thereby be written as

$$H^e(\bar{x}) = D_{kA}(\bar{x})\frac{dT}{d\bar{x}}(\bar{x}) \tag{8.33}$$

with D_{kA} according to (8.14) for heat conduction. Furthermore, (8.25) and (8.32) can be written as

$$H^e = D_{hA}\Delta T \tag{8.34}$$

with

$$D_{hA} = h_c\, A \tag{8.35}$$

for convection and

$$D_{hA} = h_r\, A \tag{8.36}$$

for radiation. The heat flows $H_1 = -H^e(0)$ and $H_2 = H^e(L)$ are referred to as *nodal flows* and are defined as positive if they are directed into the body. Since nodal temperatures and nodal flows are independent of element direction, we have here no reason to distinguish local directions from global ones, that is $\bar{\mathbf{a}}^e = \mathbf{a}^e$ and $\bar{\mathbf{f}}^e = \mathbf{f}^e$.

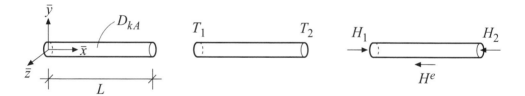

Figure 8.17   A heat-conducting element

8.2.2 Solving the Heat Conduction Equation

The general solution $T(\bar{x})$ of the differential equation for one-dimensional heat conduction (8.19) can be written as the sum of the solution $T_h(\bar{x})$ to the homogeneous differential equation and a particular solution $T_p(x)$

$$T(\bar{x}) = T_h(\bar{x}) + T_p(\bar{x}) \tag{8.37}$$

As mentioned earlier, we choose to determine the constants of the integration from the solution to the homogeneous differential equation. If the homogeneous differential equation (8.22) is divided by the conductivity D_{kA}, we obtain

$$\frac{d^2T}{d\bar{x}^2} = 0 \tag{8.38}$$

Analogous to the derivation (3.29)–(3.43) we can find a solution $T_h(\bar{x})$ to the homogeneous differential equation that expresses the temperature variation along the element as a function of the nodal temperatures of the element

$$T_h(\bar{x}) = \mathbf{N}\mathbf{a}^e = N_1(\bar{x})\,T_1 + N_2(\bar{x})\,T_2 \tag{8.39}$$

where

$$\mathbf{N} = \left[1 - \frac{\bar{x}}{L}\ \ \frac{\bar{x}}{L}\right]; \quad \mathbf{a}^e = \begin{bmatrix} T_1 \\ T_2 \end{bmatrix} \tag{8.40}$$

The general solution to the differential equation is given by (8.37). Substituting (8.39) gives

$$\boxed{T(\bar{x}) = \mathbf{N}\mathbf{a}^e + T_p(\bar{x})} \tag{8.41}$$

where the particular solution $T_p(\bar{x})$ is different for different types of supplied heat flow along the heat-conducting element. The choice to, in (8.41), express the constants of integration as functions of the nodal temperatures $\mathbf{a}^e$ gives the condition that the particular solution may not affect $\mathbf{a}^e$, that is

$$T_p(0) = 0 \tag{8.42}$$

$$T_p(L) = 0 \tag{8.43}$$

With this systematics, the general solution $T(\bar{x})$ can be understood as the sum of the temperature distribution created by the nodal temperatures $T_h(\bar{x})$ of the non-loaded element and the temperature distribution created by a supplied heat flow along an element with a temperature $T = 0$ at both its ends. Example 8.1 shows how the particular solution is obtained for a heat-conducting element with a supplied heat flow that is constant along the element.

Differentiating (8.41) gives

$$\frac{dT}{d\bar{x}} = \mathbf{B}\mathbf{a}^e + \frac{dT_p}{d\bar{x}} \tag{8.44}$$

where

$$\mathbf{B} = \frac{d\mathbf{N}}{d\bar{x}} = \frac{d\tilde{\mathbf{N}}}{d\bar{x}}\mathbf{C}^{-1} = [0\ \ 1]\begin{bmatrix} 1 & 0 \\ -\frac{1}{L} & \frac{1}{L} \end{bmatrix} = \frac{1}{L}[-1\ \ 1] \tag{8.45}$$

Substituting (8.44) into the expression for element flow (8.33) gives

$$H^e(\bar{x}) = D_{kA}\left(\mathbf{B}\mathbf{a}^e + \frac{dT_p}{dx}\right) \tag{8.46}$$

or

$$H^e(\bar{x}) = D_{kA}\mathbf{B}\mathbf{a}^e + H_p(\bar{x}) \tag{8.47}$$

where

$$H_p(\bar{x}) = D_{kA}\frac{dT_p}{d\bar{x}} \tag{8.48}$$

The definitions we have introduced for flows across the element ends (Figure 8.17) give

$$\boxed{H_1 = -H^e(0); \quad H_2 = H^e(L)} \tag{8.49}$$

Substituting (8.47) gives the nodal flows

$$H^e(0) = D_{kA}\mathbf{B}\mathbf{a}^e + H_p(0) \tag{8.50}$$

$$H^e(L) = D_{kA}\mathbf{B}\mathbf{a}^e + H_p(L) \tag{8.51}$$

With

$$\mathbf{f}_b^e = \begin{bmatrix} H_1 \\ H_2 \end{bmatrix}; \quad \mathbf{K}^e = \frac{D_{kA}}{L}\begin{bmatrix} 1 & -1 \\ -1 & 1 \end{bmatrix}; \quad \mathbf{f}_p^e = \begin{bmatrix} -H_p(0) \\ H_p(L) \end{bmatrix} \tag{8.52}$$

(8.50) and (8.51) can be written in matrix form

$$\mathbf{f}_b^e = \mathbf{K}^e\mathbf{a}^e + \mathbf{f}_p^e \tag{8.53}$$

The left-hand side of the system of equations contains the nodal flows of the element $\mathbf{f}_b^e$, that is the flows that act on both the ends of the element. On the right-hand side, these flows are divided into two parts. The product $\mathbf{K}^e\mathbf{a}^e$ gives the part of the flows that is generated by the temperatures of the end points and the vector $\mathbf{f}_p^e$ gives the part of the flows that is generated by the along the element supplied heat flow, the heat source $Q(\bar{x})$. The division of the flows into two parts is illustrated in Figure 8.18.

To prepare for a systematic handling of *loads*, we now introduce an *element load vector* $\mathbf{f}_l^e$,

$$\mathbf{f}_l^e = -\mathbf{f}_p^e = \begin{bmatrix} H_p(0) \\ -H_p(L) \end{bmatrix} \tag{8.54}$$

where the components of $\mathbf{f}_l^e$ can be interpreted as resulting flows of the heat source $Q(\bar{x})$. These resulting flows act on the nodes at the end points of the element (Figure 8.19).

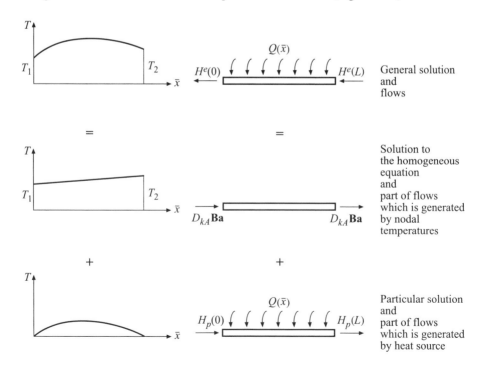

Figure 8.18   Heat-conducting element in balance

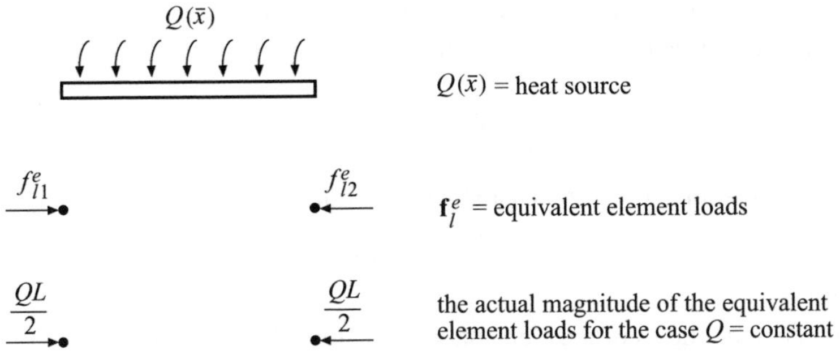

$Q(\bar{x})$

$Q(\bar{x})$ = heat source

f^e_{l1}

f^e_{l2}　　$\mathbf{f}^e_l$ = equivalent element loads

$\dfrac{QL}{2}$

$\dfrac{QL}{2}$　　the actual magnitude of the equivalent element loads for the case Q = constant

Figure 8.19　Heat source and equivalent element loads

Thereby, we can write (8.53) as

$$\mathbf{K}^e\mathbf{a}^e = \mathbf{f}^e \tag{8.55}$$

where

$$\bar{\mathbf{f}}^e = \bar{\mathbf{f}}^e_b + \bar{\mathbf{f}}^e_l \tag{8.56}$$

Equation (8.55) is the constitutive relation between nodal flows and temperatures for a heat-conducting element. The relation is referred to as the element equation for a heat-conducting element and $\mathbf{K}^e$ is the *conductivity matrix* of the element, $\mathbf{a}^e$ its *temperature vector* and $\mathbf{f}^e$ its *flow vector*. The relations leading to the relation of the heat-conducting element are summarised in Figure 8.20.

$$
\left.\begin{aligned}
H_1 &= -H^e(0) &&(8.49)\\
H_2 &= H^e(L) &&\\
H^e(\bar{x}) &= D_{kA}(\bar{x})\dfrac{dT}{d\bar{x}} &&(8.33)\\
T(\bar{x}) &= \mathbf{N}\bar{\mathbf{a}}^e + T_p(\bar{x}) &&(8.41)
\end{aligned}\right\}
\Rightarrow \mathbf{K}^e\mathbf{a}^e = \mathbf{f}^e \ (8.55)
$$

where

$$\mathbf{f}^e = \mathbf{f}^e_b + \mathbf{f}^e_l$$

$$\mathbf{K}^e = \dfrac{D_{kA}}{L}\begin{bmatrix}1 & -1\\-1 & 1\end{bmatrix}; \quad \mathbf{a}^e = \begin{bmatrix}T_1\\T_2\end{bmatrix}$$

$$\mathbf{f}^e_b = \begin{bmatrix}H_1\\H_2\end{bmatrix}; \quad \mathbf{f}^e_l = \begin{bmatrix}H_p(0)\\-H_p(L)\end{bmatrix}$$

Figure 8.20　From the cross-section level to element for one-dimensional heat-conduction

For a non-loaded element, that is for $\bar{\mathbf{f}}^e_l = \mathbf{0}$, the temperature is described by the solution to the homogeneous equation only. The case when the supplied heat is uniform is treated in Example 8.1.

For heat transport through walls, the energy flow is modelled as heat conduction in the solid materials, while the heat transport over air spaces and in the interface between the wall and adjacent air is dominated by convection and radiation. Often the two phenomena are summarised and quantified by a heat resistance R_s, where

$$R_s = \frac{1}{h_c + h_r} \tag{8.57}$$

The constitutive relation surface zone can then be written as

$$H^e = \frac{A}{R_s} \Delta T \tag{8.58}$$

Substituting the nodal flows of the element, with positive directions according to Figure 8.17 and with $\Delta T = (T_2 - T_1)$ we obtain

$$H_1 = -\frac{A}{R_s}(T_2 - T_1) \tag{8.59}$$

$$H_2 = \frac{A}{R_s}(T_2 - T_1) \tag{8.60}$$

or in matrix form

$$\mathbf{K}^e \, \mathbf{a}^e = \mathbf{f}^e \tag{8.61}$$

where $\mathbf{K}^e$ is the 'conductivity matrix' of the element, $\mathbf{a}^e$ its temperature vector and $\mathbf{f}^e$ its flow vector.

Example 8.1 A one-dimensional body with a uniformly distributed heat source

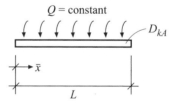

Figure 1   A heat-conducting element with uniformly distributed heat flow Q

Determine the element load vector $\mathbf{f}_l^e$ for heat-conducting element of length L with a uniformly distributed heat source Q (Figure 1).

The element load vector $\mathbf{f}_l^e$ is given by (8.54). To be able to determine $H_p(\bar{x})$, which is given by (8.48), we seek first a particular solution $T_p(\bar{x})$ to the differential equation (8.19). The particular solution is required to satisfy (8.19) and the two boundary conditions (8.42) and (8.43); see Figure 8.18. With constant Q, Equation (8.19) can, for the particular solution, be written as

$$D_{kA} \frac{d^2 T_p}{d\bar{x}^2} + Q = 0 \tag{1}$$

Integration twice gives

$$D_{kA}\frac{dT_p}{d\bar{x}} + Q\bar{x} - C_1 = 0 \tag{2}$$

$$D_{kA}T_p(\bar{x}) + Q\frac{\bar{x}^2}{2} - C_1\bar{x} - C_2 = 0 \tag{3}$$

or

$$T_p(\bar{x}) = \frac{1}{D_{kA}}\left(-Q\frac{\bar{x}^2}{2} + C_1\bar{x} + C_2\right) \tag{4}$$

The boundary conditions (8.42) and (8.43) give

$$T_p(0) = \frac{1}{D_{kA}}C_2 = 0; \quad C_2 = 0 \tag{5}$$

$$T_p(L) = \frac{1}{D_{kA}}\left(-Q\frac{L^2}{2} + C_1L + C_2\right) = 0; \quad C_1 = Q\frac{L}{2} \tag{6}$$

Substituting the constants C_1 and C_2, we obtain the particular solution

$$T_p(\bar{x}) = -\frac{Q}{D_{kA}}\left(\frac{\bar{x}^2}{2} - \frac{L\bar{x}}{2}\right) \tag{7}$$

Differentiation gives

$$\frac{dT_p}{d\bar{x}} = -\frac{Q}{D_{kA}}\left(\bar{x} - \frac{L}{2}\right) \tag{8}$$

which substituted into (8.48) gives

$$H_p(\bar{x}) = -Q\left(\bar{x} - \frac{L}{2}\right) \tag{9}$$

At the end points of the element, we have

$$H_p(0) = Q\frac{L}{2}; \quad H_p(L) = -Q\frac{L}{2} \tag{10}$$

Substituting $H_p(0)$ and $H_p(L)$ into (8.54), we obtain the element load vector

$$\bar{\mathbf{f}}_l^e = \frac{QL}{2}\begin{bmatrix}1\\1\end{bmatrix} \tag{11}$$

8.3   Networks of One-Dimensional Heat-Conducting Elements

A network consists of several one-dimensional bodies connected to each other at the nodes. The heat-conducting element we now have formulated is the basis for creation of a computational model for a network.

In the same manner as for the system of bars, we introduce a global numbering of all the temperature degrees of freedom and gather these in a global temperature vector $\mathbf{a}$,

$$\mathbf{a} = \begin{bmatrix} a_1 \\ \cdot \\ a_i \\ a_j \\ \cdot \\ a_n \end{bmatrix} \tag{8.62}$$

From the element relations for the single heat-conducting elements, we have a local numbering of the temperatures, T_1 and T_2. By continuity conditions, each of these temperatures at the element level is associated with a temperature in the global system. Continuity means that the temperature has to be continuous over a node, which yields that all connecting elements have the same nodal temperature. For an element associated with the global temperatures a_i and a_j, we obtain the following continuity conditions:

$$T_1 = a_i \tag{8.63}$$

$$T_2 = a_j \tag{8.64}$$

The continuity conditions can be written in matrix form as

$$\mathbf{a}^e = \mathbf{Ha} \tag{8.65}$$

where $\mathbf{a}^e$ is the nodal temperatures of the element (8.40), $\mathbf{a}$ is the temperature degrees of freedom of the network (8.62) and $\mathbf{H}$ is a transformation matrix with $H_{1,i} = 1$, $H_{2,j} = 1$ and all other matrix elements equal to 0; cf. (2.13).

In the discretised model of the network, external source flows can only be inserted at the nodes. It can be equivalent nodal flows from heat generated inside the heat-conducting elements, external flows which are acting directly at the nodes and boundary flows at the boundaries of the network. These flows are denoted f_i and are gathered in a global nodal flow vector $\mathbf{f}$,

$$\mathbf{f} = \begin{bmatrix} f_1 \\ \cdot \\ f_i \\ f_j \\ \cdot \\ f_n \end{bmatrix} \tag{8.66}$$

By using balance conditions, we now relate the nodal flows in the single heat-conducting elements to the network. This is done by expressing the nodal flows in a form which systematises the formulation of balance equations in the nodes of the network. From the element relation (8.55), we have the element flows expressed as nodal flows, $\mathbf{f}_b^e$. As for the spring system, we introduce an expanded nodal flow vector $\hat{\mathbf{f}}_b^e$, and here we introduce also an expanded vector for element flows $\hat{\mathbf{f}}_l^e$, each of them with equally many rows as there are degrees of freedom in the network. The expanded flow vectors can be expressed in matrix form using the same matrix $\mathbf{H}$ as was defined by the continuity conditions,

$$\hat{\mathbf{f}}_b^e = \mathbf{H}^T \mathbf{f}_b^e \tag{8.67}$$

$$\hat{\mathbf{f}}_l^e = \mathbf{H}^T \mathbf{f}_l^e \tag{8.68}$$

Substituting equations (8.65), (8.67) and (8.68) into (8.55) and (8.56) gives an element relation in expanded form

$$\hat{\mathbf{f}}_b^e = \hat{\mathbf{K}}^e \mathbf{a} - \hat{\mathbf{f}}_l^e \tag{8.69}$$

where

$$\hat{\mathbf{K}}^e = \mathbf{H}^T \mathbf{K}^e \mathbf{H} \tag{8.70}$$

The matrix $\hat{\mathbf{K}}^e$ will contain the matrix elements, which are, in $\mathbf{K}^e$, placed on the rows and columns that correspond to the global degree of freedom numbers that the element is associated with; for further details, see the discussion in Section 2.3. By this way of writing with expanded matrices, we have a formulation where flow components associated with the same global degree of freedom are on the same row in the flow vector and thus is prepared for global balance equations.

For a single degree of freedom i, the flow balance can be expressed as (Figure 8.21)

$$\sum_{e=1}^{m} \hat{f}_{b,i}^e = f_{ln,i} + f_{b,i} \tag{8.71}$$

where e denotes element number, $f_{ln,i}$ *nodal flow* (point flow acting in the node) and $f_{b,i}$ boundary flow. By establishing a balance equation for each degree of freedom, we obtain for the entire network

$$\sum_{e=1}^{m} \hat{\mathbf{f}}_b^e = \mathbf{f}_{ln} + \mathbf{f}_b \tag{8.72}$$

If the expanded element equations (8.69) are substituted into the balance relations, we obtain

$$\sum_{e=1}^{m} (\hat{\mathbf{K}}^e \mathbf{a} - \hat{\mathbf{f}}_l^e) = \mathbf{f}_{ln} + \mathbf{f}_b \tag{8.73}$$

or

$$\mathbf{K}\mathbf{a} = \mathbf{f} \tag{8.74}$$

where

$$\mathbf{K} = \sum_{e=1}^{m} \hat{\mathbf{K}}^e; \quad \mathbf{f} = \mathbf{f}_l + \mathbf{f}_b; \quad \mathbf{f}_l = \mathbf{f}_{ln} + \mathbf{f}_{lq}; \quad \mathbf{f}_{lq} = \sum_{e=1}^{m} \hat{\mathbf{f}}_l^e \tag{8.75}$$

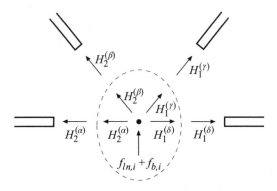

Figure 8.21   Flow balance in a nodal point

$$\hat{\mathbf{f}}_b^e = \mathbf{H}^T \mathbf{f}_b^e \quad (8.67)$$
$$\hat{\mathbf{f}}_l^e = \mathbf{H}^T \mathbf{f}_l^e \quad (8.68)$$
$$\mathbf{K}^e \mathbf{a}^e = \mathbf{f}^e \quad (8.55)$$
$$\mathbf{f}^e = \mathbf{f}_b^e + \mathbf{f}_l^e \quad (8.56)$$
$$\mathbf{a}^e = \mathbf{H}\mathbf{a} \quad (8.65)$$

$$\sum_{e=1}^{m} \hat{\mathbf{f}}_b^e = \mathbf{f}_{ln} + \mathbf{f}_b \quad (8.72)$$
$$\Rightarrow \hat{\mathbf{f}}_b^e = \hat{\mathbf{K}}^e \mathbf{a} - \hat{\mathbf{f}}_l^e \quad (8.69)$$
where
$$\hat{\mathbf{K}}^e = \mathbf{H}^T \mathbf{K}^e \mathbf{H} \quad (8.70)$$

$$\Rightarrow \mathbf{K}\mathbf{a} = \mathbf{f} \quad (8.74)$$
where
$$\mathbf{K} = \sum_{e=1}^{m} \hat{\mathbf{K}}^e$$
$$\mathbf{f} = \mathbf{f}_l + \mathbf{f}_b$$
$$\mathbf{f}_l = \mathbf{f}_{ln} + \sum_{e=1}^{m} \hat{\mathbf{f}}_l^e$$

Figure 8.22   From element for one-dimensional heat-conduction to a network of one-dimensional flows

How continuity conditions, element relations and balance equations lead to a system of equations for a network is shown in Figure 8.22.

Considering the present boundary conditions, the nodal temperatures and boundary flows can be determined from (8.74). Once the temperatures $\mathbf{a}$ have been computed, the temperatures $\mathbf{a}^e$ for an element can be determined from (8.65). The temperature distribution along the element can then be determined using (8.41) and the heat flow distribution using (8.47).

The conductivity matrix $\mathbf{K}$ and the flow vector $\mathbf{f}_l$ have here been described as sums of expanded matrices $\hat{\mathbf{K}}^e$ and vectors $\hat{\mathbf{f}}_l^e$. Normally, these expanded matrices are never actually created. The conductivity matrix $\mathbf{K}$ is instead established by creating a matrix of dimensions $n \times n$ filled with zeros and then adding the coefficients in the element matrix $\mathbf{K}^e$ for each element to the positions corresponding to the global degrees of freedom of the element. In the corresponding manner the flow vector $\mathbf{f}_l$ is created by creating a vector where first the flows acting at the nodes are inserted and then placing the equivalent nodal flows $\mathbf{f}_l^e$ into this on the rows corresponding to the global degrees of freedom of the element; cf. Section 2.3.

Example 8.2 Heat transport through a wall

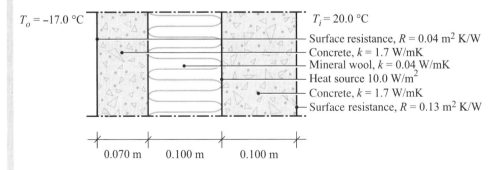

$T_o = -17.0\ °C$

$T_i = 20.0\ °C$
Surface resistance, $R = 0.04$ m^2 K/W
Concrete, $k = 1.7$ W/mK
Mineral wool, $k = 0.04$ W/mK
Heat source 10.0 W/m^2
Concrete, $k = 1.7$ W/mK
Surface resistance, $R = 0.13$ m^2 K/W

0.070 m   0.100 m   0.100 m

Figure 1   A cross-section of an external wall

The wall in Figure 1 consists of two concrete layers with mineral wool in between. The concrete layers have the conductivity $k_b = 1.7$ W/mK and the mineral wool $k_m = 0.04$ W/mK. The thicknesses of the layers are $L_1 = 0.070$ m, $L_2 = 0.100$ m and $L_3 = 0.100$ m.

The temperature on the outside of the wall is $-17.0\,°C$ and on the inside $20.0\,°C$. On the outside, the thermal surface resistance is $R = 0.04\ m^2$ K/W and on the inside $R = 0.13\ m^2$ K/W. On the boundary between the mineral wool and the interior concrete layer, there is a heat source that provides $10.0\ W/m^2$. The temperature distribution and the stationary heat flow through the wall shall be determined.

Computational model

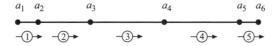

Figure 2   Computational model

We choose to study the flow per m^2 of the wall, that is we study $1\ m^2$ wall. The wall is modelled with five heat-conducting elements, denoted 1, 2, 3, 4 and 5, beginning from the outside (Figure 2). Element 1 describes the thermal surface resistance on the outside, Element 2 the exterior concrete layer, Element 3 the mineral wool, Element 4 the interior concrete layer and Element 5 the interior thermal surface resistance. The model has temperature degrees of freedom a_1, a_2, a_3, a_4, a_5 and a_6. The temperatures given on the outside and on the inside imply that degrees of freedom a_1 and a_6 shall be prescribed to be -17 and 20 $°C$, respectively. The heat source inside the wall implies that $f_{l,4} = 10\ W/m^2$.

Element matrices

For each element, an element relation $\mathbf{K}^e \mathbf{a}^e = \mathbf{f}_l^e + \mathbf{f}_b^e$ can be established. The element conductivity matrix $\mathbf{K}^e$ is given by (8.52). For the five elements in our model, we have the following:

Elements 2–4: With A, k and L known, the element conductivity matrix can be computed,

$$\bar{\mathbf{K}}^2 = \frac{kA}{L}\begin{bmatrix} 1 & -1 \\ -1 & 1 \end{bmatrix} = \frac{1.7 \cdot 1}{0.07}\begin{bmatrix} 1 & -1 \\ -1 & 1 \end{bmatrix} = 24.3\begin{bmatrix} 1 & -1 \\ -1 & 1 \end{bmatrix} \tag{1}$$

$$\bar{\mathbf{K}}^3 = \frac{0.04 \cdot 1}{0.1}\begin{bmatrix} 1 & -1 \\ -1 & 1 \end{bmatrix} = 0.4\begin{bmatrix} 1 & -1 \\ -1 & 1 \end{bmatrix} \tag{2}$$

$$\bar{\mathbf{K}}^4 = \frac{1.7 \cdot 1}{0.1}\begin{bmatrix} 1 & -1 \\ -1 & 1 \end{bmatrix} = 17.0\begin{bmatrix} 1 & -1 \\ -1 & 1 \end{bmatrix} \tag{3}$$

Elements 1 and 5: These elements describe the thermal surface resistance between the wall and the air.

$$\bar{\mathbf{K}}^1 = \frac{A}{R_s}\begin{bmatrix} 1 & -1 \\ -1 & 1 \end{bmatrix} = \frac{1}{0.04}\begin{bmatrix} 1 & -1 \\ -1 & 1 \end{bmatrix} = 25.0\begin{bmatrix} 1 & -1 \\ -1 & 1 \end{bmatrix} \tag{4}$$

$$\bar{\mathbf{K}}^5 = \frac{1}{0.13}\begin{bmatrix} 1 & -1 \\ -1 & 1 \end{bmatrix} = 7.7\begin{bmatrix} 1 & -1 \\ -1 & 1 \end{bmatrix} \tag{5}$$

Continuity conditions

The relation between the local degrees of freedom and the global degrees of freedom is described by the topology matrix:

$$
\text{topology} = \begin{bmatrix} 1 & 1 & 2 \\ 2 & 2 & 3 \\ 3 & 3 & 4 \\ 4 & 4 & 5 \\ 5 & 5 & 6 \end{bmatrix}
\tag{6}
$$

Assembling

Substituting the element conductivity matrices in a global conductivity matrix is performed using the topology information,

$$
\mathbf{K} = \begin{bmatrix}
25.0 & -25.0 & 0 & 0 & 0 & 0 \\
-25.0 & 49.3 & -24.3 & 0 & 0 & 0 \\
0 & -24.3 & 24.7 & -0.4 & 0 & 0 \\
0 & 0 & -0.4 & 17.4 & -17.0 & 0 \\
0 & 0 & 0 & -17.0 & 24.7 & -7.7 \\
0 & 0 & 0 & 0.0 & -7.7 & 7.7
\end{bmatrix}
\tag{7}
$$

The heat source of $10.0\,\text{W/m}^2$ inside the wall is placed on position 4 in the global flow vector,

$$
\mathbf{f}_l = \begin{bmatrix} 0 \\ 0 \\ 0 \\ 10.0 \\ 0 \\ 0 \end{bmatrix}
\tag{8}
$$

Boundary conditions

The temperature is prescribed in the degrees of freedom where the construction is in contact with the air outside and with the air inside, that is $a_1 = -17.0$, $a_6 = 20.0$. This is described by the boundary condition matrix

$$
\text{boundary conditions} = \begin{bmatrix} 1 & -17.0 \\ 6 & 20.0 \end{bmatrix}
\tag{9}
$$

In the degrees of freedom where the temperature is prescribed, boundary flows arise. These are unknown for now and are denoted $f_{b,1}$ and $f_{b,6}$, respectively. The temperature vector $\mathbf{a}$

and the boundary flow vector $\mathbf{f}_b$ can now be written as

$$\mathbf{a} = \begin{bmatrix} -17.0 \\ a_2 \\ a_3 \\ a_4 \\ a_5 \\ 20.0 \end{bmatrix} ; \quad \mathbf{f}_b = \begin{bmatrix} f_{b,1} \\ 0 \\ 0 \\ 0 \\ 0 \\ f_{b,6} \end{bmatrix} \tag{10}$$

Solving the system of equations

By solving the system of equations, we obtain

$$\begin{bmatrix} a_2 \\ a_3 \\ a_4 \\ a_5 \end{bmatrix} = \begin{bmatrix} -16.44 \\ -15.86 \\ 19.24 \\ 19.48 \end{bmatrix} \tag{11}$$

which means that the temperature at the surface of the wall is $-16.44\,°\mathrm{C}$ and the temperatures at the material boundaries are -15.86 and $19.24\,°\mathrm{C}$, respectively, and the temperature on the inside of the wall is $19.48\,°\mathrm{C}$. We also obtain the boundary flows

$$\begin{bmatrix} f_{b,1} \\ f_{b,6} \end{bmatrix} = \begin{bmatrix} -14.0 \\ 4.0 \end{bmatrix} \tag{12}$$

This means that we have the flow $14.0\,\mathrm{W/m^2}$ out of the wall on the outside and $4.0\,\mathrm{W/m^2}$ into the wall on the inside. We can conclude that the flows into the wall are $-14.0 + 10.0 + 4.0 = 0$, that is the external heat balance is satisfied.

Internal heat flows

With the global nodal temperatures $\mathbf{a}$ known, we can, using the continuity relations, determine the temperatures of each element

$$\mathbf{a}^1 = \begin{bmatrix} -17.00 \\ -16.44 \end{bmatrix} \tag{13}$$

$$\mathbf{a}^2 = \begin{bmatrix} -16.44 \\ -15.86 \end{bmatrix} \tag{14}$$

$$\mathbf{a}^3 = \begin{bmatrix} -15.86 \\ 19.24 \end{bmatrix} \tag{15}$$

$$\mathbf{a}^4 = \begin{bmatrix} 19.24 \\ 19.48 \end{bmatrix} \tag{16}$$

$$\mathbf{a}^5 = \begin{bmatrix} 19.48 \\ 20.00 \end{bmatrix} \tag{17}$$

The flow through each element can be computed using (8.46). For Elements 1 and 5, k/L shall be replaced by $1/R$.

$$H^{(1)} = \frac{1}{0.04} \begin{bmatrix} -1 & 1 \end{bmatrix} \begin{bmatrix} -17.00 \\ -16.44 \end{bmatrix} = 14.0 \text{ W} \tag{18}$$

$$H^{(2)} = 1.7 \begin{bmatrix} -\frac{1}{0.07} & \frac{1}{0.07} \end{bmatrix} \begin{bmatrix} -16.44 \\ -15.86 \end{bmatrix} = 14.0 \text{ W} \tag{19}$$

$$H^{(3)} = 0.04 \begin{bmatrix} -\frac{1}{0.1} & \frac{1}{0.1} \end{bmatrix} \begin{bmatrix} -15.86 \\ 19.24 \end{bmatrix} = 14.0 \text{ W} \tag{20}$$

$$H^{(4)} = 1.7 \begin{bmatrix} -\frac{1}{0.1} & \frac{1}{0.1} \end{bmatrix} \begin{bmatrix} 19.24 \\ 19.48 \end{bmatrix} = 4.0 \text{ W} \tag{21}$$

$$H^{(5)} = \frac{1}{0.13} \begin{bmatrix} -1 & 1 \end{bmatrix} \begin{bmatrix} 19.48 \\ 20.00 \end{bmatrix} = 4.0 \text{ W} \tag{22}$$

8.4 Analogies

The above-described methodology for the analysis of systems of heat-conducting bodies has several analogous applications and we briefly discuss the following areas: diffusion, laminar liquid flow in pipes, groundwater flow and electric circuits. The analogies presented have been limited to a description of steady-state (time-independent) processes.

8.4.1 Diffusion – Fick's Law

The physical concept of *diffusion* describes a process where differences in concentration decrease. The term concentration can refer to the matter itself (particles or molecules) and the motion of the matter (thermal energy). In Sections 8.1 and 8.2, in the concept of diffusion, we have studied how thermal energy is transported from hot (high concentration of molecular motion) to cold (low concentration of molecular motion). In a more limited sense, the concept of diffusion is used for description of the transport of substances where the transport is driven by differences in concentration between adjacent regions. With that meaning, *Fick's first law* was formulated in the year 1855 by the German physiologist Adolf Fick (1829–1901)

$$J = -D\frac{dc}{d\bar{x}} \tag{8.76}$$

where c denotes a *concentration* that either can be a density of mass (kg/m^3) or an amount of substance (mol/m^3), D is the *diffusivity* (m^2/s) of the material J indicates a *flux* of matter per unit of area and unit of time (kg/m^{2}s) or a *flux* of substance per unit of area and unit of time (mol/m^{2}s). One common application for buildings is vapour diffusion driven by differences in concentration of vapour.

   During a stationary (independent of time) process, the sum of all matter supplied to a delimited body is equal to zero. No matter is created nor disappears uncontrolled during the process. If we choose to consider the matter as density of mass and consider, analogous to Figure 8.13, a one-dimensional body with the length $d\bar{x}$ and the cross-sectional area $A(\bar{x})$, such a *balance of mass* results in the equation

$$\frac{dH}{d\bar{x}} - Q(\bar{x}) = 0 \tag{8.77}$$

where $H(\bar{x}) = J(\bar{x})A(\bar{x})$ denotes *mass flow* per unit of time through the body (kg/s) and $Q(\bar{x})$ denotes mass supplied to the body per length unit and time unit (kg/ms); cf. (8.10) and (8.17). The term Q is sometimes called the *source term*. Substituting (8.76) into (8.77) gives the differential equation for one-dimensional diffusion

$$D_{DA}\frac{d^2c}{d\bar{x}^2} + Q(\bar{x}) = 0; \quad 0 \leq \bar{x} \leq L \tag{8.78}$$

where D_{DA} is the *diffusivity* of the cross-section. If we assume that the diffusivity of the material D and the cross-sectional area A are constant in the interval considered $0 \leq \bar{x} \leq L$, then $D_{DA} = DA$. The derivation of the element equations for an element for diffusion follows the derivation in Section 8.2 and leads to the element relation

$$\mathbf{K}^e\mathbf{a}^e = \mathbf{f}^e_l + \mathbf{f}^e_b \tag{8.79}$$

where $\mathbf{K}^e$ is the *diffusivity matrix* of the element, $\mathbf{a}^e$ the nodal concentrations, $\mathbf{f}^e_l$ the element load vector and $\mathbf{f}^e_b$ the nodal flow vector. By assembling, a global computational model is built up in the same way as for thermal conduction in a network. For moisture diffusion through a wall (Figure 8.23), typical boundary conditions are the moisture contents c_A and c_B at the respective sides of the wall. If the moisture transport is given per m^2 wall, the area is set to $A = 1$ m^2.

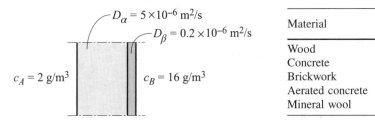

Material	Diffusivity D (m^2/s)
Wood	0.2×10^{-6}
Concrete	1.5×10^{-6}
Brickwork	4×10^{-6}
Aerated concrete	5×10^{-6}
Mineral wool	15×10^{-6}

Figure 8.23   One-dimensional diffusion.

8.4.2   *Liquid Flow in Porous Media – Darcy's Law*

Darcy's law describes how liquids flow in saturated porous media where the liquid flow is driven by differences in pressure in the region considered. A common application is *groundwater flow*. The constitutive relation was formulated in the year 1856 by the Frenchman Henri Darcy (1803–1858) after different experiments with water flowing through water saturated sand packings,

$$v_{\bar{x}} = -k\frac{dh}{d\bar{x}} \tag{8.80}$$

where h denotes *hydraulic head* (m), k is the *permeability* (m/s) of the material and $v_{\bar{x}}$ is the *velocity* (m/s) of the flowing water. For a one-dimensional body with the length $d\bar{x}$ and the cross-sectional area $A(\bar{x})$ a *flow balance* analogous to (8.17) results in the equation

$$\frac{dH}{d\bar{x}} - Q(\bar{x}) = 0 \tag{8.81}$$

where $H(\bar{x}) = v_{\bar{x}}(\bar{x})A(\bar{x})$ is *flow* of liquid through the body (m^3/s) and $Q(\bar{x})$ is an external supply of liquid per length unit of the body (m^3/sm). Substituting (8.81) in (8.80) gives the differential equation for one-dimensional saturated flow in porous media

$$D_{kA}\frac{d^2h}{d\bar{x}^2} + Q(\bar{x}) = 0; \quad 0 \le \bar{x} \le L \tag{8.82}$$

where D_{kA} is the *permeability* of the cross-section. If we assume the permeability k of the material and the cross-sectional area A to be constant in the interval considered $0 \le \bar{x} \le L$ then we have $D_{kA} = kA$.

The derivation of the element equations for a flow element follows the derivation in Section 8.2 and leads to the element relation

$$\mathbf{K}^e\mathbf{a}^e = \mathbf{f}_l^e + \mathbf{f}_b^e \tag{8.83}$$

where $\mathbf{K}^e$ is the *permeability matrix* of the element, $\mathbf{a}^e$ is the hydraulic head of the nodes, $\mathbf{f}_l^e$ is the element load vector and $\mathbf{f}_b^e$ is the nodal flow vector. By assembling, a global computational model is built up in the same manner as heat conduction in a network. Typical boundary conditions is the hydraulic heads h_A and h_B where the level of a free water surface is example of a known hydraulic head (Figure 8.24).

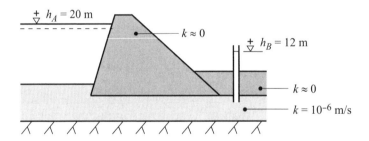

Figure 8.24   One-dimensional groundwater flow

8.4.3   Laminar Pipe Flow – Poiseuille's Law

Poiseuille's law, sometimes called Hagen–Poiseuille's law, describes an incompressible, homogeneous and viscous liquid that flows laminar through a cylindrical pipe with a constant cross-sectional area. The law was formulated and published in 1840 and 1846, respectively, by the French scientist Jean Louis Marie Poiseuille (1797–1869). Poiseuille's law expressed as a constitutive relation has the form

$$v_{\bar{x}} = -k(\bar{x})\frac{dp}{d\bar{x}} \tag{8.84}$$

where

$$k(\bar{x}) = \frac{d^2}{32\eta} \tag{8.85}$$

and where $p(\bar{x})$ is the *pressure* (N/m^2) of the liquid, $k(\bar{x})$ is conductivity created by the diameter $d(\bar{x})$ (m) of the pipe and the dynamic viscosity of the liquid η (Ns/m^2) or (kg/ms) and $v_{\bar{x}}$ is *mean velocity* along the pipe (m/s). For a one-dimensional body with the length $d\bar{x}$ and the cross-sectional area $A(\bar{x})$ a flow balance equation analogous to (8.17) results in the equation

$$\frac{dH}{d\bar{x}} - Q(\bar{x}) = 0 \tag{8.86}$$

where $H(\bar{x}) = v_{\bar{x}}(\bar{x})A(\bar{x})$ is the flow of liquid along the pipe (m³/s) and $Q(\bar{x})$ is the inflow of liquid per length unit (m³/sm). Substituting (8.84) in (8.86) gives the differential equation for laminar pipe flow

$$D_{kA}\frac{d^2p}{d\bar{x}^2} + Q(\bar{x}) = 0; \quad 0 \leq \bar{x} \leq L \tag{8.87}$$

where D_{kA} is the conductivity of the cross-section. If we assume that the material conductivity k and the cross-sectional area A are constant in the interval considered $0 \leq \bar{x} \leq L$, then we have $D_{kA} = kA$.

The derivation of the element equations for an element for pipe flow follows the derivation in Section 8.2 and leads to the element relation

$$\mathbf{K}^e\mathbf{a}^e = \mathbf{f}_l^e + \mathbf{f}_b^e \tag{8.88}$$

where $\mathbf{K}^e$ is the *conductivity matrix* of the element, $\mathbf{a}^e$ the pressure at the nodes, $\mathbf{f}_l^e$ the element load vector and $\mathbf{f}_b^e$ the nodal flow vector. By assembling, a global computational model is built up in the same way as heat conduction in a network. Typical boundary conditions are known pressures p_A and p_B (Figure 8.25). Poiseuille's law has among other things been applied in the study of the movement of the blood in arteries and veins (hemodynamics).

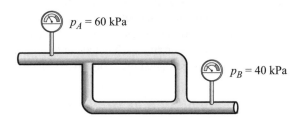

Figure 8.25   Network of pipes

8.4.4   Electricity – Ohm's Law

In the year 1827, the German physicist Georg Ohm (1789–1854) published a series of measurements on elementary electric circuits. By employing these, he was able to establish a relation between voltage and electric current. Through this pioneering work, he gave name to the constitutive relation for a material conducting electricity, *Ohm's law*

$$J = \sigma E = -\sigma\frac{dV}{d\bar{x}} \tag{8.89}$$

where V denotes *electric potential* (V), E denotes the *electric field* (V/m), σ is the *electric conductivity* of the material (A/Vm) or (S/m) and J denotes *density of current* or current per unit of area (A/m^2). A maybe more well-known form of Ohm's law we have when we go up in scale and consider a one-dimensional body conducting electricity

$$I = -\frac{1}{R}U \tag{8.90}$$

where U denotes voltage (V), R is the resistance of the body [Ω] and I denotes electric current (A).

An element conducting electricity (resistor) has as nodal variables electric potentials V_1 and V_2 and nodal flows (currents) I_1 and I_2. The voltage U over an element is the difference in electric potential $U = V_2 - V_1$; cf. (2.2). If we introduce $I^e = -I$ (cf. Section 8.2), Ohm's law (8.90) for the element can be written as

$$I^e = \frac{1}{R}\left(V_2 - V_1\right) \tag{8.91}$$

With the nodal flows of the element substituted and with positive directions according to Figure 8.17, we get

$$I_1 = -\frac{1}{R}\left(V_2 - V_1\right) \tag{8.92}$$

$$I_2 = \frac{1}{R}\left(V_2 - V_1\right) \tag{8.93}$$

or, in matrix form

$$\mathbf{K}^e\mathbf{a}^e = \mathbf{f}_b^e \tag{8.94}$$

where $\mathbf{K}^e$ is the conductivity matrix of the element, $\mathbf{a}^e$ the electric potential of the nodes and $\mathbf{f}_b^e$ the nodal flow vector. A global computational model is assembled in the same manner as heat conduction in a network. Boundary conditions in electric circuits (Figure 8.26) can be known electric potentials V_A and V_B and known currents I_C. If a node in the circuit is connected to the ground, this node has electric potential zero, that is the boundary condition $V_J = 0$.

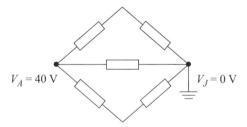

Figure 8.26   Electric circuit

8.4.5   Summary

In Table 8.1, a summary is given of constitutive relations for the different areas of applied mechanics, which have been treated.

Table 8.1 Summary of constitutive relations for different problem areas of applied mechanics

Physical problem	Material level		Cross-section level	Differential equation
Bar	Hooke's law	$\sigma_{\bar{x}} = E\frac{du}{d\bar{x}}$	$N(\bar{x}) = D_{EA}\frac{du}{d\bar{x}}$	$D_{EA}\frac{d^2u}{d\bar{x}^2} + q_{\bar{x}}(\bar{x}) = 0$
Heat conduction	Fourier's law	$q_{\bar{x}} = -k\frac{dT}{d\bar{x}}$	$H(\bar{x}) = -D_{kA}\frac{dT}{d\bar{x}}$	$D_{kA}\frac{d^2T}{d\bar{x}^2} + Q(x) = 0$
Diffusion	Fick's law	$J = -D\frac{dc}{d\bar{x}}$	$H(\bar{x}) = -D_{DA}\frac{dc}{d\bar{x}}$	$D_{DA}\frac{d^2c}{d\bar{x}^2} + Q(\bar{x}) = 0$
Groundwater flow	Darcy'slaw	$q_{\bar{x}} = -k\frac{dh}{d\bar{x}}$	$H(\bar{x}) = -D_{kA}\frac{dh}{d\bar{x}}$	$D_{kA}\frac{d^2h}{d\bar{x}^2} + Q(\bar{x}) = 0$
Laminar pipe flow	Poiseuille's law	$v_{\bar{x}} = -k\frac{dp}{d\bar{x}}$	$H(\bar{x}) = -D_{kA}\frac{dp}{d\bar{x}}$	$D_{kA}\frac{d^2p}{d\bar{x}^2} + Q(\bar{x}) = 0$
Spring			$N = k\delta$	
Electric circuit	Ohm's law	$J = -\sigma\frac{dV}{d\bar{x}}$	$I = \frac{1}{R}U$	

Exercises

8.1 In the figures, elements for modelling five different physical problems are shown. All of them can be described by the same type of element matrix

$$\mathbf{K}^e = K\begin{bmatrix} 1 & -1 \\ -1 & 1 \end{bmatrix}$$

where K denotes the stiffness or conductivity of the element. Give a value of K for each of the following elements:

(a) Spring

$k = 8000$ MN/m

(b) Bar

$E = 21$ GPa
$A = 0.04$ m^2
$L = 2$ m

(c) One-dimensional heat conduction

$k = 0.30$ W/mK
$L = 0.12$ m

(d) Heat resistance at boundary surface

$R_s = 0.04$ m^{2}K/W

(e) Electric resistance

V_1 R V_2

I_1 I_2 $R = 12\,\Omega$

8.2 Consider the wall in Example 8.2, but without internal heat source.

(a) Determine temperature distribution and heat flux through the wall if $T_o = -10\,°\mathrm{C}$ and $T_i = 20\,°\mathrm{C}$.

(b) Determine temperature distribution if $T_o = -10\,°\mathrm{C}$ and the heat flux through the wall is $q = 12\,\mathrm{W/m^2}$.

8.3

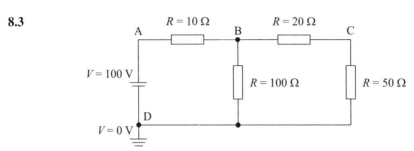

An electric circuit consists of four resistors and a voltage source of 100 V according to the figure. Establish an appropriate computational model of the circuit and determine the electric potentials and the current in the different parts of the circuit.

8.4

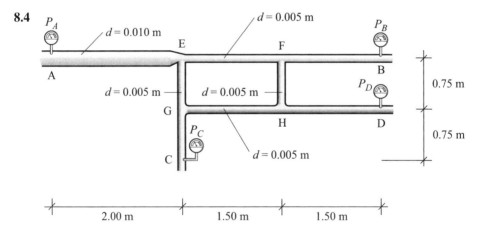

A part of a pipe network for oil is to be analysed. The length and diameter of the pipes are given in the figure. In the external limits of the network part, the pressures are known: $p_A = 200.0\,\mathrm{kPa}$, $p_B = 100.0\,\mathrm{kPa}$, $p_C = 100.0\,\mathrm{kPa}$ and $p_D = 100.0\,\mathrm{kPa}$. The oil has the density $\rho = 850\,\mathrm{kg/m^3}$ and the dynamic viscosity $\eta = 0.01\,\mathrm{Ns/m^2}$.

(a) Assume that the flow is laminar and determine the distribution of pressures and flows in the network.

(b) Check that the assumption of laminar flow is reasonable by calculating Reynolds number

$$Re = \frac{\rho v_{\bar{x}} d}{\eta}$$

for all pipes. The flow can be regarded as laminar if $Re < 2300$.

8.5

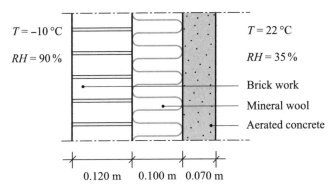

$T = -10\,°C$

$RH = 90\,\%$

$T = 22\,°C$

$RH = 35\,\%$

Brick work

Mineral wool

Aerated concrete

0.120 m   0.100 m   0.070 m

An external wall shall be checked for risk of condensation. Since water vapour condenses if the present vapour concentration is above the one at the saturation point, which depends on the temperature, both temperature and diffusion must be computed. The following thermal conductivities, thermal resistances and vapour transmissions can be assumed:

$R_{se} = 0.04\ \mathrm{m^2 K/W}$
$k_{\mathrm{brickwork}} = 0.58\ \mathrm{W/mK}$ $\qquad$ $D_{\mathrm{brickwork}} = 50 \times 10^{-7}\ \mathrm{m^2/s}$
$k_{\mathrm{mineral\ wool}} = 0.040\ \mathrm{W/mK}$ $\qquad$ $D_{\mathrm{mineral\ wool}} = 175 \times 10^{-7}\ \mathrm{m^2/s}$
$k_{\mathrm{aerated\ concrete}} = 0.15\ \mathrm{W/mK}$ $\qquad$ $D_{\mathrm{aerated\ concrete}} = 50 \times 10^{-7}\ \mathrm{m^2/s}$
$R_{si} = 0.13\ \mathrm{m^2 K/W}$

(a) Compute the temperature distribution in the wall.
(b) Compute the vapour concentration in the wall and compare at the material boundaries to the vapour concentration at saturation point given by the expressions

$$c_s = 625.67 \frac{\left(1.098 + \frac{T}{100}\right)^{8.02}}{273.15 + T}\ (\mathrm{g/m^3}); \quad 0 \le T \le 30\ (°C)$$

$$c_s = 10.16 \frac{\left(1.486 + \frac{T}{100}\right)^{12.3}}{273.15 + T}\ (\mathrm{g/m^3}); \quad -20 \le T \le 0\ (°C)$$

In the figure, relative humidity $RH = \frac{c}{c_s}$ in the air at the outside and at the inside of the wall is given.

(c) If the computed vapour concentration exceeds the vapour concentration at saturation point at any point, perform a new computation where the vapour concentration at that point is set equal to the vapour concentration at saturation point. Determine the amount of water condensed inside the wall during a week.

9

Geometrical Non-Linearity

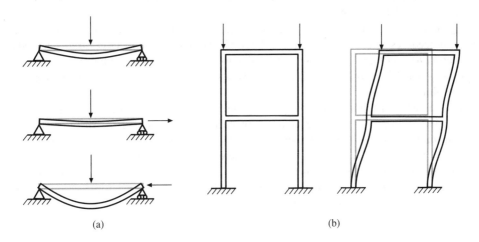

(a)   (b)

Figure 9.1 The effect of axial forces on the stiffness of elements and structures

We have so far established equilibrium equations for cross-section lamellas and for structures in their undeformed position. This assumption has given us linear systems of equations, which means that displacements and section forces are directly proportional to the level of the external load and the system of equations can be solved in a single computation. For small forces and small displacements, this is a reasonable assumption. If the forces and/or displacements successively increase in size, the assumption gives a deteriorating description of the mode of action of the structure.

Figure 9.1(a) shows a simply supported beam with a point load at the mid-point. The figure shows how the deflection caused by the point load is affected by a tensile and a compressive *axial force* $Q_{\bar{x}}$, respectively. When $Q_{\bar{x}}$ is a tensile force the deflection is reduced and when $Q_{\bar{x}}$ is a compressive force the deflection is increased. In the linear computational model, $Q_{\bar{x}}$ acts

Structural Mechanics: Modelling and Analysis of Frames and Trusses, First Edition.
Karl-Gunnar Olsson and Ola Dahlblom.
© 2016 John Wiley & Sons, Ltd. Published 2016 by John Wiley & Sons, Ltd.

on the initially straight beam and does not affect the deflection. With a more accurate model in which equilibria instead are established in a deformed position (a deformed geometry), we can capture the phenomenon. The drawback is that the calculation procedure becomes non-linear and that a solution must be determined by an iterative process. A calculation procedure that is based on establishment of equilibria for the bodies and structures in their deformed position is called *geometrically non-linear*.

We may in Figure 9.1 make the interpretation that the axial force $Q_{\bar{x}}$ affects the stiffness of the beam. A tensile axial force $Q_{\bar{x}}$ increases the stiffness (decreases the deflection) and a compressive axial force $Q_{\bar{x}}$ reduces the stiffness (increases the deflection). This is an interpretation that suits well with the structure of the system of equations, which a geometrically non-linear computational model generates. The internal axial forces will be a part of the stiffness matrix of the structure.

If large internal compressive forces occur in a slender structure, the structure may completely lose its stiffness and become *unstable*, as in Figure 9.1(b). Usually, the term *buckling* is used to designate such an unstable state.

In general, in structural mechanics, we distinguish between two types of non-linear calculation procedures – those that are due to large displacements and/or large strains are called *geometrically non-linear* and those that are due to a non-linear material behaviour are called *materially non-linear*. In Chapter 10, we discuss a simple and very useful procedure for material non-linearity.

We also distinguish between two types of geometrically non-linear calculation procedures: those who belong to the change of the geometric configuration of a structure are called *second-order theory*, and those associated with the definition of strain of the material points (cf. Chapter 3) are called *third-order theory*. Here, we discuss only second-order theory. In Section 9.1, geometrically non-linear calculation procedures are described in general terms. Thereafter, geometrically non-linear computational models are derived in Sections 9.2 and 9.3 for trusses and frames, respectively.

9.1 Methods of Calculation

There are two types of analyses where computational models that include the effect of internal axial forces are of special interest:

- computation of deformations and internal forces when the effect of internal axial forces on the stiffness of a structure is considered;
- computation of the risk of buckling (instability) caused by large internal compressive forces in a structure.

The starting point for both analyses is the system of equations of the structure

$$\mathbf{K}(\mathbf{Q}_{\bar{x}})\mathbf{a} = \mathbf{f} \tag{9.1}$$

where $\mathbf{K}(\mathbf{Q}_{\bar{x}})$ is the stiffness matrix of the structure, which is dependent on the internal axial forces $\mathbf{Q}_{\bar{x}}$.

In the first type of analysis, the deformations and internal forces are computed taking into consideration that internal axial forces affect the stiffness. For a known external load $\mathbf{f}$ the system of equations (9.1) is solved in an iterative calculation procedure. Since the internal axial forces $\mathbf{Q}_{\bar{x}}$ initially are unknown, the calculation begins with the stiffness $\mathbf{K}_0 = \mathbf{K}(\mathbf{Q}_{\bar{x},0})$ for $\mathbf{Q}_{\bar{x},0} = \mathbf{0}$, that is without considering the axial forces. From the system of equations

$$\mathbf{K}_0 \mathbf{a}_0 = \mathbf{f} \tag{9.2}$$

the displacement vector $\mathbf{a}_0$ of the first iteration can be determined and the corresponding internal forces (normal forces, shear forces and moments) can be calculated. With that, we have determined a set of section forces from which the axial forces $\mathbf{Q}_{\bar{x},1}$ of the first iteration can be determined. We can now establish a new system of equations

$$\mathbf{K}_1 \mathbf{a}_1 = \mathbf{f} \tag{9.3}$$

where $\mathbf{K}_1 = \mathbf{K}(\mathbf{Q}_{\bar{x},1})$. From this system of equations, a new displacement vector $\mathbf{a}_1$ can be determined and the internal axial forces $\mathbf{Q}_{\bar{x},2}$ computed. The procedure is repeated until some error norm $\|\mathbf{Q}_{x,n} - \mathbf{Q}_{x,n-1}\|$ is sufficiently small. The displacement vector $\mathbf{a}_n$ and the corresponding section forces are then the final approximate solution. The computed stiffness matrix $\mathbf{K}_n = \mathbf{K}(\mathbf{Q}_{\bar{x},n})$ is called the secant stiffness matrix of the system. In Figure 9.2, the method of calculation described here is illustrated in a load–displacement diagram.

Large axial forces can result in a total loss of stiffness for the entire structure. In the other type of analysis it is, from a given load level, estimated how much the load can increase before the structure has lost all its stiffness and becomes *unstable*.

Consider a frame loaded by a set of external loads $P_1, P_2, \ldots, P_n$ (Figure 9.3). The external loads give rise to internal axial forces $Q_1, Q_2, \ldots, Q_m$. We let the external loads and the

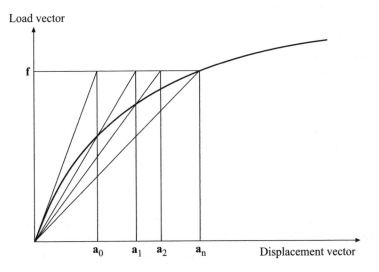

Figure 9.2   Procedure for solving geometrically non-linear problems

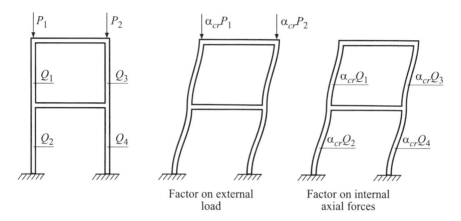

Factor on external
load

Factor on internal
axial forces

Figure 9.3 Reference state and unstable state

corresponding internal forces be a reference state for the frame. The reference state can for example be a design load case.

An increase of the load can be formulated by multiplying either the external loads with a factor $\alpha > 1$, αP_1, αP_2, $\ldots$, αP_n, or the internal axial forces with a factor $\alpha > 1$, αQ_1, αQ_2, $\ldots$, αQ_m. For large internal compressive forces, an increased load implies that the stiffness of the frame is reduced. When $\alpha = \alpha_{cr}$ (cr is an abbreviation of critical), all stiffness is lost. For the reference state, we have $\alpha = 1$ and the value α_{cr} then becomes a measure of how much the load can be increased before the frame becomes unstable. The scaling factor α_{cr} is therefore referred to as the *buckling safety* of the frame.

The buckling safety can be expressed either in terms of the external load or in terms of the internal axial forces. The results differ usually only slightly, but for frames the latter way gives simpler calculations and the concept of buckling safety can be extended to safety against instability for a general change in stiffness.[1]

There are different ways to examine whether a frame is unstable. If all internal stiffness is lost, we have

$$\det \mathbf{K} = 0 \tag{9.4}$$

where $\mathbf{K}$ is the stiffness matrix reduced with consideration taken to prescribed displacements; cf. (5.11). Instability can also be interpreted as if the structure is deformed without any additional external load acting on it, which can be written as

$$\mathbf{K}\boldsymbol{\varphi} = 0 \tag{9.5}$$

where $\boldsymbol{\varphi}$ is a set of possible displacement vectors, that is possible deformation patterns. The shape of these patterns can be determined, but the magnitude is arbitrary.

If we first apply the factor α on the external load, αP_1, αP_2, $\ldots$, αP_n, the buckling safety α_{cr} is determined by a stepwise increase of α from the reference state $\alpha = 1$. For every step,

---

[1] We can for example study the effect of reduced modulus of elasticity for increasing temperature or the effect of a reduced moment of inertia due to a reduced cross-section as a result of fire.

det $\mathbf{K}$ is checked. When det $\mathbf{K} = 0$ (with some tolerance for numerical inaccuracies), we have instability, that is $\alpha = \alpha_{cr}$.

An alternative way to determine α_{cr} is to apply the factor on the stiffness matrix. For a successively increasing load, the stiffness of the frame can be written as

$$\mathbf{K} = \mathbf{K}_0 + \alpha(\mathbf{K}_a - \mathbf{K}_0) \tag{9.6}$$

where $\mathbf{K}_0$ is the stiffness matrix from Chapter 4, which is independent of the magnitude of the load and where $\mathbf{K}_a$ is the stiffness matrix of the frame with the effect of internal axial forces considered. The difference $(\mathbf{K}_a - \mathbf{K}_0)$ is then a measure of the effect of the axial forces on the stiffness. If we choose an approximate element formulation (9.93), this difference can be written as $(\mathbf{K}_a - \mathbf{K}_0) = \mathbf{K}_\sigma$, and we get

$$\mathbf{K} = \mathbf{K}_0 + \alpha\mathbf{K}_\sigma \tag{9.7}$$

For the reference state of the external load $\alpha = 1$, we have from (9.6) that $\mathbf{K} = \mathbf{K}_a$. We now seek a value of α for which the structure is unstable. At instability, substituting (9.6) into (9.5) gives

$$(\mathbf{K}_0 + \alpha(\mathbf{K}_a - \mathbf{K}_0))\boldsymbol{\varphi} = \mathbf{0} \tag{9.8}$$

which can be reformulated to the generalised eigenvalue problem

$$(\mathbf{K}_a - \lambda\mathbf{K}_0)\boldsymbol{\varphi} = \mathbf{0} \tag{9.9}$$

where

$$\lambda = \frac{\alpha - 1}{\alpha} \tag{9.10}$$

If we solve the generalised eigenvalue problem, we obtain a set of eigenvalues λ_i. From these eigenvalues, the buckling safety α_{cr} can be determined as

$$\alpha_{cr} = \min(\alpha_i) \tag{9.11}$$

where

$$\alpha_i = \frac{1}{1 - \lambda_i} \tag{9.12}$$

The number of values α_i obtained is the same as the number of degrees of freedom in the reduced system of equations, but only the lowest of these is relevant as a measure of buckling safety.

9.2 Trusses with Geometrical Non-Linearity Considered

We here consider trusses (systems of bars) and in the analysis include how the stiffness of the truss is affected if the equilibria of the nodes are established in the deformed state of the truss. In the deformed state, the normal forces N^α and N^β in bar element (α) and (β), respectively, have two components, $Q_{\bar{x}}$ and $Q_{\bar{y}}$. Figure 9.4 shows how the normal forces are decomposed into these two components at a node. Both $Q_{\bar{x}}$ and $Q_{\bar{y}}$, therefore, have to be considered in the derivation of the element relations of the bar element, first in the local coordinate system $(\bar{x}, \bar{y})$ and then transformed into a global coordinate system (x, y).

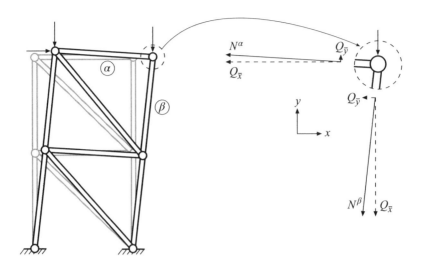

Figure 9.4   Equilibrium of a node in the deformed state of the truss

9.2.1   The Differential Equation for Bar Action

We start at the bar action cross-section level and consider a slice $d\bar{x}$ of a bar, rotated a small angle θ from the initial state (Figure 9.5). Since bar action only yields axial deformation all cross-section lamellas will have rotated the same angle θ. It is assumed that no distributed load acts along the bar.

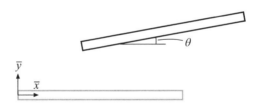

Figure 9.5   Displaced and deformed bar element and cross-section lamella $d\bar{x}$

The Kinematics and Constitutive Relation of the Cross-Section Level

In Chapter 3, we established a relation (3.19) between the deformation measure $u(\bar{x})$ of bar action and the deformation measure $\varepsilon_{\bar{x}}(\bar{x})$ of the cross-section level

$$\varepsilon_{\bar{x}}(\bar{x}) = \frac{du}{d\bar{x}} \tag{9.13}$$

and a relation (3.17) between the generalised strain $\varepsilon_{\bar{x}}(\bar{x})$ and the normal force $N(\bar{x})$.

$$N(\bar{x}) = D_{EA}(\bar{x})\,\varepsilon_{\bar{x}}(\bar{x}) \tag{9.14}$$

These relations are valid for bar action even when geometrical non-linearity is considered.

Equilibrium

Here, we consider the case without external load, that is $q_{\bar{x}} = 0$. The slice of the bar has rotated an angle $\theta(\bar{x})$ compared with the initial direction and we decompose the normal force $N(\bar{x})$ into components $Q_{\bar{x}}(\bar{x})$ and $Q_{\bar{y}}(\bar{x})$ parallel to the $\bar{x}$- and $\bar{y}$-axis, respectively; see Figure 9.6

$$Q_{\bar{x}}(\bar{x}) = N(\bar{x}) \cos \theta(\bar{x}) \tag{9.15}$$

$$Q_{\bar{y}}(\bar{x}) = N(\bar{x}) \sin \theta(\bar{x}) \tag{9.16}$$

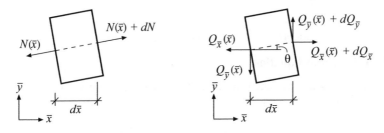

Figure 9.6 A slice of the bar rotated a small angle θ

Since the angle $\theta(\bar{x})$ is small ($\sin \theta \approx \tan \theta \approx \theta$; $\cos \theta \approx 1$), expressions (9.15) and (9.16) can be written as

$$Q_{\bar{x}}(\bar{x}) = N(\bar{x}) \tag{9.17}$$

$$Q_{\bar{y}}(\bar{x}) = N(\bar{x})\theta(\bar{x}) \tag{9.18}$$

Equilibrium in the $\bar{x}$- and $\bar{y}$-direction give

$$-Q_{\bar{x}}(\bar{x}) + (Q_{\bar{x}}(\bar{x}) + dQ_{\bar{x}}) = 0 \tag{9.19}$$

$$-Q_{\bar{y}}(\bar{x}) + (Q_{\bar{y}}(\bar{x}) + dQ_{\bar{y}}) = 0 \tag{9.20}$$

which can be rewritten as

$$dQ_{\bar{x}} = 0 \tag{9.21}$$

$$dQ_{\bar{y}} = 0 \tag{9.22}$$

Substituting (9.14), (9.13) and (9.17) into the equilibrium relation (9.21) and using the assumption that the stiffness D_{EA} is constant along the bar yield

$$D_{EA} \frac{du}{d\bar{x}} = \text{constant} \tag{9.23}$$

Substituting (9.18) into (9.22) and integrating give

$$N(\bar{x})\theta(\bar{x}) = \text{constant} \tag{9.24}$$

9.2.2 Bar Element

For the bar element, we introduce four degrees of freedom. Using these, we can express $Q_{\bar{x}}$ and $Q_{\bar{y}}$ at both ends of the bar (Figure 9.7).

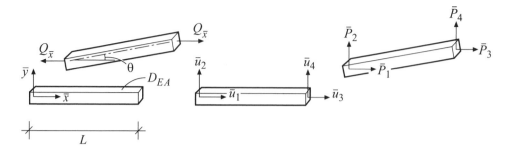

Figure 9.7   Bar element and degrees of freedom

Solving the Differential Equation

From Chapter 3, we know that the solution of (9.23) can be expressed in terms of the nodal displacements, denoted here by $\bar{u}_1$ and $\bar{u}_3$,

$$u(\bar{x}) = \mathbf{N}\bar{\mathbf{a}}^e = \begin{bmatrix} 1 - \frac{\bar{x}}{L} & \frac{\bar{x}}{L} \end{bmatrix} \begin{bmatrix} \bar{u}_1 \\ \bar{u}_3 \end{bmatrix} \tag{9.25}$$

Using (9.13) and substituting into (9.14) enable us to express the normal force using nodal displacements

$$N = D_{EA} \begin{bmatrix} -\frac{1}{L} & \frac{1}{L} \end{bmatrix} \begin{bmatrix} \bar{u}_1 \\ \bar{u}_3 \end{bmatrix} \tag{9.26}$$

Substitution into (9.17) gives the force component in the $\bar{x}$-direction, $Q_{\bar{x}}$

$$Q_{\bar{x}} = D_{EA} \begin{bmatrix} -\frac{1}{L} & \frac{1}{L} \end{bmatrix} \begin{bmatrix} \bar{u}_1 \\ \bar{u}_3 \end{bmatrix} \tag{9.27}$$

For pure bar action, the angle θ is constant along the element and can be expressed as a function of the nodal displacements, $\bar{u}_2$ and $\bar{u}_4$, and the length L of the bar; see Figure 9.8

$$\theta \approx \sin\theta = \frac{\bar{u}_4 - \bar{u}_2}{L} \tag{9.28}$$

With θ being constant, we obtain from (9.24) that N is constant. When considering this and substituting (9.28) and (9.17) into (9.18), we obtain the force component in the $\bar{y}$-direction, $Q_{\bar{y}}$

$$Q_{\bar{y}} = Q_{\bar{x}} \begin{bmatrix} -\frac{1}{L} & \frac{1}{L} \end{bmatrix} \begin{bmatrix} \bar{u}_2 \\ \bar{u}_4 \end{bmatrix} \tag{9.29}$$

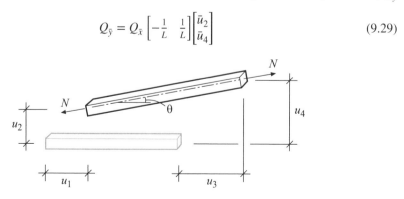

Figure 9.8   Displaced bar element

In Figure 9.7, we have defined the forces that act at the nodes of the element, $\bar{P}_1$, $\bar{P}_2$, $\bar{P}_3$ and $\bar{P}_4$ as positive in the $\bar{x}$- and $\bar{y}$-direction, respectively. This implies that

$$\bar{P}_1 = -Q_{\bar{x}}(0) \tag{9.30}$$

$$\bar{P}_2 = -Q_{\bar{y}}(0) \tag{9.31}$$

$$\bar{P}_3 = Q_{\bar{x}}(L) \tag{9.32}$$

$$\bar{P}_4 = Q_{\bar{y}}(L) \tag{9.33}$$

which with the substitution of (9.27) and (9.29) gives

$$\boxed{\bar{\mathbf{K}}^e \, \bar{\mathbf{a}}^e = \bar{\mathbf{f}}^e} \tag{9.34}$$

where

$$\bar{\mathbf{K}}^e = \bar{\mathbf{K}}_0^e + \bar{\mathbf{K}}_\sigma^e \tag{9.35}$$

and

$$\bar{\mathbf{K}}_0^e = \frac{D_{EA}}{L} \begin{bmatrix} 1 & 0 & -1 & 0 \\ 0 & 0 & 0 & 0 \\ -1 & 0 & 1 & 0 \\ 0 & 0 & 0 & 0 \end{bmatrix} ; \quad \bar{\mathbf{K}}_\sigma^e = \frac{Q_{\bar{x}}}{L} \begin{bmatrix} 0 & 0 & 0 & 0 \\ 0 & 1 & 0 & -1 \\ 0 & 0 & 0 & 0 \\ 0 & -1 & 0 & 1 \end{bmatrix} \tag{9.36}$$

$$\bar{\mathbf{a}}^e = \begin{bmatrix} \bar{u}_1 \\ \bar{u}_2 \\ \bar{u}_3 \\ \bar{u}_4 \end{bmatrix} ; \quad \bar{\mathbf{f}}_b^e = \begin{bmatrix} \bar{P}_1 \\ \bar{P}_2 \\ \bar{P}_3 \\ \bar{P}_4 \end{bmatrix} \tag{9.37}$$

In element relation (9.34), the element stiffness matrix $\bar{\mathbf{K}}^e$ is composed of two parts: one element stiffness matrix $\bar{\mathbf{K}}_0^e$, which is identical to the one we have derived in Chapter 3, and one element stiffness matrix $\bar{\mathbf{K}}_\sigma^e$, which describes the effect of the axial force on the stiffness. For a positive (tensile) axial force, the stiffness increases. For a negative (compressive) axial force, the stiffness decreases.

From Local to Global Coordinates

To be able to model a truss, the element relations have to be transformed to a global coordinate system. From coordinate transformations, described in Chapter 3, we have the transformation relation

$$\mathbf{K}^e = \mathbf{G}^T \bar{\mathbf{K}}^e \mathbf{G} \tag{9.38}$$

where the transformation matrix is given by

$$\mathbf{G} = \begin{bmatrix} n_{x\bar{x}} & n_{y\bar{x}} & 0 & 0 \\ n_{x\bar{y}} & n_{y\bar{y}} & 0 & 0 \\ 0 & 0 & n_{x\bar{x}} & n_{y\bar{x}} \\ 0 & 0 & n_{x\bar{y}} & n_{y\bar{y}} \end{bmatrix} \tag{9.39}$$

9.2.3   Trusses

The system of equations for a truss is established by assembling the element stiffness matrices of the bar elements to a global stiffness matrix and defining a load vector and boundary conditions. The stiffness matrix depends on the axial forces in the elements. Since the axial forces initially are unknown and thus is a result from the calculations, the solution has to be determined in an iterative process. First, the global stiffness matrix is established with an assumption that all axial forces are zero. Then the system of equations is solved and from the computed displacements, the axial forces are computed. After that, the stiffness matrix is recomputed, now with the computed values of the axial forces inserted, and a new solution can be determined. The procedure is repeated until the computed axial forces correspond sufficiently well to the ones assumed when calculating the stiffness matrix, that is until the solution has converged.

Example 9.1   Truss analysis with geometrical non-linearity considered
The truss in Figure 1 consists of two bars hinged at A and C. For the truss, $E = 10.0$ GPa, $A_1 = 4.0 \times 10^{-2}$ m^2, $A_2 = 1.0 \times 10^{-2}$ m^2, $L = 1.6$ m, $P = 10.0$ MN and $F = 0.2$ MN. The displacements in B and the axial forces in the two bars is to be determined considering geometrical non-linearity.

The element stiffness matrices are given in local coordinates by (9.35). With the substitution of the values of known quantities, we have

$$\bar{\mathbf{K}}^1 = \begin{bmatrix} 250 & 0 & -250 & 0 \\ 0 & 0 & 0 & 0 \\ -250 & 0 & 250 & 0 \\ 0 & 0 & 0 & 0 \end{bmatrix} 10^6 + Q_{\bar{x}}^{(1)} \begin{bmatrix} 0 & 0 & 0 & 0 \\ 0 & 0.625 & 0 & -0.625 \\ 0 & 0 & 0 & 0 \\ 0 & -0.625 & 0 & 0.625 \end{bmatrix} \tag{1}$$

$$\bar{\mathbf{K}}^2 = \begin{bmatrix} 50 & 0 & -50 & 0 \\ 0 & 0 & 0 & 0 \\ -50 & 0 & 50 & 0 \\ 0 & 0 & 0 & 0 \end{bmatrix} 10^6 + Q_{\bar{x}}^{(2)} \begin{bmatrix} 0 & 0 & 0 & 0 \\ 0 & 0.5 & 0 & -0.5 \\ 0 & 0 & 0 & 0 \\ 0 & -0.5 & 0 & 0.5 \end{bmatrix} \tag{2}$$

For Element 1, the local coordinate system coincides with the global one, which gives

$$\mathbf{K}^1 = \bar{\mathbf{K}}^1 \tag{3}$$

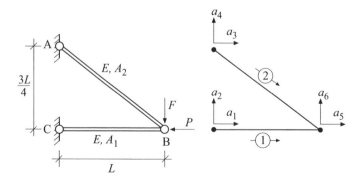

Figure 1   A truss and a computational model with degrees of freedom

For Element 2 we have the direction cosines $n_{x\bar{x}} = 0.8$, $n_{y\bar{x}} = -0.6$, $n_{x\bar{y}} = 0.6$ and $n_{y\bar{y}} = 0.8$. Substitution into (9.38) this yields

$$\mathbf{K}^2 = \begin{bmatrix} 32 & -24 & -32 & 24 \\ -24 & 18 & 24 & -18 \\ -32 & 24 & 32 & -24 \\ 24 & -18 & -24 & 18 \end{bmatrix} 10^6 + Q_{\bar{x}}^{(2)} \begin{bmatrix} 0.18 & 0.24 & -0.18 & -0.24 \\ 0.24 & 0.32 & -0.24 & -0.32 \\ -0.18 & -0.24 & 0.18 & 0.24 \\ -0.24 & -0.32 & 0.24 & 0.32 \end{bmatrix} \quad (4)$$

Assembling the element stiffness matrices gives a global stiffness matrix $\mathbf{K}$ and substituting the loads P and F into the load vector $\mathbf{f}$, we obtain a system of equations $\mathbf{Ka} = \mathbf{f}$ for the truss. Taking the boundary conditions $a_1 = 0$, $a_2 = 0$, $a_3 = 0$ and $a_4 = 0$ into consideration, the system of equations can be reduced to

$$\begin{bmatrix} 282 \times 10^6 + 0.18Q_{\bar{x}}^{(2)} & -24 \times 10^6 + 0.24Q_{\bar{x}}^{(2)} \\ -24 \times 10^6 + 0.24Q_{\bar{x}}^{(2)} & 18 \times 10^6 + 0.625Q_{\bar{x}}^{(1)} + 0.32Q_{\bar{x}}^{(2)} \end{bmatrix} \begin{bmatrix} a_5 \\ a_6 \end{bmatrix} = \begin{bmatrix} -10 \times 10^6 \\ -0.2 \times 10^6 \end{bmatrix} \quad (5)$$

The stiffness matrix depends on the axial forces $Q_{\bar{x}}^{(1)}$ and $Q_{\bar{x}}^{(2)}$, which are unknown at the beginning of the computation. To obtain a solution, we use an iterative procedure. We start by solving the system of equations with $Q_{\bar{x}}^{(1)} = 0$ and $Q_{\bar{x}}^{(2)} = 0$, which gives

$$\begin{bmatrix} a_5 \\ a_6 \end{bmatrix} = \begin{bmatrix} -41.067 \\ -65.867 \end{bmatrix} 10^{-3} \quad (6)$$

Transformation of the computed displacements to local coordinates and substitution into (9.26) give the axial forces

$$Q_{\bar{x}}^{(1)} = -10.2667 \times 10^6 \quad (7)$$

$$Q_{\bar{x}}^{(2)} = 0.3333 \times 10^6 \quad (8)$$

The new values of $Q_{\bar{x}}^{(1)}$ and $Q_{\bar{x}}^{(2)}$ are substituted into the system of equations, then the displacements and thereafter the axial forces are recomputed. The procedure is repeated until the axial forces computed in a step correspond sufficiently well to the values used to compute the element stiffness matrices in that step. In this example, the solution has converged after six iterations; see Table 1.

Table 1 Computed displacements and axial forces

Iteration number	a_5 (mm)	a_6 (mm)	$Q_{\bar{x}}^{(1)}$ (MN)	$Q_{\bar{x}}^{(2)}$ (MN)
1	-41.067	-65.867	-10.2667	0.3333
2	-44.653	-108.477	-11.1632	1.4682
3	-44.569	-109.106	-11.1421	1.4905
4	-44.544	-108.843	-11.1359	1.4835
5	-44.544	-108.833	-11.1360	1.4833
6	-44.544	-108.835	-11.1360	1.4833

The vertical displacement at B is 1.6 times larger than in the linear computation (iteration 1) and the axial force in Element 2 is 4.5 times larger.

So far, we have only studied the effect of geometrical non-linearity that is associated with bar action. For Example 9.1, this means that we have not considered local instability (local buckling) for individual bars, a phenomenon associated with beam action. Often local instability is crucial to the load capacity of the truss and it is, therefore, advisable to let a geometrically non-linear analysis for a truss include both bar and beam action.

9.3 Frames with Geometrical Non-Linearity Considered

With a procedure corresponding to the one for the truss, the starting point for the geometrically non-linear frame is that the normal and shear forces of the beam element have two components, $Q_{\bar{x}}$ and $Q_{\bar{y}}$, at the nodes; cf. Figure 9.4. These components are included as section forces in the derivation of the beam element, first in a local coordinate system $(\bar{x}, \bar{y})$, and then transformed into a global coordinate system (x, y).

9.3.1 The Differential Equation for Beam Action

We consider a slice $d\bar{x}$ of a beam, which has rotated a small angle θ compared with the initial state; Figure 9.9. In the $\bar{y}$-direction, the load $q_{\bar{y}}(\bar{x})$ acts on the beam. It is assumed that no distributed load acts in the $\bar{x}$-direction. Normal force $N(\bar{x})$ and shear force $V(\bar{x})$ are directed perpendicular and parallel, respectively, to the cross-section of the beam.

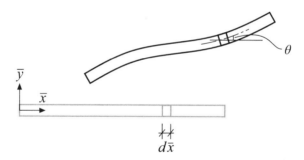

Figure 9.9   A deformed beam element

The Kinematics and Constitutive Relation of the Cross-Section

From Chapters 3 and 4, we have the kinematic relations (3.10), (4.15) and (4.16) between the cross-section level and the bar level and the beam level, respectively,

$$\varepsilon_{\bar{x}}(\bar{x}) = \frac{du}{d\bar{x}} \tag{9.40}$$

$$\frac{dv}{d\bar{x}} = \tan(\theta(\bar{x})) = \theta(\bar{x}) \tag{9.41}$$

$$\frac{d^2v}{d\bar{x}^2} = \frac{d\theta}{d\bar{x}} = \kappa(\bar{x}) \tag{9.42}$$

where $u(\bar{x})$, $v(\bar{x})$ and $\theta(\bar{x})$ are the displacements and rotation of the system line and where $\varepsilon_{\bar{x}}$ and $\kappa(\bar{x})$ are the strain and curvature, respectively, of the cross-section. Moreover, we have the constitutive relations (3.16) and (4.12)

$$N(\bar{x}) = D_{EA}(\bar{x})\,\varepsilon_{\bar{x}}(\bar{x}) \tag{9.43}$$

$$M(\bar{x}) = D_{EI}(\bar{x})\,\kappa(\bar{x}) \tag{9.44}$$

between the force measures $N(\bar{x})$ and $M(\bar{x})$ of the cross-section and its deformation measures $\varepsilon_{\bar{x}}(\bar{x})$ and $\kappa_{\bar{x}}(\bar{x})$.

Equilibrium

At the deformed cross-section, normal force $N(\bar{x})$ and shear force $V(\bar{x})$ act perpendicular and parallel, respectively, to the cross-sectional surface. When we establish equilibria in the deformed state, we use the directions of the coordinates $\bar{x}$ and $\bar{y}$, thus we obtain the section forces in these directions as axial force $Q_{\bar{x}}(\bar{x})$ and transverse force $Q_{\bar{y}}(\bar{x})$ (Figure 9.10). $Q_{\bar{x}}(\bar{x})$ and $Q_{\bar{y}}(\bar{x})$ can also be expressed in normal force $N(\bar{x})$ and shear force $V(\bar{x})$

$$Q_{\bar{x}}(\bar{x}) = N(\bar{x})\cos\theta(\bar{x}) - V(\bar{x})\sin\theta(\bar{x}) \tag{9.45}$$

$$Q_{\bar{y}}(\bar{x}) = N(\bar{x})\sin\theta(\bar{x}) + V(\bar{x})\cos\theta(\bar{x}) \tag{9.46}$$

or

$$\begin{bmatrix} Q_{\bar{x}}(\bar{x}) \\ Q_{\bar{y}}(\bar{x}) \end{bmatrix} = \begin{bmatrix} \cos\theta(\bar{x}) & -\sin\theta(\bar{x}) \\ \sin\theta(\bar{x}) & \cos\theta(\bar{x}) \end{bmatrix} \begin{bmatrix} N(\bar{x}) \\ V(\bar{x}) \end{bmatrix} \tag{9.47}$$

Inversely, we have

$$\begin{bmatrix} N(\bar{x}) \\ V(\bar{x}) \end{bmatrix} = \begin{bmatrix} \cos\theta(\bar{x}) & \sin\theta(\bar{x}) \\ -\sin\theta(\bar{x}) & \cos\theta(\bar{x}) \end{bmatrix} \begin{bmatrix} Q_{\bar{x}}(\bar{x}) \\ Q_{\bar{y}}(\bar{x}) \end{bmatrix} \tag{9.48}$$

Since the angle θ is small ($\sin\theta \approx \tan\theta \approx \theta$, $\cos\theta \approx 1$), the expressions (9.47) and (9.48) can be written as

$$\begin{bmatrix} Q_{\bar{x}}(\bar{x}) \\ Q_{\bar{y}}(\bar{x}) \end{bmatrix} = \begin{bmatrix} 1 & -\theta(\bar{x}) \\ \theta(\bar{x}) & 1 \end{bmatrix} \begin{bmatrix} N(\bar{x}) \\ V(\bar{x}) \end{bmatrix} \tag{9.49}$$

$$\begin{bmatrix} N(\bar{x}) \\ V(\bar{x}) \end{bmatrix} = \begin{bmatrix} 1 & \theta(\bar{x}) \\ -\theta(\bar{x}) & 1 \end{bmatrix} \begin{bmatrix} Q_{\bar{x}}(\bar{x}) \\ Q_{\bar{y}}(\bar{x}) \end{bmatrix} \tag{9.50}$$

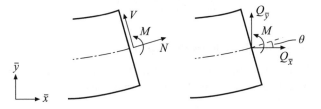

Figure 9.10 Section forces in the deformed cross-section and expressed in the $\bar{x}\bar{y}$-system

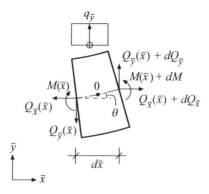

Figure 9.11 The forces which act on a slice of the beam

The forces which, in the local $\bar{x}\bar{y}$-system, act on a slice $d\bar{x}$ are shown in Figure 9.11. Equilibrium in the $\bar{x}$-direction gives

$$- Q_{\bar{x}}(\bar{x}) + (Q_{\bar{x}}(\bar{x}) + dQ_{\bar{x}}) = 0 \tag{9.51}$$

or

$$\frac{dQ_{\bar{x}}}{d\bar{x}} = 0 \tag{9.52}$$

Equilibrium in the $\bar{y}$-direction gives

$$- Q_{\bar{y}}(\bar{x}) + (Q_{\bar{y}}(\bar{x}) + dQ_{\bar{y}}) + q_{\bar{y}}(\bar{x})d\bar{x} = 0 \tag{9.53}$$

or

$$\frac{dQ_{\bar{y}}}{d\bar{x}} + q_{\bar{y}}(\bar{x}) = 0 \tag{9.54}$$

Moment equilibrium about a point at the right end of the considered slice gives

$$- M(\bar{x}) + (M(\bar{x}) + dM) + Q_{\bar{y}}(\bar{x})d\bar{x} - Q_{\bar{x}}(\bar{x})d\bar{x}\theta(\bar{x}) + q_{\bar{y}}(\bar{x})d\bar{x}\frac{d\bar{x}}{2} = 0 \tag{9.55}$$

or

$$\frac{dM}{d\bar{x}} + Q_{\bar{y}}(\bar{x}) - Q_{\bar{x}}(\bar{x})\theta(\bar{x}) = 0 \tag{9.56}$$

If we combine expressions (9.54), (9.56) and (9.41) and observe that $Q_{\bar{x}}$ is constant according to (9.52), we obtain

$$\frac{d^2M}{d\bar{x}^2} - Q_{\bar{x}}\frac{d^2v}{d\bar{x}^2} - q_{\bar{y}}(\bar{x}) = 0 \tag{9.57}$$

Differential Equations for Bar and Beam Action

Substituting (9.43) and (9.40) into (9.52) and also (9.44) and (9.42) into (9.57) along with the assumption that the stiffnesses D_{EA} and D_{EI} are constant give

$$D_{EA}\frac{d^2u}{d\bar{x}^2} = 0 \tag{9.58}$$

$$D_{EI}\frac{d^4 v}{d\bar{x}^4} - Q_{\bar{x}}\frac{d^2 v}{d\bar{x}^2} - q_{\bar{y}}(\bar{x}) = 0 \qquad (9.59)$$

These two differential equations describe bar and beam action with equilibria established in the deformed state. The expressions show that the axial force $Q_{\bar{x}}$ locally only affects the bending stiffness and not the axial stiffness.

9.3.2 Beam Element

For the beam element, we have to introduce six degrees of freedom by which we can express the components of the normal and shear force of the deformed element.

Solving the Differential Equation for Bar Action

From Chapter 3, we know that the solution to (9.58) can be expressed in terms of the nodal displacements, here denoted $\bar{u}_1$ and $\bar{u}_4$ according to Figure 9.12,

$$u(\bar{x}) = \mathbf{N}\bar{\mathbf{a}}^e = \begin{bmatrix} 1 - \frac{\bar{x}}{L} & \frac{\bar{x}}{L} \end{bmatrix}\begin{bmatrix} \bar{u}_1 \\ \bar{u}_4 \end{bmatrix} \qquad (9.60)$$

Then, by substituting (9.40) into (9.43), the normal force can be expressed in terms of nodal displacements

$$N = D_{EA}\begin{bmatrix} -\frac{1}{L} & \frac{1}{L} \end{bmatrix}\begin{bmatrix} \bar{u}_1 \\ \bar{u}_4 \end{bmatrix} \qquad (9.61)$$

From (9.50), we have

$$Q_{\bar{x}}(\bar{x}) = N(\bar{x}) + \theta(\bar{x})Q_{\bar{y}}(\bar{x}) \qquad (9.62)$$

But since a geometrically non-linear analysis is only relevant for large normal forces and since the kinematics of beam action presume small rotations, the term $\theta(\bar{x})Q_{\bar{y}}(\bar{x})$ can be neglected, that is

$$Q_{\bar{x}}(\bar{x}) = N(\bar{x}) \qquad (9.63)$$

Substitution of (9.61) gives

$$Q_{\bar{x}} = D_{EA}\begin{bmatrix} -\frac{1}{L} & \frac{1}{L} \end{bmatrix}\begin{bmatrix} \bar{u}_1 \\ \bar{u}_4 \end{bmatrix} \qquad (9.64)$$

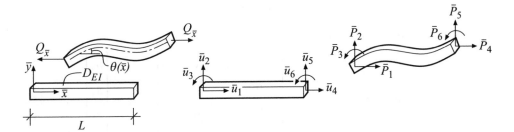

Figure 9.12 A beam element with six degrees of freedom

In Figure 9.12, we have defined the forces that act at the nodes of the element. In the $\bar{x}$-direction, we have $\bar{P}_1$ and $\bar{P}_4$ that is

$$\bar{P}_1 = -Q_{\bar{x}}(0) \tag{9.65}$$

$$\bar{P}_4 = Q_{\bar{x}}(L) \tag{9.66}$$

Substituting (9.64), the element relations for bar action can be expressed as

$$\bar{\mathbf{f}}_b^e = \bar{\mathbf{K}}_0^e \bar{\mathbf{a}}^e \tag{9.67}$$

where

$$\bar{\mathbf{K}}_0^e = \frac{D_{EA}}{L} \begin{bmatrix} 1 & -1 \\ -1 & 1 \end{bmatrix} ; \quad \bar{\mathbf{a}}^e = \begin{bmatrix} \bar{u}_1 \\ \bar{u}_4 \end{bmatrix} ; \quad \bar{\mathbf{f}}_b^e = \begin{bmatrix} \bar{P}_1 \\ \bar{P}_4 \end{bmatrix} \tag{9.68}$$

Approximate Solution of the Differential Equation for Beam Action

For a beam element, an exact solution can be obtained (see later), but analogous to the solution for a beam on a transverse flexible support (Chapter 6), we introduce an approximate solution instead. The differential equation (9.59) can be written as

$$D_{EI} \frac{d^4 v}{d\bar{x}^4} + p_{\bar{y}}(\bar{x}) - q_{\bar{y}}(\bar{x}) = 0 \tag{9.69}$$

where the effect of the axial force $Q_{\bar{x}}$ on the beam has been interpreted as an effect of a distributed load

$$p_{\bar{y}}(\bar{x}) = -Q_{\bar{x}} \frac{d^2 v}{d\bar{x}^2} \tag{9.70}$$

If we assume that the transverse displacement varies as a third-degree polynomial, the load $p_{\bar{y}}(\bar{x})$ in the differential equation can be written as

$$p_{\bar{y}}(\bar{x}) = -Q_{\bar{x}} \frac{d^2 \mathbf{N}}{d\bar{x}^2} \bar{\mathbf{a}}^e \tag{9.71}$$

where $\mathbf{N}\bar{\mathbf{a}}^e$ is the third-degree polynomial expressed in terms of the nodal displacements of the beam, according to (4.47). The homogeneous equation corresponding to (9.69) is

$$D_{EI} \frac{d^4 v_h}{d\bar{x}^4} = 0 \tag{9.72}$$

As we have seen in Chapter 4, the solution of (9.72) is given from (4.47)

$$v_h(\bar{x}) = \mathbf{N}\bar{\mathbf{a}}^e \tag{9.73}$$

where according to (4.48) and (4.44)

$$\mathbf{N} = \bar{\mathbf{N}}\mathbf{C}^{-1} = \begin{bmatrix} 1 & \bar{x} & \bar{x}^2 & \bar{x}^3 \end{bmatrix} \begin{bmatrix} 1 & 0 & 0 & 0 \\ 0 & 1 & 0 & 0 \\ -\frac{3}{L^2} & -\frac{2}{L} & \frac{3}{L^2} & -\frac{1}{L} \\ \frac{2}{L^3} & \frac{1}{L^2} & -\frac{2}{L^3} & \frac{1}{L^2} \end{bmatrix} ; \quad \bar{\mathbf{a}}^e = \begin{bmatrix} \bar{u}_2 \\ \bar{u}_3 \\ \bar{u}_5 \\ \bar{u}_6 \end{bmatrix} \tag{9.74}$$

The general solution to (9.69) is

$$v(\bar{x}) = v_h(\bar{x}) + v_p(\bar{x})$$
(9.75)

Integrating (9.69) four times with the substitution of (9.71) gives for the case with a constant load $q_{\bar{y}}$

$$v_p(\bar{x}) = \frac{Q_{\bar{x}}}{D_{EI}} \begin{bmatrix} 0 & 0 & \frac{\bar{x}^4}{12} & \frac{\bar{x}^5}{20} \end{bmatrix} \mathbf{C}^{-1}\bar{\mathbf{a}}^e + \frac{1}{D_{EI}}\left(q_{\bar{y}}\frac{\bar{x}^4}{24} + C_1\frac{\bar{x}^3}{6} + C_2\frac{\bar{x}^2}{2} + C_3\bar{x} + C_4 \right)$$
(9.76)

The boundary conditions (4.56)–(4.59) give the constants of integration

$$C_1 = -Q_{\bar{x}}\begin{bmatrix} 0 & 0 & L & \frac{9L^2}{10} \end{bmatrix} \mathbf{C}^{-1}\bar{\mathbf{a}}^e - q_{\bar{y}}\frac{L}{2}$$
(9.77)

$$C_2 = Q_{\bar{x}}\begin{bmatrix} 0 & 0 & \frac{L^2}{6} & \frac{L^3}{5} \end{bmatrix} \mathbf{C}^{-1}\bar{\mathbf{a}}^e + q_{\bar{y}}\frac{L^2}{12}$$
(9.78)

$$C_3 = 0$$
(9.79)

$$C_4 = 0$$
(9.80)

Substituting C_1, C_2, C_3 and C_4, the particular solution can be written as

$$v_p(\bar{x}) = \frac{Q_{\bar{x}}}{D_{EI}} \begin{bmatrix} 0 & 0 & \left(\frac{\bar{x}^4}{12} - \frac{L\bar{x}^3}{6} + \frac{L^2\bar{x}^2}{12}\right) & \left(\frac{\bar{x}^5}{20} - \frac{3L^2\bar{x}^3}{20} + \frac{L^3\bar{x}^2}{10}\right) \end{bmatrix} \mathbf{C}^{-1}\bar{\mathbf{a}}^e$$

$$+ \frac{q_{\bar{y}}}{D_{EI}}\left(\frac{\bar{x}^4}{24} - \frac{L\bar{x}^3}{12} + \frac{L^2\bar{x}^2}{24} \right)$$
(9.81)

$$\frac{d^2v_p}{d\bar{x}^2} = \frac{Q_{\bar{x}}}{D_{EI}} \begin{bmatrix} 0 & 0 & \left(\bar{x}^2 - L\bar{x} + \frac{L^2}{6}\right) & \left(\bar{x}^3 - \frac{9L^2\bar{x}}{10} + \frac{L^3}{5}\right) \end{bmatrix} \mathbf{C}^{-1}\bar{\mathbf{a}}^e$$

$$+ \frac{q_{\bar{y}}}{D_{EI}}\left(\frac{\bar{x}^2}{2} - \frac{L\bar{x}}{2} + \frac{L^2}{12} \right)$$
(9.82)

$$\frac{d^3v_p}{d\bar{x}^3} = \frac{Q_{\bar{x}}}{D_{EI}} \begin{bmatrix} 0 & 0 & (2\bar{x} - L) & \left(3\bar{x}^2 - \frac{9L^2}{10}\right) \end{bmatrix} \mathbf{C}^{-1}\bar{\mathbf{a}}^e + \frac{q_{\bar{y}}}{D_{EI}}\left(\bar{x} - \frac{L}{2}\right)$$
(9.83)

Equation (9.44) together with (9.42) and (9.75) gives an expression for the moment as a function of $\bar{x}$

$$M(\bar{x}) = D_{EI}\mathbf{B}\bar{\mathbf{a}}^e + M_p(\bar{x})$$
(9.84)

where

$$M_p(\bar{x}) = D_{EI}\frac{d^2v_p}{d\bar{x}^2}$$
(9.85)

and with the substitution of (9.82)

$$
M_p(\bar{x}) = Q_{\bar{x}}
\begin{bmatrix}
\dfrac{2\bar{x}^3}{L^3} - \dfrac{3\bar{x}^2}{L^2} + \dfrac{6\bar{x}}{5L} - \dfrac{1}{10} \\[2mm]
\dfrac{\bar{x}^3}{L^2} - \dfrac{2\bar{x}^2}{L} + \dfrac{11\bar{x}}{10} - \dfrac{2L}{15} \\[2mm]
-\dfrac{2\bar{x}^3}{L^3} + \dfrac{3\bar{x}^2}{L^2} - \dfrac{6\bar{x}}{5L} + \dfrac{1}{10} \\[2mm]
\dfrac{\bar{x}^3}{L^2} - \dfrac{\bar{x}^2}{L} + \dfrac{\bar{x}}{10} + \dfrac{L}{30}
\end{bmatrix}^T
\begin{bmatrix}
\bar{u}_2 \\ \bar{u}_3 \\ \bar{u}_5 \\ \bar{u}_6
\end{bmatrix}
+ q_{\bar{y}}\left(\dfrac{\bar{x}^2}{2} - \dfrac{L\bar{x}}{2} + \dfrac{L^2}{12}\right)
\tag{9.86}
$$

Equation (9.50) together with (9.56), (9.84) and (9.85) gives an expression for the shear force as a function of $\bar{x}$

$$
V(\bar{x}) = -\frac{dM}{d\bar{x}} = -D_{EI}\frac{d\mathbf{B}}{d\bar{x}}\bar{\mathbf{a}}^e + V_p(\bar{x})
\tag{9.87}
$$

where

$$
V_p(\bar{x}) = -D_{EI}\frac{d^3 v_p}{d\bar{x}^3}
\tag{9.88}
$$

Substituting (9.83)

$$
V_p(\bar{x}) = Q_{\bar{x}}
\begin{bmatrix}
-\dfrac{6\bar{x}^2}{L^3} + \dfrac{6\bar{x}}{L^2} - \dfrac{6}{5L} \\[2mm]
-\dfrac{3\bar{x}^2}{L^2} + \dfrac{4\bar{x}}{L} - \dfrac{11}{10} \\[2mm]
\dfrac{6\bar{x}^2}{L^3} - \dfrac{6\bar{x}}{L^2} + \dfrac{6}{5L} \\[2mm]
-\dfrac{3\bar{x}^2}{L^2} + \dfrac{2\bar{x}}{L} - \dfrac{1}{10}
\end{bmatrix}^T
\begin{bmatrix}
\bar{u}_2 \\ \bar{u}_3 \\ \bar{u}_5 \\ \bar{u}_6
\end{bmatrix}
- q_{\bar{y}}\left(\bar{x} - \dfrac{L}{2}\right)
\tag{9.89}
$$

From (9.50), we have

$$
Q_{\bar{y}}(\bar{x}) = V(\bar{x}) + Q_{\bar{x}}(\bar{x})\theta(\bar{x})
\tag{9.90}
$$

and for $\bar{x} = 0$ we obtain the nodal force $\bar{P}_2$ as

$$
\bar{P}_2 = -Q_{\bar{y}}(0) = -V(0) - Q_{\bar{x}}(0)\theta(0)
\tag{9.91}
$$

or

$$
\bar{P}_2 = D_{EI}
\begin{bmatrix}
\dfrac{12}{L^3} \\[2mm]
\dfrac{6}{L^2} \\[2mm]
-\dfrac{12}{L^3} \\[2mm]
\dfrac{6}{L^2}
\end{bmatrix}^T
\begin{bmatrix}
\bar{u}_2 \\ \bar{u}_3 \\ \bar{u}_5 \\ \bar{u}_6
\end{bmatrix}
+ Q_{\bar{x}}
\begin{bmatrix}
\dfrac{6}{5L} \\[2mm]
\dfrac{11}{10} \\[2mm]
-\dfrac{6}{5L} \\[2mm]
\dfrac{1}{10}
\end{bmatrix}^T
\begin{bmatrix}
\bar{u}_2 \\ \bar{u}_3 \\ \bar{u}_5 \\ \bar{u}_6
\end{bmatrix}
+ Q_{\bar{x}}
\begin{bmatrix}
0 \\ -1 \\ 0 \\ 0
\end{bmatrix}^T
\begin{bmatrix}
\bar{u}_2 \\ \bar{u}_3 \\ \bar{u}_5 \\ \bar{u}_6
\end{bmatrix}
- q_{\bar{y}}\dfrac{L}{2}
\tag{9.92}
$$

In the same manner, we can determine the nodal forces $\bar{P}_3 = -M(0)$, $\bar{P}_5 = Q_{\bar{y}}(L)$ and $\bar{P}_6 = M(L)$. If we finally gather the nodal forces in a nodal force vector $\bar{\mathbf{f}}_b^e$, we obtain the element relations for beam action with geometrical non-linearity considered

$$
\boxed{\bar{\mathbf{K}}^e \bar{\mathbf{a}}^e = \bar{\mathbf{f}}^e}
\tag{9.93}
$$

where

$$\bar{\mathbf{K}}^e = \bar{\mathbf{K}}_0^e + \bar{\mathbf{K}}_\sigma^e; \quad \bar{\mathbf{f}}^e = \bar{\mathbf{f}}_b^e + \bar{\mathbf{f}}_l^e \tag{9.94}$$

$$\bar{\mathbf{K}}_0^e = \frac{D_{EI}}{L^3} \begin{bmatrix} 12 & 6L & -12 & 6L \\ 6L & 4L^2 & -6L & 2L^2 \\ -12 & -6L & 12 & -6L \\ 6L & 2L^2 & -6L & 4L^2 \end{bmatrix} \tag{9.95}$$

$$\bar{\mathbf{K}}_\sigma^e = \frac{Q_{\bar{x}}}{30L} \begin{bmatrix} 36 & 3L & -36 & 3L \\ 3L & 4L^2 & -3L & -L^2 \\ -36 & -3L & 36 & -3L \\ 3L & -L^2 & -3L & 4L^2 \end{bmatrix} \tag{9.96}$$

$$\bar{\mathbf{a}}^e = \begin{bmatrix} \bar{u}_2 \\ \bar{u}_3 \\ \bar{u}_5 \\ \bar{u}_6 \end{bmatrix}; \quad \bar{\mathbf{f}}_b^e = \begin{bmatrix} \bar{P}_2 \\ \bar{P}_3 \\ \bar{P}_5 \\ \bar{P}_6 \end{bmatrix}; \quad \bar{\mathbf{f}}_l^e = q_{\bar{y}} \begin{bmatrix} \frac{L}{2} \\ \frac{L^2}{12} \\ \frac{L}{2} \\ -\frac{L^2}{12} \end{bmatrix} \tag{9.97}$$

where the element stiffness matrix $\bar{\mathbf{K}}^e$ consists of two parts: one element stiffness matrix $\bar{\mathbf{K}}_0^e$, which is identical to the one we derived in Chapter 4, and one element stiffness matrix $\bar{\mathbf{K}}_\sigma^e$, which describes the effect of the axial force $Q_{\bar{x}}$ on the stiffness.

Exact Solution of the Differential Equation for Beam Action

The differential equation for beam action with regard to geometric non-linearity is above solved approximately. As a complement, an exact solution is given below without derivation. The homogeneous equation corresponding to (9.59) is

$$D_{EI} \frac{d^4 v_h}{d\bar{x}^4} - Q_{\bar{x}} \frac{d^2 v}{d\bar{x}^2} = 0 \tag{9.98}$$

For $Q_{\bar{x}} < 0$, the solution can be written as

$$v_h(\bar{x}) = \alpha_1 + \alpha_2 \bar{x} + \alpha_3 \cos \lambda \bar{x} + \alpha_4 \sin \lambda \bar{x} \tag{9.99}$$

where

$$\lambda = \sqrt{-\frac{Q_{\bar{x}}}{D_{EI}}} \tag{9.100}$$

and for $Q_{\bar{x}} > 0$ the solution can be written as

$$v_h(\bar{x}) = \alpha_1 + \alpha_2 \bar{x} + \alpha_3 \cosh \lambda \bar{x} + \alpha_4 \sinh \lambda \bar{x} \tag{9.101}$$

where

$$\lambda = \sqrt{\frac{Q_{\bar{x}}}{D_{EI}}} \tag{9.102}$$

These solutions give the element stiffness matrix

$$\bar{\mathbf{K}}^e = \frac{D_{EI}}{L^3} \begin{bmatrix} 12\phi_5 & 6L\phi_2 & -12\phi_5 & 6L\phi_2 \\ 6L\phi_2 & 4L^2\phi_3 & -6L\phi_2 & 2L^2\phi_4 \\ -12\phi_5 & -6L\phi_2 & 12\phi_5 & -6L\phi_2 \\ 6L\phi_2 & 2L^2\phi_4 & -6L\phi_2 & 4L^2\phi_3 \end{bmatrix} \tag{9.103}$$

where for $Q_{\bar{x}} < 0$

$$\phi_2 = \frac{\lambda^2 L^2}{12(1 - \phi_1)}; \quad \phi_3 = \frac{\phi_1}{4} + \frac{3\phi_2}{4}; \quad \phi_4 = -\frac{\phi_1}{2} + \frac{3\phi_2}{2}; \quad \phi_5 = \phi_1\phi_2 \tag{9.104}$$

with

$$\phi_1 = \frac{\lambda L}{2} \cot \frac{\lambda L}{2} \tag{9.105}$$

and for $Q_{\bar{x}} > 0$

$$\phi_2 = -\frac{\lambda^2 L^2}{12(1 - \phi_1)}; \quad \phi_3 = \frac{\phi_1}{4} + \frac{3\phi_2}{4}; \quad \phi_4 = -\frac{\phi_1}{2} + \frac{3\phi_2}{2}; \quad \phi_5 = \phi_1\phi_2 \tag{9.106}$$

with

$$\phi_1 = \frac{\lambda L}{2} \coth \frac{\lambda L}{2} \tag{9.107}$$

For the case constant load $q_{\bar{y}}$, the element load vector can be written as

$$\bar{\mathbf{f}}_l^e = q_{\bar{y}} \begin{bmatrix} \dfrac{L}{2} \\ \dfrac{L^2}{12}\psi \\ \dfrac{L}{2} \\ -\dfrac{L^2}{12}\psi \end{bmatrix} \tag{9.108}$$

where for $Q_{\bar{x}} < 0$

$$\psi = \frac{6}{\lambda L} \left(\frac{2}{\lambda L} - \frac{1 + \cos \lambda L}{\sin \lambda L} \right) \tag{9.109}$$

and for $Q_{\bar{x}} > 0$

$$\psi = -\frac{6}{\lambda L} \left(\frac{2}{\lambda L} - \frac{1 + \cosh \lambda L}{\sinh \lambda L} \right) \tag{9.110}$$

Geometrically Non-Linear Beam Element with Six Degrees of Freedom

We obtain the element relations for the geometrically non-linear element with six degrees of freedom if we combine the relations (9.67) and (9.93) in one system of equations

$$\boxed{\bar{\mathbf{K}}^e \bar{\mathbf{a}}^e = \bar{\mathbf{f}}^e} \tag{9.111}$$

where

$$\bar{\mathbf{K}}^e = \bar{\mathbf{K}}_0^e + \bar{\mathbf{K}}_\sigma^e; \quad \bar{\mathbf{f}}^e = \bar{\mathbf{f}}_b^e + \bar{\mathbf{f}}_l^e \tag{9.112}$$

$$
\bar{\mathbf{K}}_0^e =
\begin{bmatrix}
\dfrac{D_{EA}}{L} & 0 & 0 & -\dfrac{D_{EA}}{L} & 0 & 0 \\[2mm]
0 & \dfrac{12D_{EI}}{L^3} & \dfrac{6D_{EI}}{L^2} & 0 & -\dfrac{12D_{EI}}{L^3} & \dfrac{6D_{EI}}{L^2} \\[2mm]
0 & \dfrac{6D_{EI}}{L^2} & \dfrac{4D_{EI}}{L} & 0 & -\dfrac{6D_{EI}}{L^2} & \dfrac{2D_{EI}}{L} \\[2mm]
-\dfrac{D_{EA}}{L} & 0 & 0 & \dfrac{D_{EA}}{L} & 0 & 0 \\[2mm]
0 & \dfrac{-12D_{EI}}{L^3} & -\dfrac{6D_{EI}}{L^2} & 0 & \dfrac{12D_{EI}}{L^3} & -\dfrac{6D_{EI}}{L^2} \\[2mm]
0 & \dfrac{6D_{EI}}{L^2} & \dfrac{2D_{EI}}{L} & 0 & -\dfrac{6D_{EI}}{L^2} & \dfrac{4D_{EI}}{L}
\end{bmatrix}
\tag{9.113}
$$

$$
\bar{\mathbf{K}}_\sigma^e = Q_{\bar{x}}
\begin{bmatrix}
0 & 0 & 0 & 0 & 0 & 0 \\[2mm]
0 & \dfrac{6}{5L} & \dfrac{1}{10} & 0 & -\dfrac{6}{5L} & \dfrac{1}{10} \\[2mm]
0 & \dfrac{1}{10} & \dfrac{2L}{15} & 0 & -\dfrac{1}{10} & -\dfrac{L}{30} \\[2mm]
0 & 0 & 0 & 0 & 0 & 0 \\[2mm]
0 & -\dfrac{6}{5L} & -\dfrac{1}{10} & 0 & \dfrac{6}{5L} & -\dfrac{1}{10} \\[2mm]
0 & \dfrac{1}{10} & -\dfrac{L}{30} & 0 & -\dfrac{1}{10} & \dfrac{2L}{15}
\end{bmatrix}
\tag{9.114}
$$

$$
\bar{\mathbf{a}}^e =
\begin{bmatrix}
\bar{u}_1 \\ \bar{u}_2 \\ \bar{u}_3 \\ \bar{u}_4 \\ \bar{u}_5 \\ \bar{u}_6
\end{bmatrix} ; \quad
\bar{\mathbf{f}}_b^e =
\begin{bmatrix}
\bar{P}_1 \\ \bar{P}_2 \\ \bar{P}_3 \\ \bar{P}_4 \\ \bar{P}_5 \\ \bar{P}_6
\end{bmatrix} ; \quad
\bar{\mathbf{f}}_l^e = q_{\bar{y}}
\begin{bmatrix}
0 \\[2mm] \dfrac{L}{2} \\[2mm] \dfrac{L^2}{12} \\[2mm] 0 \\[2mm] \dfrac{L}{2} \\[2mm] -\dfrac{L^2}{12}
\end{bmatrix}
\tag{9.115}
$$

The element relations for pure bar action (9.34) can be derived from (9.111) if we let $q_{\bar{y}} = 0$ and condense degree of freedom $\bar{u}_3$ and $\bar{u}_6$. Note that in the element relations for bar action (9.34) the axial stiffness of the element is not affected by the axial force $Q_{\bar{x}}$. The bar system as a whole is, however, affected since the term $\dfrac{Q_{\bar{x}}}{L}$ affects the stiffness of the system perpendicular to a bar element. In the element relations for the beam element (9.111), an axial force gives both this effect and a local effect on the bending stiffness of the element.

From Local to Global Coordinates

To be able to model a frame, we have to transform the element relation to a global coordinate system. From coordinate transformations, described in Chapter 4, we have the transformation relations

$$
\mathbf{K}^e = \mathbf{G}^T \bar{\mathbf{K}}^e \mathbf{G}; \quad \mathbf{f}_l^e = \mathbf{G}^T \bar{\mathbf{f}}_l^e
\tag{9.116}
$$

where the transformation matrix is given by

$$
\mathbf{G} =
\begin{bmatrix}
n_{x\bar{x}} & n_{y\bar{x}} & 0 & 0 & 0 & 0 \\
n_{x\bar{y}} & n_{y\bar{y}} & 0 & 0 & 0 & 0 \\
0 & 0 & 1 & 0 & 0 & 0 \\
0 & 0 & 0 & n_{x\bar{x}} & n_{y\bar{x}} & 0 \\
0 & 0 & 0 & n_{x\bar{y}} & n_{y\bar{y}} & 0 \\
0 & 0 & 0 & 0 & 0 & 1
\end{bmatrix}
\tag{9.117}
$$

Example 9.2 Comparison between exact and approximate solution

By performing comparative computations with the exact and the approximative formulation, we can gain understanding of conditions that must be fulfilled for the approximate solution to be applicable, that is to get a reasonable accuracy. Consider a beam, hinged at both ends and loaded by a moment M_0 at its left end (Figure 1). The buckling load for a hinged beam is $Q_{\bar{x}} = \pi^2 \frac{D_{EI}}{L^2}$ according to the second of Euler's buckling cases. This relation can be reformulated to a quotient between the effect of the axial force on the stiffness $\frac{Q_{\bar{x}}}{L}$ and the bending stiffness $\frac{D_{EI}}{L^3}$, that is $\frac{Q_{\bar{x}}}{D_{EI}}L^2 = \pi^2$ or $\sqrt{\frac{Q_{\bar{x}}}{D_{EI}}}L = \pi$. Computations have been performed for the values 0, 1.5, 2.0 and 2.5 of $\sqrt{\frac{Q_{\bar{x}}}{D_{EI}}}L$. In Figures 2–4, the results are shown. Solid lines represent exact solutions and dashed lines the approximative ones. Largest differences are obtained for the largest axial force. For $\sqrt{\frac{Q_{\bar{x}}}{D_{EI}}}L = 1.5$, the difference between the exact and the approximate solution is negligible.

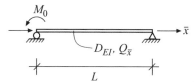

Figure 1 A beam with an axial force loaded by a moment M_0

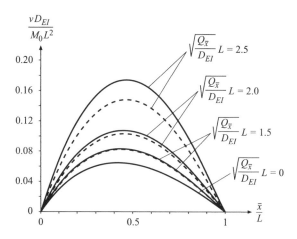

Figure 2 Displacement $v(\bar{x})$

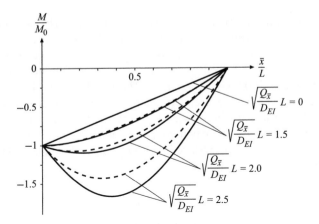

Figure 3   Moment $M(\bar{x})$

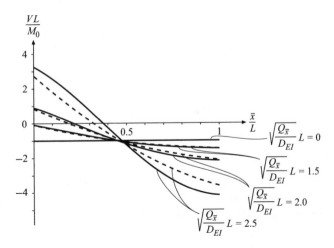

Figure 4   Shear force $V(\bar{x})$

9.3.3   Frames

Example 9.3 A frame analysis with geometrical non-linearity considered
Consider a frame composed of three beams with the cross-sectional areas $A_1 = 2.0 \times 10^{-3}$ m^2, $A_2 = 2.0 \times 10^{-3}$ m^2 and $A_3 = 6.0 \times 10^{-3}$ m^2, the moments of inertia $I_1 = 1.6 \times 10^{-5}$ m^4, $I_2 = 1.6 \times 10^{-5}$ m^4 and $I_3 = 5.4 \times 10^{-5}$ m^4 and with the modulus of elasticity $E = 200.0$ GPa (Figure 1). The lengths of the beams are $L_1 = 4.0$ m, $L_2 = 4.0$ m and $L_3 = 6.0$ m, respectively. Along the horizontal beam, the frame is loaded by a uniformly distributed load $q_0 = 50$ kN/m and at the upper left corner a horizontal point load $P = 10$ kN acts. At the bottom, the structure is rigidly supported at the left end and hinged at the right end. The frame in this example is the same as the one in Example 4.2, but with a five times larger load.

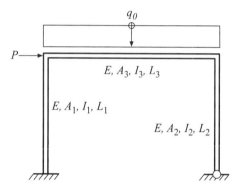

Figure 1   A frame structure

Computational model

The frame is modelled with three beam elements, denoted 1, 2 and 3; see Figure 2. The system has the degrees of freedom $a_1, a_2, \dots, a_{12}$. The displacements a_1, a_2, a_{10} and a_{11} as well as the rotation a_3 are prescribed to be zero.

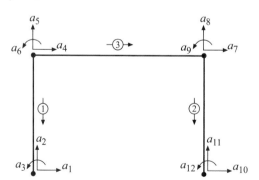

Figure 2   Computational model

Computational procedure

The element stiffness matrices are functions of the axial force $Q_{\bar{x}}$ acting on respective element. Since these forces are a result of the computation, the computational procedure is iterative. The element stiffness matrices are in the first iteration computed with $Q_{\bar{x}} = 0$, that is in a linear computation. The element stiffness matrices are functions of the axial force $Q_{\bar{x}}$ acting on respective element. Since these forces are a result from the computation, the computational procedure is iterative. The element stiffness matrices are in the first iteration computed with $Q_{\bar{x}} = 0$, that is in a linear computation. From the computation, values of $Q_{\bar{x}}$ in the three elements are obtained and these are used to determine new element stiffness matrices whereupon a new solution is found. The procedure is repeated until the computed axial forces with a sufficient accuracy agree with the presumed ones. Computed values are found in Table 1. In this case, the solution has converged after four iterations. The computed displacements and moment distribution are shown with solid lines in Figures 3 and 4, respectively, while the results according to linear theory are shown with dashed lines.

Table 1 Computed displacements and axial forces

Iteration number	a_4 (mm)	a_5 (mm)	a_6 (−)	$Q_{\bar{x}}^{(1)}$ (kN)	$Q_{\bar{x}}^{(2)}$ (kN)	$Q_{\bar{x}}^{(3)}$ (kN)
1	37.6785	−1.4370	-2.6867×10^{-2}	−143.704	−156.296	−19.634
2	45.1286	−1.4241	-2.8102×10^{-2}	−142.413	−157.587	−18.187
3	45.1366	−1.4242	-2.8097×10^{-2}	−142.417	−157.583	−18.163
4	45.1364	−1.4242	-2.8097×10^{-2}	−142.417	−157.583	−18.163

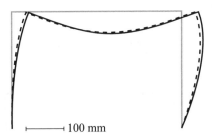

├─────────┤ 100 mm

Figure 3   Computed displacements

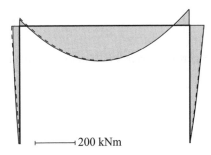

├─────────┤ 200 kNm

Figure 4   Computed moment distribution

Compared with a linear computation, the horizontal displacement is about 20 % larger in the geometrically non-linear solution.

Example 9.4 Buckling safety for a frame

The frame that was treated in Example 9.3 is here analysed with respect to buckling safety. Figure 1 shows how det $\mathbf{K}$ varies with the value of α. By solving the generalised eigenvalue problem and computing the value of the buckling safety, we obtain $\alpha_{cr} = 6.89$. The corresponding shape of the deformation is given by the eigenvector corresponding to this value and is shown in Figure 2.

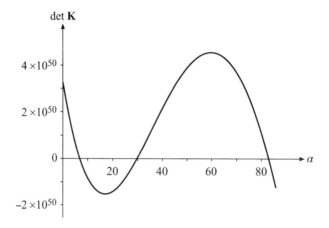

Figure 1   Variation of det $\mathbf{K}$

Figure 2   Computed deformation shape at instability

9.4 Three-Dimensional Geometric Non-Linearity

Earlier, we have studied how the internal axial tensile and compressive forces affect the stiffness of a two-dimensional structure. If we, as discussed in Chapter 7, combine two-dimensional beam action in the $\bar{x}\bar{y}$-plane with two-dimensional beam action in the $\bar{x}\bar{z}$-plane, we can extend the theory to be valid for three-dimensional structures. We can thereby determine the safety against instability for three-dimensional structures according to internal axial forces $Q_{\bar{x}}$.

In addition to this instability phenomenon, there are two important forms of instability. Both the phenomena belong to a two-dimensional loading situation (the $\bar{x}\bar{y}$-plane), but generate deformations in the third dimension (the $\bar{z}$-direction). The first phenomenon is called *lateral buckling* and is associated with the bending moment of beam action. Lateral buckling may occur in beams which at their supports are prevented from torsion. In Figure 9.13, this is shown by the fact that the simply supported beam also has two horizontal struts at its supports. If we, in the $\bar{x}\bar{y}$-plane, consider the bending moment as a couple of forces, the compressive force tends to deform the upper flange of a bent beam in the $\bar{x}\bar{z}$-plane. Because the beam, at the upper flange, has horizontal supports in the z-direction, the tensile force at the lower flange of the beam prevents the lateral buckling of it. This is expressed by the beam rotation in the $\bar{y}\bar{z}$-plane.

In Figure 9.14, the same beam is shown but without the upper horizontal strut in the $\bar{z}$-direction. The force patterns that the load causes will then contain a vertical pressure that tends to get the beam to 'overturn' as a rigid body. We call this phenomenon *overturning*.

Lateral buckling is usually avoided by transversely stabilising the compressed flanges of the beams. Overturning is avoided for example by web stiffeners. The theory and methodology for modelling and analysing lateral buckling and overturning are outside the scope of this textbook.

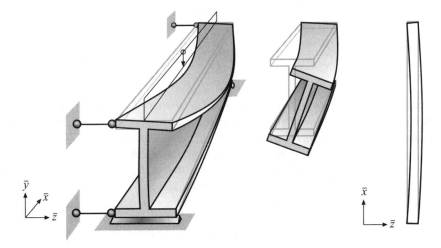

Figure 9.13 Lateral buckling

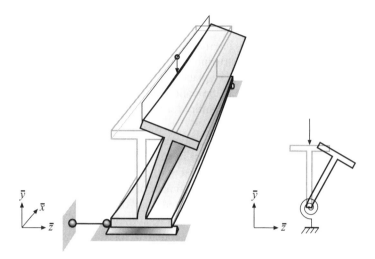

Figure 9.14   Overturning

Exercises

9.1

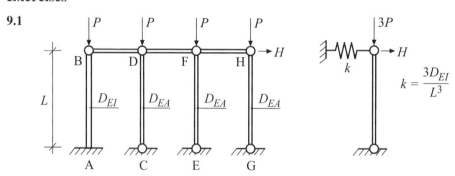

The figure to the left shows a model of a bearing structure. The bearing structure consists of a column-beam system which is connected to a stabilising stairwell. Each column is loaded by a vertical force P. In addition, the entire bearing structure is loaded by a horizontal force H. The figure to the right shows a simplified model of the bearing structure. A bar with the load $3P$ and the cross-sectional stiffness $3D_{EA}$ represents the three hinged columns CD, EF and GH, and a spring with stiffness $k = \frac{3D_{EI}}{L^3}$ represents the cantilevered beam AB. (Example 5.3 shows how a cantilever beam can be represented by a spring with equivalent stiffness.) The model is based on the assumption that the hinged columns have a bending stiffness which is large enough for neglection of geometrical non-linearity due to beam action. Establish a computational model consisting of a geometrically non-linear bar element and an elastic spring. For the case $D_{EI} = 1$, $L = 1$, $P = 0.5$ and $H = 0.15$, determine the force in the spring and the horizontal displacement at B. Compare the force in the spring with the magnitude of the external horizontal force H and comment on the result.

9.2   Consider the structure in Example 9.1. Compute the displacements a_5 and a_6 and also
the forces $Q_{\bar{x}}^{(1)}$ and $Q_{\bar{x}}^{(2)}$ for the case $P = 5.0$ MN and $F = 0.2$ MN in
(a) the linear case
(b) the non-linear case, with one computational step.
Compare with the results from the example.

9.3   Consider the truss in Example 9.1. Use CALFEM to determine the buckling safety.

9.4   Derive the element relation for a bar, (9.34), by starting from the element relation for a
beam when considering geometrical non-linearity, (9.111), and condense out the rotational degrees of freedom.

9.5

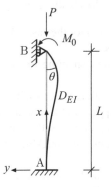

The column according to the figure is loaded by an axial compressive force
$P = 0.4\frac{\pi^2 D_{EI}}{L^2}$.
(a) Determine by a linear analysis the rotation at B and the sectional moments at A, B
and $x = \frac{L}{2}$.
(b) Determine by a geometrically non-linear analysis the rotation at B and the sectional
moments at A, B and $x = \frac{L}{2}$.
(c) Compute how much the load shall be increased for the column to become unstable.
Compare the computed result with Euler's third case of column buckling.

9.6

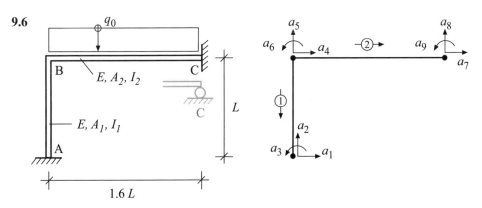

The frame of Exercise 4.8 but here loaded only by a vertical distributed load $q_0 =$ 100 kN/m is to be analysed considering geometric non-linearity. Determine the horizontal displacement at B and the support moment at A for the following four cases:
(a) linear analysis, clamped at C
(b) geometrical non-linear analysis, clamped at C
(c) linear analysis, roller support at C
(d) geometrical non-linear analysis, roller support at C.

9.7 Consider the frame in Exercise 9.6 for the case where the frame has a roller support at C.
(a) Use Euler's cases of column buckling to find an estimation of an interval for the buckling safety α_{cr}.
(b) Use CALFEM to compute the buckling safety α_{cr}.

9.8 Consider the frame in Exercise 9.6 for the case where the frame is clamped at C and with a distributed load $q_0 = 500$ kN/m.
(a) Use Euler's cases of column buckling to find an estimation of an interval for the buckling safety α_{cr}.
(b) If a structure is modelled with few elements it may, due to the assumed deformation shape of the elements, happen that the model is unable to capture the deformation mode occurring at buckling and then the computed buckling safety may be incorrect. Use CALFEM to compute the buckling safety α_{cr} using a computational model according to Exercise 9.6.
(c) Modify the computational model so that part AB is modelled with two elements of length $0.5L$ and use CALFEM to compute the buckling safety α_{cr}.

10

Material Non-Linearity

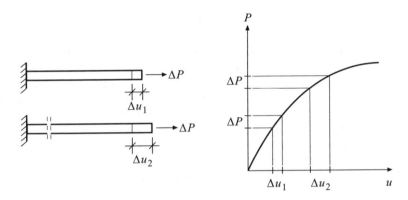

Figure 10.1   The effect of a non-linear material relation on the stiffness

Until now, we have assumed that Hooke's law defines the relation between stress and strain at the material level, that is the stress is proportional to the strain. Such a material relation is referred to as a *linear material relation*; cf. Figure 3.9. Since most structural materials under moderate loading act in accordance with Hooke's law, this assumption is often reasonable. To obtain a more exact description of how a structure behaves at high load levels, which cause material degradation, we extend the description of the constitutive relations of the materials (Figure 10.1).

A general description of the material behaviour is obtained if we load the material up to its failure load and record its force versus deformation behaviour in a stress–strain diagram (Figure 10.2). A typical stress–strain graph consists of three parts: the initial stiffness, the strength and the softening of the material. The modulus of elasticity, E, refers in most cases to the *initial stiffness* of the material. The maximal (ultimate) stress that a material can carry is referred to as its *strength* and is denoted σ_u (u is an abbreviation of ultimate). With modern test methods, it is possible to run the test in displacement control and thereby follow the material behaviour also after the material strength has been reached. The behaviour that the material then exhibits determines its *softening behaviour*. If the strength and softening behaviour of

Structural Mechanics: Modelling and Analysis of Frames and Trusses, First Edition.
Karl-Gunnar Olsson and Ola Dahlblom.
© 2016 John Wiley & Sons, Ltd. Published 2016 by John Wiley & Sons, Ltd.

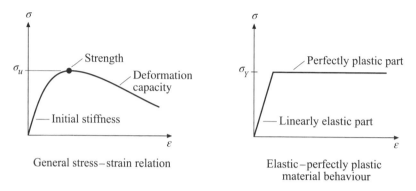

Figure 10.2   Stress–strain diagrams

the materials of a structure are known, the load-carrying capacity as well as the behaviour at failure of the structure can be simulated.

An important special case for the stress–strain graph is an *elastic–perfectly plastic material behaviour* (Figure 10.2). This idealisation of the stress–strain graph means that the material first is linearly elastic and then perfectly plastic. The latter implies that the material is deformed without any increase in the stress, which can be interpreted as the material yields. The stress level is called yield stress and is denoted σ_Y (where Y is an abbreviation for yield). With this stress–strain graph, simulations of load-carrying capacity and behaviour at failure can yield accurate results for structures of e.g. steel, reinforced concrete and aluminium.

When a structure is loaded above a certain load level, zones where the stress approaches the material strength will arise locally in the structure. In these zones, the internal structure of the material is damaged. One says that the material *yields*, which implies that its stiffness decreases or, for the case of perfect plasticity, completely vanishes. This affects the stiffness matrix $\mathbf{K}$ of the structure; we obtain a stiffness matrix that depends on the magnitude of the internal forces. These are in turn determined by the present displacements $\mathbf{a}$, which initially are unknown in the computation, that is $\mathbf{K} = \mathbf{K}(a)$. Therefore, an incremental calculation procedure has to be formulated. This procedure is related to the gradual degradation of the material and the associated stiffness reduction and is therefore referred to as a *material non-linear* calculation. For analyses of trusses and frames, the idealised material description elastic–perfectly plastic material behaviour enables us to use relatively uncomplicated non-linear methods of solution.

Section 10.1 describes one of the most common calculation procedures for material non-linearity. In Section 10.2, a more thorough description of the idealisation elastic–perfectly plastic material is given, and Sections 10.3 and 10.4 demonstrate how to use the described calculation procedure to analyse a truss and a frame, respectively.

10.1   Calculation Procedures

Analogous to geometrical non-linearity, there are two important types of analyses, which take into account that the materials of a structure yield:

- calculation of deformations and internal forces for a gradually increasing internal load
- limit load analyses where the maximal load-carrying capacity of a structure can be estimated under the assumption of an elastic–perfectly plastic material behaviour.

Next, we discuss the first type of analysis. The basis for this analysis is the system of equations

$$\mathbf{K(a)a} = \mathbf{f} \tag{10.1}$$

of the structure, where the stiffness matrix $\mathbf{K(a)}$ depends on the present displacements $\mathbf{a}$.

A robust method for considering the gradual yielding of a material is to perform the calculation in small steps using, so called, *incremental formulation*. This means that the external loads are divided into a finite number of increments $\Delta \mathbf{f}^i$, according to Figure 10.3,

$$\mathbf{f} = \sum_{i=1}^{n} \Delta \mathbf{f}^i \tag{10.2}$$

where $\mathbf{f}$ is the force vector and n is the number of increments. If the problem contains prescribed displacements, these are divided into increments in the same manner.

There are different strategies for performing a step in the incremental calculation procedure. Here, one of the most straightforward strategies will be discussed. This is referred to as the *forward Euler* method. Assume that we have reached state i where we have known external loads $\mathbf{f}^i$, known displacements $\mathbf{a}^i$ and a known stiffness $\mathbf{K}^i$ of the structure. In the next calculation step, the structure is loaded by a load increment $\Delta \mathbf{f}^i$ such that the applied load now is $\mathbf{f}^{i+1} = \mathbf{f}^i + \Delta \mathbf{f}^i$. To determine the corresponding displacements $\mathbf{a}^{i+1} = \mathbf{a}^i + \Delta \mathbf{a}^i$, we calculate the displacement increment $\Delta \mathbf{a}^i$ by solving the system of equations

$$\mathbf{K}^i \Delta \mathbf{a}^i = \Delta \mathbf{f}^i \tag{10.3}$$

With forward Euler method, we take $\mathbf{K}^i$ to be the known stiffness of the structure in state i. This is commonly referred to as the *tangent stiffness* and denoted $\mathbf{K}_T^i$, since it, in a model with only one degree of freedom represents the tangent stiffness in the present state. The tangent stiffness $\mathbf{K}_T^i$ is calculated by considering the magnitude of the internal forces, normal forces N_i and bending moments M_i, in the present state. If it is possible that these change during

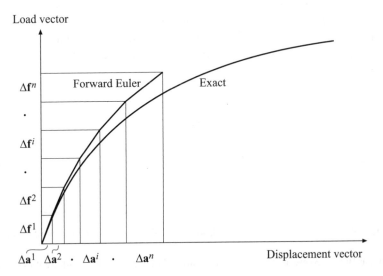

Figure 10.3 Calculation procedure for incrementally solving materially non-linear problems

the calculation step, the calculated displacement increment $\Delta \mathbf{a}^i$ can deviate from the exact solution. For yielding materials, the assumed stiffness is always larger than the exact stiffness, since the section forces that determine $\mathbf{K}_T^i$ always are underestimated. This implies that each calculation step adds a deviation from the exact solution, as shown in Figure 10.3. For forward Euler, the total deviation can be kept down by using small load increments.

To summarise, a calculation procedure for material non-linearity can be described by the following points:

For increment $i = 1$ to n:

- Establish the tangent element stiffness matrices $\mathbf{K}_T^{e,i}$ at the state i.
- Assemble all $\mathbf{K}_T^{e,i}$ into a global tangent stiffness matrix $\mathbf{K}_T^i$.
- Solve the system of equations $\mathbf{K}_T^i \Delta \mathbf{a}^i = \Delta \mathbf{f}^i$.
- Determine the element displacement increments $\Delta \mathbf{a}^{e,i}$.
- Calculate the stresses/section forces at state $i + 1$.
- Calculate the total displacements $\mathbf{a}^{i+1} = \mathbf{a}^i + \Delta \mathbf{a}^i$.
- Calculate the total external forces $\mathbf{f}^{i+1} = \mathbf{f}^i + \Delta \mathbf{f}^i$.

10.2   Elastic–Perfectly Plastic Material

When a material yields, an internal degradation of the material takes place. A stress–strain graph which deviates from an initially straight line can be an indication of material degradation. Figure 10.4 shows examples of typical stress–strain graphs for different materials whose behaviour is characterised by substantial yielding at high stress levels.

To find out how the material yielding affects the behaviour of a structure at high load levels, a simplified stress–strain graph can be very useful. Figure 10.4 shows how the approximation of *elastic–perfectly plastic material* can be done (dashed lines). In the stress–strain graphs, σ_Y denotes the *yield stress* of the material. For an elastic–perfectly plastic material, the stress cannot exceed this level. It is said that the material yields, which means that its deformation increases without any increase in the forces on it.

With the idealisation of elastic–perfectly plastic behaviour, there is no limit to how stretched the material can be. This infinite deformation capacity is of course not realistic; every material will eventually fail. Therefore, it is common to add to the assumption of elastic–perfectly plastic material a measure of the *deformation capacity*, that is the maximal strain that cannot be exceeded.

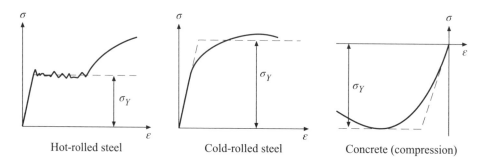

Figure 10.4   Experimental and idealised stress–strain diagrams for different materials

10.3   Trusses with Material Non-Linearity Considered

The behaviour of a truss at high load levels can be simulated if the model of calculation allows the material in the bars to yield. Moreover, with such a computational model, the failure load can be accurately determined. In Section 10.2, it was shown that several of the most common building materials can be represented by an elastic–perfectly plastic material model. For a homogeneous cross-section, the stress σ is constant across the cross-sectional surface of a bar and the normal force N can according to (3.14) be written as

$$N = \int_A \sigma dA = \sigma A \tag{10.4}$$

The yield stress σ_Y is reached simultaneously across the entire cross-section and corresponds to a *yield force* $N_Y = \sigma_Y A$. For a bar with no external axial load and with a constant cross-sectional area, the stress is the same throughout the entire bar. Thus, yielding will be initiated simultaneously in the entire bar.

The calculations become very simple if the analysis of a truss is performed with an elastic–perfectly plastic material behaviour and by using the forward Euler method. A bar behaves either elastically or fully plastic. The criterion for when plastic deformations occur in the bar is $N = N_Y$. If $N < N_Y$, the bar behaves elastically. There are two different tangent stiffnesses E_T for the material, one for the initial linear-elastic range of action ($\sigma < \sigma_Y$) where $E_T = E$ and another one for the plastic range ($\sigma = \sigma_Y$) where $E_T = 0$. The incremental element equation for a bar with elastic–perfectly plastic material can then be written in the local coordinate system as

$$\frac{E_T A}{L} \begin{bmatrix} 1 & -1 \\ -1 & 1 \end{bmatrix} \begin{bmatrix} \Delta u_1 \\ \Delta u_2 \end{bmatrix} = \begin{bmatrix} \Delta P_1 \\ \Delta P_2 \end{bmatrix} \tag{10.5}$$

where

$$E_T = E \quad \text{for} \quad N < N_Y$$

$$E_T = 0 \quad \text{for} \quad N = N_Y \tag{10.6}$$

How to apply the incremental calculation procedure described in Section 10.1 to the analysis of a truss by using these relations is shown in Example 10.1.

Example 10.1 Truss analysis with material non-linearity considered

The truss from Example 3.1 is analysed here with an assumption of a non-linear material behaviour. Each bar is fixed by a hinge in one end and connected to the other bars in a joint in its other end, as shown in Figure 1. The lengths of the bars are $L_1 = 1.6$ m, $L_2 = 1.2$ m and $L_3 = \sqrt{1.6^2 + 1.2^2} = 2.0$ m, respectively, and the cross-sectional areas are $A_1 = 6.0 \times 10^{-4}$ m^2, $A_2 = 3.0 \times 10^{-4}$ m^2 and $A_3 = 10.0 \times 10^{-4}$ m^2, respectively. The material in the bars is assumed to be elastic–perfectly plastic with the modulus of elasticity $E = 200.0$ GPa and the yield stress $\sigma_Y = 400.0$ MPa. The truss is loaded by a downwards directed force P in the free joint and we seek the maximal load P_u that the truss can carry before it becomes a mechanism due to yielding in the bars and collapses.

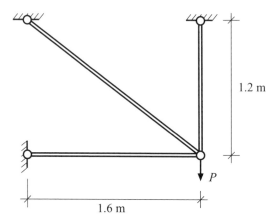

Figure 1   A plane truss consisting of three bars

Computational model

The truss model is built up by three bar elements and is given eight degrees of freedom according to Figure 2. In degree of freedom a_1, a_2, a_3, a_4, a_7 and a_8, the displacement is prescribed to be zero.

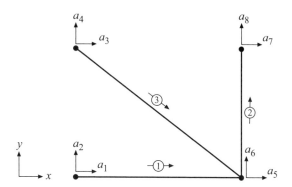

Figure 2   Computational model

Calculation procedure

The external load is applied gradually with small load increment ΔP_i. For each new load increment, a system of equations is established where the assembled element stiffness matrices depend on the present material stiffnesses E_T. The corresponding displacement increments are computed and the element section forces determined. Then, the displacements and external loads of the truss are updated. This is in accordance with the calculation procedure for forward Euler described in Section 10.1.

Forward Euler for one load increment

We choose to use a load increment of $\Delta P_i = 4$ kN and start the loading. The element stiffness matrices $\mathbf{K}_T^e$ for the three elements are (cf. Example 3.1)

$$\mathbf{K}_T^1 = \frac{E_T^{(1)} \cdot 6.0 \times 10^{-4}}{1.6} \begin{bmatrix} 1 & 0 & -1 & 0 \\ 0 & 0 & 0 & 0 \\ -1 & 0 & 1 & 0 \\ 0 & 0 & 0 & 0 \end{bmatrix} \tag{1}$$

$$\mathbf{K}_T^2 = \frac{E_T^{(2)} \cdot 3.0 \times 10^{-4}}{1.2} \begin{bmatrix} 0 & 0 & 0 & 0 \\ 0 & 1 & 0 & -1 \\ 0 & 0 & 0 & 0 \\ 0 & -1 & 0 & 1 \end{bmatrix} \tag{2}$$

$$\mathbf{K}_T^3 = \frac{E_T^{(3)} \cdot 10.0 \times 10^{-4}}{2.0} \begin{bmatrix} 0.64 & -0.48 & -0.64 & 0.48 \\ -0.48 & 0.36 & 0.48 & -0.36 \\ -0.64 & 0.48 & 0.64 & -0.48 \\ 0.48 & -0.36 & -0.48 & 0.36 \end{bmatrix} \tag{3}$$

where the modulus of elasticity E_T is

$$E_T = 200.0 \text{ GPa for } N < N_Y \tag{4}$$

$$E_T = 0 \quad \text{for} \quad N = N_Y \tag{5}$$

When the element matrices have been assembled and the system of equations has been reduced considering prescribed displacements, it becomes

$$10^{-5} \begin{bmatrix} 37.5E_T^{(1)} + 32E_T^{(3)} & -24E_T^{(3)} \\ -24E_T^{(3)} & 25E_T^{(2)} + 18E_T^{(3)} \end{bmatrix} \begin{bmatrix} \Delta a_5 \\ \Delta a_6 \end{bmatrix} = \begin{bmatrix} 0 \\ -\Delta P_i \end{bmatrix} 10^3 \tag{6}$$

which initially, with $E_T = 200$ GPa for all the bars and known quantities inserted, gives

$$10^6 \begin{bmatrix} 139 & -48 \\ -48 & 86 \end{bmatrix} \begin{bmatrix} \Delta a_5 \\ \Delta a_6 \end{bmatrix} = \begin{bmatrix} 0 \\ -4 \end{bmatrix} 10^3 \tag{7}$$

which has the solution

$$\begin{bmatrix} \Delta a_5 \\ \Delta a_6 \end{bmatrix} = \begin{bmatrix} -19.90 \\ -57.62 \end{bmatrix} 10^{-6} \tag{8}$$

For Element 1, we have initially the element displacement increments

$$\Delta \mathbf{a}^1 = \mathbf{G} \Delta \mathbf{a}^1 = \begin{bmatrix} 1 & 0 & 0 & 0 \\ 0 & 0 & 1 & 0 \end{bmatrix} \begin{bmatrix} 0 \\ 0 \\ -19.90 \\ -57.62 \end{bmatrix} 10^{-6} = \begin{bmatrix} 0 \\ -19.90 \end{bmatrix} 10^{-6} \tag{9}$$

which give the normal force increment

$$\Delta N^{(1)} = EA_1 \mathbf{B} \Delta \mathbf{a}^1$$

$$= 200.0 \times 10^9 \cdot 6.0 \times 10^{-4} \frac{1}{1.6} \begin{bmatrix} -1 & 1 \end{bmatrix} \begin{bmatrix} 0 \\ -19.90 \end{bmatrix} 10^{-6}$$

$$= -1.492 \times 10^3 \tag{10}$$

For Elements 2 and 3, we can in the same manner determine the initial values of $\Delta N^{(2)}$ and $\Delta N^{(3)}$

$$\Delta N^{(2)} = 2.881 \times 10^3 \tag{11}$$
$$\Delta N^{(3)} = 1.865 \times 10^3 \tag{12}$$

In the end of each load increment, the element normal forces are updated

$$N^{e,i+1} = N^{e,i} + \Delta N^e \tag{13}$$

as well as the displacements of the truss and the external load

$$\mathbf{a}^{i+1} = \mathbf{a}^i + \Delta \mathbf{a}^i \tag{14}$$

$$\mathbf{f}^{i+1} = \mathbf{f}^i + \Delta \mathbf{f}^i \tag{15}$$

Check for yielding and collapse

To determine the material stiffnesses E_T in a new calculation step, the present normal force N is compared with the yield force N_Y for each element. The three bar elements have the yield forces

$$N_Y^{(1)} = \sigma_Y A_1 = 240 \text{ kN} \tag{16}$$

$$N_Y^{(2)} = \sigma_Y A_2 = 120 \text{ kN} \tag{17}$$

$$N_Y^{(3)} = \sigma_Y A_3 = 400 \text{ kN} \tag{18}$$

During the first 41 load increments, none of the normal forces reaches the corresponding yield force; thus, the initial stiffness matrix

$$\mathbf{K}_T = 10^6 \begin{bmatrix} 139 & -48 \\ -48 & 86 \end{bmatrix} \tag{19}$$

can be used throughout all these increments.

At load increment 42, we obtain $N^{(2),42} = 120.99$ kN, which exceeds the yield force of 120 kN. A new and reduced stiffness matrix $\mathbf{K}_T$ is calculated

$$\mathbf{K}_T = 10^6 \begin{bmatrix} 139 & -48 \\ -48 & 36 \end{bmatrix} \tag{20}$$

where the material stiffness for Element 2 has been reduced to $E_T^{(2)} = 0$. Thereafter, we first check whether the plastic flow in bar Element 2 leads to a collapse of the truss, but since $\det \mathbf{K}_T > 0$ there is no collapse yet and we can continue the calculation, where we choose to keep the load increment $\Delta P_i = 4$ kN.

After load increment 76, we obtain $N^{(1),76} = -244.00$ kN, which exceeds the yield force of 240 kN. A new reduced stiffness matrix $\mathbf{K}_T$ for the truss is calculated

$$\mathbf{K}_T = 10^6 \begin{bmatrix} 64 & -48 \\ -48 & 36 \end{bmatrix} \tag{21}$$

where also the material stiffness $E_T^{(1)} = 0$. The determinant of the reduced stiffness matrix $\det \mathbf{K}_T = 0$, that is we have a mechanism, the truss has collapsed and the calculation is terminated.

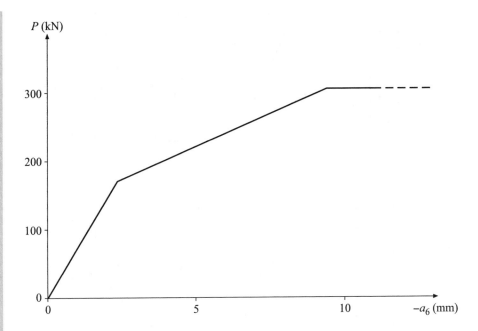

Figure 3   Calculated force P versus displacement of degree of freedom 6

Figure 3 shows the load–displacement relation of degree of freedom 6. Note that the load when the truss collapses is 1.8 times larger than the load when yielding first occurs in one element of the truss.

10.4   Frames with Material Non-Linearity Considered

In a truss, the bars carry load by pure bar action. If the external load acts only at the hinges of the truss, there is a constant normal force in each bar. Moreover, if the cross-sections are homogeneous and the cross-sectional areas constant along the bars, the computational model gives simultaneous yielding of the entire bar. The fact that yielding actually will be located at the weakest zone along the bar will not affect the results of the calculation.

In a frame with pure beam action or with combined bar and beam action, yielding will, in practise as well as in the computational model, be localised to certain zones. These will arise where the bending moment has its maxima, which may be at fixed locations such as where point loads act, at clamps and/or at the corners of the frame, but the maxima can also appear at locations which depend on where yielding already has occurred. When a zone has attained fully developed plastic flow, we say that a *plastic hinge* has arisen. A plastic hinge arises progressively as the plastic flow extends over an increasingly large part of the cross-section during the increase of the magnitude of the internal forces, normal force and moment. The yield stress is first reached at the outermost part of the beam cross-section and then, the plastic zone grows until plastic flow occurs in the entire cross-section (Figure 10.5).

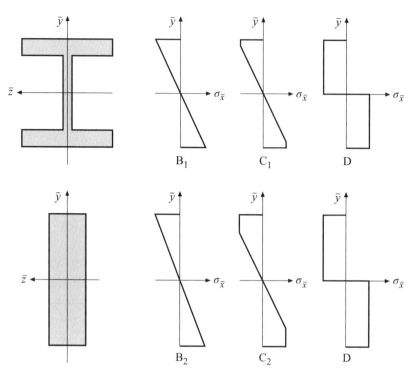

Figure 10.5 The development of a plastic hinge

Based on the elastic–perfectly plastic stress–strain relation at material level (Figure 10.2), we can establish the corresponding diagram, an M–κ diagram, at the cross-section level, where according to (4.7) and (4.8)

$$\varepsilon_{\bar{x}}(\bar{x}, \bar{y}, \bar{z}) = -\kappa(\bar{x})\,\bar{y} \tag{10.7}$$

$$M(\bar{x}) = -\int_A \sigma_{\bar{x}}(\bar{x}, \bar{y}, \bar{z})\,\bar{y}\,dA \tag{10.8}$$

Such a diagram can, among other things, show how different cross-sectional shapes affect the yielding progress. Figure 10.6 shows two different yielding progresses for pure beam action and homogeneous cross-sections, one for a rectangular cross-section (dashed line) and one for an H-shaped cross-section (solid line). Initially, we have an elastic behaviour from A to B, described by the constitutive relation of the cross-section level according to Equation (4.12)

$$M = D_{EI}\,\frac{d^2v}{dx^2} = D_{EI}\,\kappa \tag{10.9}$$

At B, yielding is initiated in the outermost parts of the cross-section. From B, the plastic zone grows and at D it is a fully developed plastic hinge. The moment–curvature relations in Figure 10.6 show that this development is faster for an H-shaped cross-section than for a rectangular one. The maximal bending moment the cross-section has capacity for is referred to as its yield moment and is denoted M_Y. By considering a beam loaded by a point load at

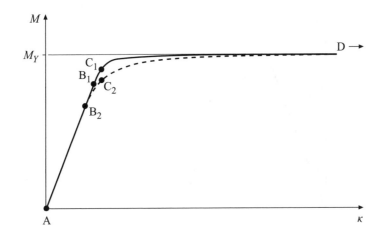

Figure 10.6 Moment–curvature relations at cross-section level

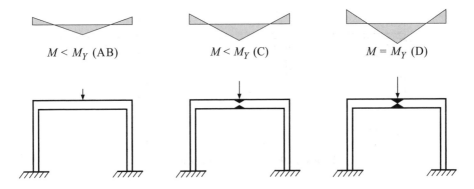

Figure 10.7 The development of a plastic hinge

its midpoint (Figure 10.7), we can follow how a plastic zone is initiated and grows to a fully developed plastic hinge.

To simplify the materially non-linear analysis of frames, it is often assumed that an elastic–perfectly plastic moment–curvature relation can be applied at the cross-section level (Figure 10.8). This assumption implies that the computational model describes a progress which is slightly stiffer than the one obtained from the original moment–curvature relations in Figure 10.6.

When a frame is gradually loaded, the section forces will increase elastically up to the load level where the first plastic hinge develops. With an elastic–perfectly plastic moment–curvature relation at the cross-section level, this plastic hinge will appear instantaneously. At the plastic hinge, the cross-section will continue to deform without any increase in the bending moment. The criterion for plastic deformations to arise across a cross-section is $M = M_Y$. As long as $M < M_Y$, the cross-section behaves elastically.

Different strategies can be employed for the analysis of a frame with the forward Euler method and elastic–perfectly plastic beam lamellae. The strategy presented here is based on

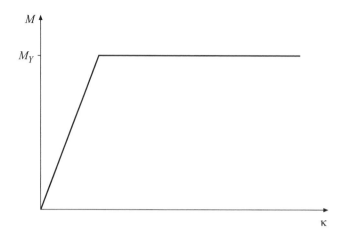

Figure 10.8   An elastic–perfectly plastic moment–curvature relation at the cross-section level

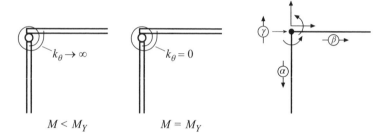

$M < M_Y$ $M = M_Y$

Figure 10.9   Modelling of a plastic hinge

that locations where yielding can be expected are predefined in the computational model. At these locations, a rotational spring is inserted between the adjoining beam elements (Figure 10.9). A rigid connection between the beam elements can then be modelled by assigning an infinite stiffness k_θ to the inserted springs. In practise, it is appropriate to choose a k_θ that is in an order of magnitude of 10^4 to 10^6 times the rotational stiffness of the adjoining beam elements. This rotational stiffness is $\frac{4D_{EI}}{L}$; cf. $K_{3,3}$ in (4.87). When yielding occurs, the plastic hinge is modelled by letting the stiffness of the corresponding rotational spring be zero, $k_\theta = 0$.

Analogous to the bar in a truss, (10.5) and (10.6), an incremental element relation for a rotational spring can be established.

$$k_\theta \begin{bmatrix} 1 & -1 \\ -1 & 1 \end{bmatrix} \begin{bmatrix} \Delta u_1 \\ \Delta u_2 \end{bmatrix} = \begin{bmatrix} \Delta P_1 \\ \Delta P_2 \end{bmatrix} \tag{10.10}$$

where

$$k_\theta = 10^6 \cdot \frac{4D_{EI}}{L} \quad \text{for} \quad M < M_Y$$

$$k_\theta = 0 \quad \text{for} \quad M = M_Y \tag{10.11}$$

How to apply the incremental calculation procedure described in Section 10.1 to the analysis of a frame by using these relations is shown in Example 10.2.

Example 10.2 Frame analysis with material non-linearity considered

The load-carrying capacity of the frame in Example 4.2 will be determined with material non-linearity considered. The frame is rigidly fixed in the lower left end and hinged in the lower right end, according to Figure 1. The lengths of the beams are $L_1 = 4.0$ m, $L_2 = 4.0$ m and $L_3 = 6.0$ m, respectively, the cross-sectional areas are $A_1 = 2.0 \times 10^{-3}$ m^2, $A_2 = 2.0 \times 10^{-3}$ m^2 and $A_3 = 6.0 \times 10^{-3}$ m^2, respectively, and the moments of inertia $I_1 = 1.6 \times 10^{-5}$ m^4, $I_2 = 1.6 \times 10^{-5}$ m^4 and $I_3 = 5.4 \times 10^{-5}$ m^4, respectively. The beams are made of an elastic–perfectly plastic material with elastic modulus $E = 200$ GPa and their yield moment are $M_{Y,1} = 50$ kNm, $M_{Y,2} = 50$ kNm and $M_{Y,3} = 100$ kNm, respectively. The uniformly distributed load q_0 in Example 4.2 is here replaced by a downwards directed point load P_1 acting at the midpoint of the horizontal beam. In the upper left corner, a horizontally directed point load $P_2 = P_1/10$ acts. We seek the maximal load $P_{1,u}$ that the frame can carry before it becomes a mechanism and collapses.

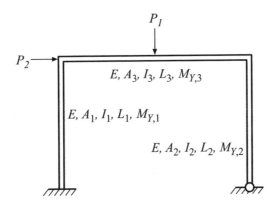

Figure 1   A frame with an elastic–perfectly plastic material

Computational model

The computational model is built up of four beam elements denoted 1–4 according to Figure 2. To enable modelling of the plastic hinges of the frame, the four rigid connections are replaced by rotational springs, denoted 5–8. Initially, these springs are given a very large stiffness, $k_\theta = 10^6 \cdot \frac{4DEI}{L} = 3.2 \times 10^6$ MNm^2, such that their effect on the calculation results is negligible. Once the yield moment is reached at an adjacent beam end, the stiffness of the rotational spring is set to zero, $k_\theta = 0$. The beam element with the lowest yield moment determines when yielding occurs in the rotational spring, thus for the four springs $M_Y^{(5)} = 50$ kNm, $M_Y^{(6)} = 50$ kNm, $M_Y^{(7)} = 100$ kNm and $M_Y^{(8)} = 50$ kNm.

A total of 19 degrees of freedom is required for the materially non-linear computational model. Of these are a_1, a_2, a_3, a_{17} and a_{18} prescribed to be zero. With element and degree of freedom numbering according to Figure 2, we obtain the following two topology matrices for the frame

$$
\text{topology} = \begin{bmatrix} 1 & 5 & 6 & 7 & 1 & 2 & 4 \\ 2 & 13 & 14 & 16 & 17 & 18 & 19 \\ 3 & 5 & 6 & 8 & 9 & 10 & 11 \\ 4 & 9 & 10 & 12 & 13 & 14 & 15 \end{bmatrix} \tag{1}
$$

$$
\text{topology} = \begin{bmatrix} 5 & 3 & 4 \\ 6 & 7 & 8 \\ 7 & 11 & 12 \\ 8 & 16 & 15 \end{bmatrix} \tag{2}
$$

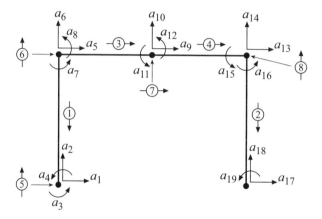

Figure 2   Computational model

Calculation procedure

The external load is applied gradually with small load increments ΔP_i. For each load increment, a system of equations is established, where the assembled element stiffness matrices are determined by the present spring stiffnesses k_θ. Corresponding displacement increment is calculated, after which the element section forces as well as the displacements and external load of the frame are updated according to the calculation procedure for the forward Euler method described in Section 10.1.

Forward Euler for one load increment

We choose a load increment of $\Delta P_1 = 1.0$ kN and begin the loading of the frame. The incremental load vector becomes

$$\Delta \mathbf{f}_l = \begin{bmatrix} 0 \\ 0 \\ 0 \\ 0 \\ 0.1 \\ 0 \\ 0 \\ 0 \\ 0 \\ -1.0 \\ 0 \\ 0 \\ 0 \\ 0 \\ 0 \\ 0 \\ 0 \\ 0 \\ 0 \end{bmatrix} 10^3 \tag{3}$$

The element stiffness matrices $\mathbf{K}_T^e$ for all the beam and spring elements are established according to (4.108) and (2.7). These are assembled into a global tangent stiffness matrix $\mathbf{K}_T$, which is reduced, considering the boundary conditions. Thereafter, the incremental displacements $\Delta \mathbf{a}$ are calculated by solving the system of equations

$$\mathbf{K}_T \Delta \mathbf{a} = \Delta \mathbf{f}_l \tag{4}$$

From the calculated incremental displacements, incremental section moments are determined by using (4.69) and (2.3). The incremental moments in the rotational springs are initially calculated to

$$\Delta M^{(5)} = -0.0958 \text{ kNm} \tag{5}$$

$$\Delta M^{(6)} = -0.1334 \text{ kNm} \tag{6}$$

$$\Delta M^{(7)} = 1.2145 \text{ kNm} \tag{7}$$

$$\Delta M^{(8)} = 0.4376 \text{ kNm} \tag{8}$$

At the end of each load increment, the element section forces are updated

$$N^{e,i+1} = N^{e,i} + \Delta N^e \tag{9}$$

$$V^{e,i+1} = V^{e,i} + \Delta V^e \tag{10}$$

$$M^{e,i+1} = M^{e,i} + \Delta M^e \tag{11}$$

as well as the displacements and loading of the frame

$$\mathbf{a}^{i+1} = \mathbf{a}^i + \Delta \mathbf{a}^i \qquad (12)$$

$$\mathbf{f}^{i+1} = \mathbf{f}^i + \Delta \mathbf{f}^i \qquad (13)$$

Check for yielding and collapse

For each load increment, we check that updated bending moments $M^{e,i+1}$ do not exceed the yield moment in any element. After load increment 83 we have the loads $P_1 = 83 \cdot 1.0 = 83$ kN and $P_2 = 83 \cdot 0.1 = 8.3$ kN. The moments in the rotational springs are

$$M^{(5)} = 83 \cdot (-0.0958) = -7.95 \text{ kNm} \qquad (14)$$

$$M^{(6)} = 83 \cdot (-0.1334) = -11.07 \text{ kNm} \qquad (15)$$

$$M^{(7)} = 83 \cdot 1.2145 = 100.80 \text{ kNm} \qquad (16)$$

$$M^{(8)} = 83 \cdot 0.4376 = 36.32 \text{ kNm} \qquad (17)$$

In Element 7, the moment has been calculated to 100.80 kNm, thus it has exceeded the yield moment of 100 kNm. To check whether the yielding in Element 7 leads to a collapse of the frame, we calculate a new reduced stiffness matrix $\mathbf{K}_T$ with the stiffness $k_\theta^{(7)} = 0$. Since also now $\det \mathbf{K}_T > 0$ the calculation can continue with new load increments.

We proceed with the same load increment $\Delta P_1 = 1.0$ kN which, with the updated stiffness matrix $\mathbf{K}_T$, gives new incremental moments in the rotational springs.

$$\Delta M^{(5)} = 0.1914 \text{ kNm} \qquad (18)$$

$$\Delta M^{(6)} = -1.2043 \text{ kNm} \qquad (19)$$

$$\Delta M^{(7)} = 0 \qquad (20)$$

$$\Delta M^{(8)} = 1.7957 \text{ kNm} \qquad (21)$$

After 91 increments, the loads are $P_1 = 91 \cdot 1.0 = 91$ kN and $P_2 = 91 \cdot 0.1 = 9.1$ kN. The moments in the rotational springs are then

$$M^{(5)} = -7.95 + (91 - 83) \cdot 0.1914 = -6.42 \text{ kNm} \qquad (22)$$

$$M^{(6)} = -11.07 + (91 - 83) \cdot (-1.2043) = -20.71 \text{ kNm} \qquad (23)$$

$$M^{(7)} = 100.80 + (91 - 83) \cdot 0 = 100.80 \text{ kNm} \qquad (24)$$

$$M^{(8)} = 36.32 + (91 - 83) \cdot 1.7957 = 50.69 \text{ kNm} \qquad (25)$$

In Element 8, we now have a bending moment of 50.69 kNm which exceeds the yield moment of 50 kNm. To check whether the yielding in Element 8 leads to a collapse of the frame, we calculate a new reduced stiffness matrix $\mathbf{K}_T$ with the stiffness $k_\theta^{(8)} = 0$. Since $\det \mathbf{K}_T > 0$, the calculation can once again proceed with new load increments.

We keep the load increment of $\Delta P_1 = 1.0$ kN which with the updated stiffness matrix $\mathbf{K}_T$ gives new incremental moments in the rotational springs

$$\Delta M^{(5)} = -3.4000 \text{ kNm} \tag{26}$$

$$\Delta M^{(6)} = -3.0000 \text{ kNm} \tag{27}$$

$$\Delta M^{(7)} = 0 \tag{28}$$

$$\Delta M^{(8)} = 0 \tag{29}$$

After 101 increments, the loads are $P_1 = 101 \cdot 1.0 = 101$ kN and $P_2 = 101 \cdot 0.1 = 10.1$ kN. The moments in the rotational springs are then

$$M^{(5)} = -6.42 + (101 - 91) \cdot (-3.4000) = -40.42 \text{ kNm} \tag{30}$$

$$M^{(6)} = -20.71 + (101 - 91) \cdot (-3.0000) = -50.71 \text{ kNm} \tag{31}$$

$$M^{(7)} = 100.80 + (101 - 91) \cdot 0 = 100.80 \text{ kNm} \tag{32}$$

$$M^{(8)} = 50.69 + (101 - 91) \cdot 0 = 50.69 \text{ kNm} \tag{33}$$

In Element 6, we obtain a bending moment of 50.71 kNm, which exceeds the yielding moment of 50 kNm. To check whether the yielding in Element 6 leads to a collapse of the frame, we calculate a new reduced stiffness matrix $\mathbf{K}_T$ with the stiffness $k_\theta^{(6)} = 0$. The determinant of the reduced stiffness matrix $\det \mathbf{K}_T = 0$, that is we have a mechanism, the frame has collapsed and the calculation is finished.

The calculated displacements after 101 increments are shown in Figure 3 and the moment distribution after 83, 91 and 101 increments is shown in Figure 4. The load P_1 versus displacement of the midpoint of the horizontal beam is shown in the diagram in Figure 5.

The calculated maximal load is approximately 25% larger than the load where the first plastic hinge arises.

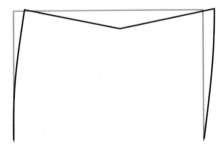

Figure 3   Calculated displacements after increment 101

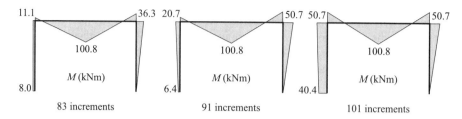

83 increments   91 increments   101 increments

Figure 4   Calculated moment distribution after 83, 91 and 101 increments

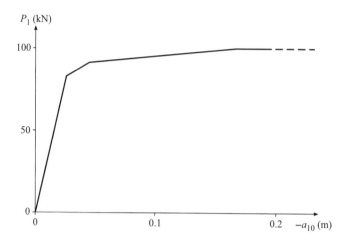

Figure 5   Calculated relation of the load P_1 versus the vertical displacement of the midpoint of the horizontal beam

Exercises

10.1

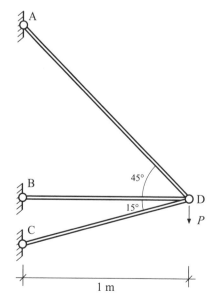

A truss with dimensions as shown consists of three bars and loaded with a force P, which is initially 0 and then increases until it reaches the ultimate load, at which the truss collapses. The material of the bars is elastic–perfectly plastic with $E = 200.0$ GPa and $\sigma_Y = 200.0$ MPa in both tension and compression. All bars have the cross-sectional area $A = 4.0 \times 10^{-4}$ m^2. Determine a load–displacement relationship for vertical displacement at D and the load. Calculate the maximal load P_u that the truss can carry before a mechanism arises and the structure collapses.

10.2

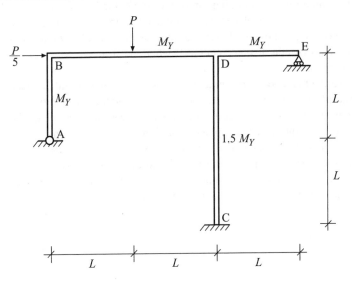

The frame in the figure is to be analysed considering material non-linearity. To prepare for a rational calculation procedure rotational springs with high initial stiffness are placed at the positions where plastic hinges may occur.

(a) Show in a figure an appropriate computational model with elements and degrees of freedom.
(b) Give the topology of the elements.
(c) Give the degrees of freedom which should be prescribed.
(d) How many plastic hinges are formed at least before a mechanism arises?
(e) How many plastic hinges are formed at the most before a mechanism arises?

10.3

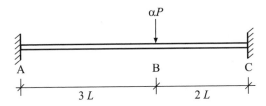

The figure shows a clamped beam made of an elastic–perfectly plastic material with $E = 10,000$ and $M_Y = 1$. The beam is loaded with a point load αP where $P = 1$. For $A = 100$, $I = 1$, and $L = 1$, determine the ultimate load $\alpha_{cr}P$ for which the beam collapses. Describe the load–displacement relationship for vertical displacement at B and give the rotation at C.

10.4 The frame in Exercise 10.2 is constructed of beams made of an elastic–perfectly plastic material with the properties $E = 12\,\text{GPa}$, $M_Y = 100\,\text{kNm}$ and with the plastic rotational capacity $\phi_{cr} = 30 \times 10^{-3}$. For all beams $A = 0.12\,\text{m}^2$ and $I = 1.6 \times 10^{-3}\,\text{m}^4$, and for the in the figure given measures of length $L = 1.5\,\text{m}$. Start from a state where the frame is loaded by a vertical load P and a horizontal load $P/5$ according to the figure and where $P = 100\,\text{kN}$. At this condition, we have the load factor $\alpha = 1$. From this state the load is increased gradually.

(a) Determine the moment diagram of the frame, the load factor α and the magnitude of the horizontal displacement at B when the first plastic hinge arises.

(b) Determine the moment diagram of the frame, the load factor α_{cr} and the magnitude of the horizontal displacement at B when the structure becomes a mechanism and collapses.

(c) Describe the load–displacement relationship for horizontal displacement at B.

(d) The plastic rotation for a plastic hinge is obtained as the difference in rotation between the two connecting rotational degrees of freedom of the plastic hinge. Determine the maximum plastic rotation that arises in any of the plastic hinges and check that the specified rotation capacity is not exceeded.

(e) Assume that we consider the material to be linearly elastic and do not allow any plastic flow. What yield moment M_Y would the frame have to be designed for if it is exposed to the load $\alpha_{cr}P$ according to (b)?

Appendix A

Notations

$\mathbf{f}$ force vector

$\mathbf{f}^e$ element force vector

$\hat{\mathbf{f}}^e$ expanded element force vector

$\bar{\mathbf{f}}^e$ element force vector in local coordinates

$\mathbf{f}_b$ boundary force vector

$\mathbf{f}_b^e$ nodal force vector

$\hat{\mathbf{f}}_b^e$ expanded nodal force vector

$\bar{\mathbf{f}}_b^e$ nodal force vector in local coordinates

$\mathbf{f}_l$ load vector

$\mathbf{f}_{ln}$ nodal load vector

$\mathbf{f}_{lq}$ equivalent nodal load vector

$\mathbf{f}_l^e$ element load vector

$\hat{\mathbf{f}}_l^e$ expanded element load vector

$\bar{\mathbf{f}}_l^e$ element load vector in local coordinates

$\mathbf{a}$ displacement vector

$\mathbf{a}^e$ element displacement vector

$\bar{\mathbf{a}}^e$ element displacement vector in local coordinates

---

Structural Mechanics: Modelling and Analysis of Frames and Trusses, First Edition.
Karl-Gunnar Olsson and Ola Dahlblom.
© 2016 John Wiley & Sons, Ltd. Published 2016 by John Wiley & Sons, Ltd.

K   stiffness matrix

$\mathbf{K}^e$ element stiffness matrix

$\hat{\mathbf{K}}^e$ expanded element stiffness matrix

$\bar{\mathbf{K}}^e$ element stiffness matrix in local coordinates

$\boldsymbol{\alpha}$ vector with constants of integration

C   matrix with constants

$\bar{\mathbf{N}}$ matrix with variables

N   matrix with variables arranged to form shape functions

B   differentiated **N**-matrix

G   matrix for transformation between different coordinate systems

H   matrix for transformation between different sets of degrees of freedom

Appendix B

Answers to the Exercises

1.1 (a) $\begin{bmatrix} 2 & 1 & 3 \\ 5 & 8 & 2 \end{bmatrix}$

    (b) $\begin{bmatrix} -10 & 0 \\ -16 & 4 \end{bmatrix}$

    (c) $\begin{bmatrix} 4 & 8 & 0 \\ -4 & -6 & 2 \\ 16 & 28 & -4 \end{bmatrix}$

    (d) $\begin{bmatrix} 11 & 2 & 8 \\ 36 & 16 & 20 \end{bmatrix}$

    (e) 28

1.2 (a) 4×6
    (b) 3×1
    (c) it is not possible to perform this operation
    (d) 4×8
    (e) 6×4

1.3 $\begin{bmatrix} a_1 \\ a_2 \\ a_3 \end{bmatrix} = \begin{bmatrix} 1 \\ -2 \\ 2 \end{bmatrix}$

1.4 (a) $a_3 = 2, f_1 = -4, f_2 = -6$
    (b) $a_2 = 3, a_3 = 2, f_1 = -10$
    (c) $a_2 = -1, a_4 = 1, f_1 = -8, f_3 = -1, f_5 = 6$

1.5 (a) $\mathbf{A} + \mathbf{B} = \mathbf{B} + \mathbf{A}$
    (b) $\mathbf{AB} \neq \mathbf{BA}$

Structural Mechanics: Modelling and Analysis of Frames and Trusses, First Edition.
Karl-Gunnar Olsson and Ola Dahlblom.
© 2016 John Wiley & Sons, Ltd. Published 2016 by John Wiley & Sons, Ltd.

(c) $(\mathbf{AB})^T = \mathbf{B}^T \mathbf{A}^T, (\mathbf{BA})^T \neq \mathbf{B}^T \mathbf{A}^T$

(d) $\mathbf{CD}$ 4×4-matrix, $\mathbf{DC}$ 1×1-matrix

(e) $\mathbf{C}^T \mathbf{AC}$ 1×1-matrix

(f) $\det \mathbf{A} = 416, \mathbf{AA}^{-1} = \mathbf{I}$

1.6 (a) $\begin{bmatrix} a_1 \\ a_2 \\ a_3 \\ a_4 \end{bmatrix} = \begin{bmatrix} 4 \\ 6 \\ 1 \\ -4 \end{bmatrix}$

(b) infinitely many solutions, $\det \mathbf{K} = 0, \mathbf{f} = \mathbf{0}$

(c) there is no unique solution, $\det \mathbf{K} = 0, \mathbf{f} \neq \mathbf{0}$

1.7 (a) $\lambda_1 = 1, \lambda_2 = 11$

(b) $\mathbf{a}_1 = t_1 \begin{bmatrix} 1 \\ 3 \end{bmatrix}, \mathbf{a}_2 = t_2 \begin{bmatrix} 3 \\ -1 \end{bmatrix}$

2.1 $a_2 = 0.4\frac{F}{k}, \quad a_3 = 0.6\frac{F}{k}, \quad f_1 = -0.4F, \quad f_4 = -0.6F, \quad N^{(1)} = 0.4F, \quad N^{(2)} = 0.4F,$
$N^{(3)} = -0.6F$

2.2 $a_2 = 0.4, a_3 = 0.6, f_1 = -0.4, f_4 = -0.6, N^{(1)} = 0.4, N^{(2)} = 0.4, N^{(3)} = -0.6$

2.3 (a) $\mathbf{K}^1 = \begin{bmatrix} 3 & -3 \\ -3 & 3 \end{bmatrix}; \quad \mathbf{K}^2 = \begin{bmatrix} 6 & -6 \\ -6 & 6 \end{bmatrix}; \quad \mathbf{K}^3 = \begin{bmatrix} 7 & -7 \\ -7 & 7 \end{bmatrix};$ topology $= \begin{bmatrix} 1 & 1 & 2 \\ 2 & 2 & 3 \\ 3 & 3 & 4 \end{bmatrix};$

$\mathbf{K} = \begin{bmatrix} 3 & -3 & 0 & 0 \\ -3 & 9 & -6 & 0 \\ 0 & -6 & 13 & -7 \\ 0 & 0 & -7 & 7 \end{bmatrix}$

(b) $\det \mathbf{K} = 0$, there is no unique solution. Boundary conditions not defined, rigid body motion is not prevented

(c) one

(d) $a_2 = 0.1543, a_3 = 0.1481, f_1 = -0.4630, f_4 = -1.0370$

(e) $a_2 = 0.0356, a_3 = 0.0433, f_1 = -0.0467, f_4 = 0.0467$

2.4 topology $= \begin{bmatrix} 1 & 1 & 2 \\ 2 & 2 & 3 \\ 3 & 3 & 4 \\ 4 & 3 & 4 \\ 5 & 2 & 4 \\ 6 & 4 & 5 \end{bmatrix}$

$\mathbf{K} = \begin{bmatrix} k_1 & -k_1 & 0 & 0 & 0 \\ -k_1 & k_1 + k_2 + k_5 & -k_2 & -k_5 & 0 \\ 0 & -k_2 & k_2 + k_3 + k_4 & -k_3 - k_4 & 0 \\ 0 & -k_5 & -k_3 - k_4 & k_3 + k_4 + k_5 + k_6 & -k_6 \\ 0 & 0 & 0 & -k_6 & k_6 \end{bmatrix}$

3.1 (a) $y_0 = 0.200$ m

(b) $y_0 = 0.191$ m

(c) $y_0 = 0.222$ m

3.2 (a) $D_{EA} = 3.6$ GN, $D_{EA} = 3.83$ GN, $D_{EA} = 3.6$ GN

(b) $N = 3.6$ MN, $N = 3.83$ MN, $N = 3.6$ MN

3.3 (a)

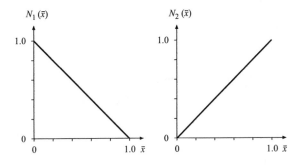

(b)

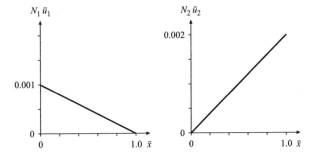

(c)

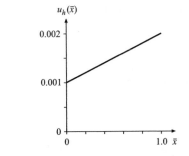

(d) $N(\bar{x}) = 1.0 \times 10^6$

3.4 (a) $u_p(\bar{x}) = \dfrac{q_0 L}{6 D_{EA}} \left(\bar{x} - \dfrac{\bar{x}^3}{L^2} \right)$

(b)

$$\frac{u_p(\bar{x})}{q_0 L^2/D_{EA}}$$

0.10

0.05

0

0   1.0 $\frac{\bar{x}}{L}$

(c) $N_p(\bar{x}) = \frac{q_0 L}{6}\left(1 - \frac{3\bar{x}^2}{L^2}\right)$

$$\frac{N_p(\bar{x})}{q_0 L}$$

0.2

1.0 $\frac{\bar{x}}{L}$

−0.2

−0.4

(d) $\bar{\mathbf{f}}_l^e = \frac{q_0 L}{6}\begin{bmatrix}1\\2\end{bmatrix}$

3.5 $u_B = \dfrac{q_0 L^2}{3 D_{EA}}$

3.6 $u(\bar{x}) = 7.5 \times 10^{-4}\bar{x} - 6.25 \times 10^{-5}\bar{x}^3,\ N(\bar{x}) = 3.0 \times 10^5 - 7.5 \times 10^4\bar{x}^2$

3.7 (a) $n_{x\bar{x}} = \dfrac{\sqrt{3}}{2} = 0.866,\ n_{y\bar{x}} = 0.5$

      (b) $n_{x\bar{x}} = \dfrac{1}{\sqrt{10}} = 0.316,\ n_{y\bar{x}} = \dfrac{3}{\sqrt{10}} = 0.948$

      (c) $n_{x\bar{x}} = -\dfrac{1}{\sqrt{10}} = -0.316,\ n_{y\bar{x}} = \dfrac{3}{\sqrt{10}} = 0.948$

3.8 $\mathbf{G} = \begin{bmatrix}0.894 & 0.447 & 0 & 0\\ 0 & 0 & 0.894 & 0.447\end{bmatrix}$

3.9 $K_{11}^e: K_{7,7},\ K_{24}^e: K_{8,16},\ K_{32}^e: K_{15,8}$

3.10 (a) $\bar{u}_1^{(7)} = -4.128$ mm, $\bar{u}_2^{(7)} = -5.628$ mm
       (b) $\delta = -1.500$ mm, $N = -120$ kN, $\sigma = -120$ MPa

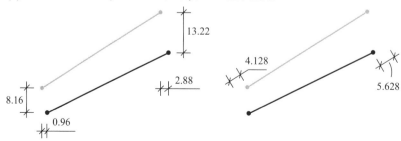

Note that the displacements have been drawn in exaggerated scale.

3.11 (a)
$$\mathbf{K}^1 = \begin{bmatrix} 3.60 & 4.80 & -3.60 & -4.80 \\ 4.80 & 6.40 & -4.80 & -6.40 \\ -3.60 & -4.80 & 3.60 & 4.80 \\ -4.80 & -6.40 & 4.80 & 6.40 \end{bmatrix} 10^6$$

(b)
$$\mathbf{K}^2 = \begin{bmatrix} 0 & 0 & 0 & 0 \\ 0 & 12.50 & 0 & -12.50 \\ 0 & 0 & 0 & 0 \\ 0 & -12.50 & 0 & 12.50 \end{bmatrix} 10^6; \quad \mathbf{K}^3 = \begin{bmatrix} 3.60 & -4.80 & -3.60 & 4.80 \\ -4.80 & 6.40 & 4.80 & -6.40 \\ -3.60 & 4.80 & 3.60 & -4.80 \\ 4.80 & -6.40 & -4.80 & 6.40 \end{bmatrix} 10^6$$

(c)
$$\text{topology} = \begin{bmatrix} 1 & 7 & 8 & 5 & 6 \\ 2 & 7 & 8 & 3 & 4 \\ 3 & 1 & 2 & 7 & 8 \end{bmatrix}$$

(d)
$$\mathbf{K} = \begin{bmatrix} 3.60 & -4.80 & 0 & 0 & 0 & 0 & -3.60 & 4.80 \\ -4.80 & 6.40 & 0 & 0 & 0 & 0 & 4.80 & -6.40 \\ 0 & 0 & 0 & 0 & 0 & 0 & 0 & 0 \\ 0 & 0 & 0 & 12.50 & 0 & 0 & 0 & -12.50 \\ 0 & 0 & 0 & 0 & 3.60 & 4.80 & -3.60 & -4.80 \\ 0 & 0 & 0 & 0 & 4.80 & 6.40 & -4.80 & -6.40 \\ -3.60 & 4.80 & 0 & 0 & -3.60 & -4.80 & 7.20 & 0 \\ 4.80 & -6.40 & 0 & -12.50 & -4.80 & -6.40 & 0 & 25.30 \end{bmatrix} 10^6$$

(e)
$$\text{boundary conditions} = \begin{bmatrix} 1 & 0 \\ 2 & 0 \\ 3 & 0 \\ 4 & 0 \\ 5 & 0 \\ 6 & 0 \end{bmatrix}; \quad \mathbf{f}_l = \begin{bmatrix} 0 \\ 0 \\ 0 \\ 0 \\ 0 \\ 0 \\ 40 \\ -30 \end{bmatrix} 10^3$$

(f) $a_7 = 5.556$ mm, $a_8 = -1.186$ mm, $f_{b,1} = -25.69$ kN, $f_{b,2} = 34.26$ kN, $f_{b,3} = 0$,
$f_{b,4} = 14.82$ kN, $f_{b,5} = -14.31$ kN, $f_{b,6} = -19.08$ kN

(g) $N^{(1)} = -23.85$ kN, $N^{(2)} = 14.82$ kN, $N^{(3)} = 42.82$ kN

(h)
$$\mathbf{a}^1 = \begin{bmatrix} 5.556 \\ -1.186 \\ 0 \\ 0 \end{bmatrix} 10^{-3}; \quad \mathbf{a}^2 = \begin{bmatrix} 5.556 \\ -1.186 \\ 0 \\ 0 \end{bmatrix} 10^{-3}; \quad \mathbf{a}^3 = \begin{bmatrix} 0 \\ 0 \\ 5.556 \\ -1.186 \end{bmatrix} 10^{-3}$$

4.1 (a) $D_{EI} = 48.0$ MNm^2, $D_{EI} = 52.8$ MNm^2, $D_{EI} = 46.1$ MNm^2

(b) $M = 48.0$ kNm, $M = 52.8$ kNm, $M = 46.1$ kNm

4.2

(a)

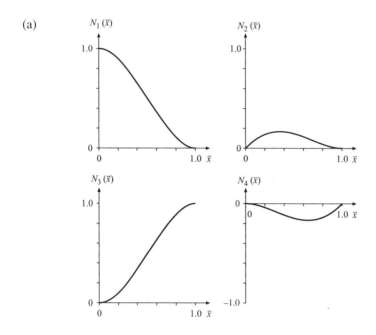

(b)

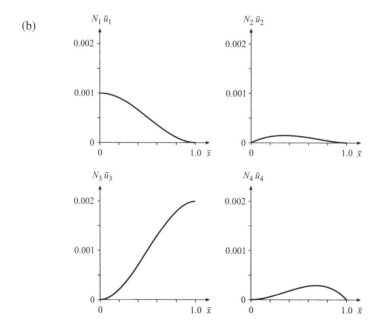

(c)

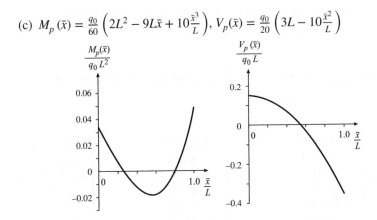

(d) $V(\bar{x}) = 180$ kN, $M(\bar{x}) = 60 - 180\bar{x}$ kNm

(e) $V(0) = 180$ kN, $V(L) = 180$ kN, $M(0) = 60$ kNm, $M(L) = -120$ kNm

4.3   (a) $v_p(\bar{x}) = \dfrac{q_0}{120 D_{EI}} \left(2L^2\bar{x}^2 - 3L\bar{x}^3 + \dfrac{\bar{x}^5}{L} \right)$

(b)

$$\frac{v_p(\bar{x})}{q_0\, L^4/D_{EI}}$$

(c) $M_p(\bar{x}) = \dfrac{q_0}{60} \left(2L^2 - 9L\bar{x} + 10\dfrac{\bar{x}^3}{L} \right)$, $V_p(\bar{x}) = \dfrac{q_0}{20} \left(3L - 10\dfrac{\bar{x}^2}{L} \right)$

$$\frac{M_p(\bar{x})}{q_0\, L^2}$$

$$\frac{V_p(\bar{x})}{q_0\, L}$$

4.4 $a_4 = -\dfrac{q_0 L^3}{48 EI}$

4.5 $v(\bar{x}) = 4.167 \times 10^{-5}\bar{x}^4 - 1.042 \times 10^{-4}\bar{x}^3 + 6.25 \times 10^{-5}\bar{x}^2$, $M(\bar{x}) = 500\bar{x}^2 - 625\bar{x} + 125$, $V(\bar{x}) = -1000\bar{x} + 625$

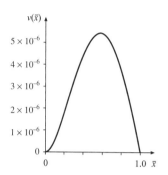

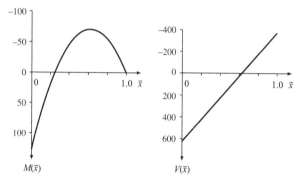

4.6   (a) $n_{x\bar{x}} = \dfrac{\sqrt{3}}{2} = 0.866$, $n_{y\bar{x}} = 0.5$, $n_{x\bar{y}} = -0.5$, $n_{y\bar{y}} = \dfrac{\sqrt{3}}{2} = 0.866$

      (b) $n_{x\bar{x}} = \dfrac{1}{\sqrt{10}} = 0.316$, $n_{y\bar{x}} = \dfrac{3}{\sqrt{10}} = 0.948$, $n_{x\bar{y}} = -\dfrac{3}{\sqrt{10}} = -0.948$,

        $n_{y\bar{y}} = \dfrac{1}{\sqrt{10}} = 0.316$

      (c) $n_{x\bar{x}} = -\dfrac{1}{\sqrt{10}} = -0.316$, $n_{y\bar{x}} = \dfrac{3}{\sqrt{10}} = 0.948$, $n_{x\bar{y}} = -\dfrac{3}{\sqrt{10}} = -0.948$,

        $n_{y\bar{y}} = -\dfrac{1}{\sqrt{10}} = -0.316$

4.7   (a)

$$\mathbf{a}^1 = \begin{bmatrix} 0.0000 \\ 0.0000 \\ 0.0000 \\ 8.7296 \\ -0.6960 \\ -13.8747 \end{bmatrix} 10^{-3}; \quad \mathbf{a}^2 = \begin{bmatrix} 0.0000 \\ 0.0000 \\ 0.0000 \\ 8.6325 \\ -0.7440 \\ -11.8603 \end{bmatrix} 10^{-3}; \quad \mathbf{a}^3 = \begin{bmatrix} 8.7296 \\ -0.6960 \\ -13.8747 \\ 8.6325 \\ -0.7440 \\ -11.8603 \end{bmatrix} 10^{-3}$$

(b)

$$\bar{\mathbf{a}}^1 = \begin{bmatrix} 0.0000 \\ 0.0000 \\ 0.0000 \\ -0.6960 \\ -8.7296 \\ -13.8747 \end{bmatrix} 10^{-3}$$

(c) $N^{(1)}(0) = -34.801$ kN, $V^{(1)}(0) = 5.706$ kN, $M^{(1)}(0) = 5.862$ kNm,

$N^{(1)}(L) = -34.801$ kN, $V^{(1)}(L) = 5.706$ kN, $M^{(1)}(L) = -16.962$ kNm

4.8 (a)

$$\bar{\mathbf{K}}^1 = \begin{bmatrix} 210 & 0 & 0 & -210 & 0 & 0 \\ 0 & 0.896 & 1.344 & 0 & -0.896 & 1.344 \\ 0 & 1.344 & 2.688 & 0 & -1.344 & 1.344 \\ -210 & 0 & 0 & 210 & 0 & 0 \\ 0 & -0.896 & -1.344 & 0 & 0.896 & -1.344 \\ 0 & 1.344 & 1.344 & 0 & -1.344 & 2.688 \end{bmatrix} 10^6;$$

$$\bar{\mathbf{f}}_l^1 = \begin{bmatrix} 0 \\ 15 \\ 7.5 \\ 0 \\ 15 \\ -7.5 \end{bmatrix} 10^3$$

$$\bar{\mathbf{K}}^2 = \begin{bmatrix} 210 & 0 & 0 & -210 & 0 & 0 \\ 0 & 0.4375 & 1.05 & 0 & -0.4375 & 1.05 \\ 0 & 1.05 & 3.36 & 0 & -1.05 & 1.68 \\ -210 & 0 & 0 & 210 & 0 & 0 \\ 0 & -0.4375 & -1.05 & 0 & 0.4375 & -1.05 \\ 0 & 1.05 & 1.68 & 0 & -1.05 & 3.36 \end{bmatrix} 10^6;$$

$$\bar{\mathbf{f}}_l^2 = \begin{bmatrix} 0 \\ -48.0 \\ -38.4 \\ 0 \\ -48.0 \\ 38.4 \end{bmatrix} 10^3$$

(b)

$$\mathbf{K}^1 = \begin{bmatrix} 0.896 & 0 & 1.344 & -0.896 & 0 & 1.344 \\ 0 & 210 & 0 & 0 & -210 & 0 \\ 1.344 & 0 & 2.688 & -1.344 & 0 & 1.344 \\ -0.896 & 0 & -1.344 & 0.896 & 0 & -1.344 \\ 0 & -210 & 0 & 0 & 210 & 0 \\ 1.344 & 0 & 1.344 & -1.344 & 0 & 2.688 \end{bmatrix} 10^6;$$

$$\mathbf{f}_l^1 = \begin{bmatrix} 15.0 \\ 0 \\ 7.5 \\ 15.0 \\ 0 \\ -7.5 \end{bmatrix} 10^3$$

$$\mathbf{K}^2 = \begin{bmatrix} 210 & 0 & 0 & -210 & 0 & 0 \\ 0 & 0.4375 & 1.05 & 0 & -0.4375 & 1.05 \\ 0 & 1.05 & 3.36 & 0 & -1.05 & 1.68 \\ -210 & 0 & 0 & 210 & 0 & 0 \\ 0 & -0.4375 & -1.05 & 0 & 0.4375 & -1.05 \\ 0 & 1.05 & 1.68 & 0 & -1.05 & 3.36 \end{bmatrix} 10^6;$$

$$\mathbf{f}_l^2 = \begin{bmatrix} 0 \\ -48.0 \\ -38.4 \\ 0 \\ -48.0 \\ 38.4 \end{bmatrix} 10^3$$

(c)

$$\mathbf{K} = \begin{bmatrix} 0.896 & 0 & -1.344 & -0.896 & 0 & -1.344 & 0 & 0 & 0 \\ 0 & 210 & 0 & 0 & -210 & 0 & 0 & 0 & 0 \\ -1.344 & 0 & 2.688 & 1.344 & 0 & 1.344 & 0 & 0 & 0 \\ -0.896 & 0 & 1.344 & 210.896 & 0 & 1.344 & -210 & 0 & 0 \\ 0 & -210 & 0 & 0 & 210.4375 & 1.05 & 0 & -0.4375 & 1.05 \\ -1.344 & 0 & 1.344 & 1.344 & 1.05 & 6.048 & 0 & -1.05 & 1.68 \\ 0 & 0 & 0 & -210 & 0 & 0 & 210 & 0 & 0 \\ 0 & 0 & 0 & 0 & -0.4375 & -1.05 & 0 & 0.4375 & -1.05 \\ 0 & 0 & 0 & 0 & 1.05 & 1.68 & 0 & -1.05 & 3.36 \end{bmatrix} 10^6$$

$$\mathbf{f}_l = \begin{bmatrix} 15 \\ 0 \\ -7.5 \\ 15 \\ -48 \\ -15.9 \\ 0 \\ -48 \\ 38.4 \end{bmatrix} 10^3$$

(d)

$$\text{boundary conditions} = \begin{bmatrix} 1 & 0 \\ 2 & 0 \\ 3 & 0 \\ 7 & 0 \\ 8 & 0 \\ 9 & 0 \end{bmatrix}$$

(e) $a_4 = 0.0878$ mm, $a_5 = -0.2151$ mm, $a_6 = -2.611 \times 10^{-3}$, $f_{b,1} = -11.57$ kN, $f_{b,2} = 45.16$ kN, $f_{b,3} = 4.109$ kNm, $f_{b,7} = -18.43$ kN, $f_{b,8} = 50.84$ kN, $f_{b,9} = -43.013$ kNm.

(f)

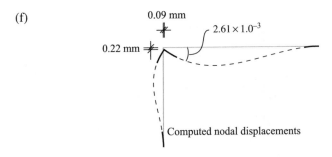

Computed nodal displacements

(g)

$$\bar{\mathbf{a}}^1 = \begin{bmatrix} 0.2151 \\ 0.0878 \\ -2.6111 \\ 0 \\ 0 \\ 0 \end{bmatrix} 10^{-3}; \quad \bar{\mathbf{f}}_b^1 = \begin{bmatrix} 45.164 \\ -18.431 \\ -14.401 \\ -45.164 \\ -11.569 \\ 4.109 \end{bmatrix} 10^3$$

$$\bar{\mathbf{a}}^2 = \begin{bmatrix} 0.0878 \\ -0.2151 \\ -2.6111 \\ 0 \\ 0 \\ 0 \end{bmatrix} 10^{-3}; \quad \bar{\mathbf{f}}_b^2 = \begin{bmatrix} 18.431 \\ 45.164 \\ 29.401 \\ -18.431 \\ 50.836 \\ -43.013 \end{bmatrix} 10^3$$

(h) $M^{(1)}(\bar{x}) = (14.401 - 18.431\bar{x} + 5\bar{x}^2) \times 10^3$
$V^{(1)}(\bar{x}) = (18.431 - 10\bar{x}) \times 10^3$
$M^{(2)}(\bar{x}) = (-29.401 + 45.164\bar{x} - 10\bar{x}^2) \times 10^3$
$V^{(2)}(\bar{x}) = (-45.164 + 20\bar{x}) \times 10^3$

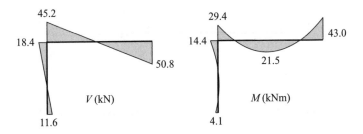

4.10 (a) $a_4 = -0.195$ mm, $a_5 = -3.505$ mm, $a_6 = 4.241 \times 10^{-3}$
(b) $f_{b,3} = 86.310$ kNm
(c) $\sigma = 77.90$ MPa

4.11   (a) $f_{b,1} = 63$ kN, $f_{b,3} = 69$ kN, $f_{b,5} = -12$ kN

(b) $f_{b,1} = 29$ kN, $f_{b,3} = -73$ kN, $f_{b,5} = 44$ kN

(c) $f_{b,1} = 92$ kN, $f_{b,3} = -4$ kN, $f_{b,5} = 32$ kN

4.12

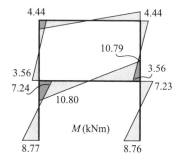

4.13   Maximum stress in bar -70.7 MPa (the leaning bar to the left), maximum bending moment in the beam 22.1 kNm, bending moment 4 m from the left support -7.9 kNm, bending moment 8 m from the left support -3.9 kNm.

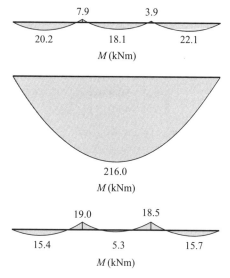

The difference in the support moment occurs because the axial stiffness of the beam is finite.

5.1   (a) $a_{13} = 0$, $a_{15} = 0$

(b) Symmetric: $a_{13} = 0$, $a_{15} = 0$, Anti-symmetric: $a_{14} = 0$

5.2 $a_2 = 0$, $a_{11} = 0$, $a_{13} = 0$ (note that a_1 shall not be prescribed)

5.3 Horizontal displacement at E: 4.97 mm, vertical displacement at E: −1.27 mm.

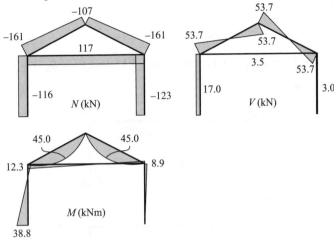

5.4 $a_4 = -0.1031$ mm, $a_5 = -0.0006$ mm, $a_6 = 0.0183$, $a_7 = -0.1305$ mm, $a_9 = -0.0480$, $f_{b,1} = 11.283$ kN, $f_{b,2} = 4.956$ kN, $f_{b,3} = -18.979$ kNm, $f_{b,8} = 18.804$ kN

5.5 Vertical deflection at the right end: 0.237 m (downwards). Support forces: 2.8 MN to the right, 0.05 MN upwards, 0.715 MNm moment.

5.6 (a) $a_3 = 5.2083 \times 10^{-3}$, $a_4 = 0$

(b) $$\begin{bmatrix} 12 & 6 & -12 & 6 \\ 6 & 4 & -6 & 2 \\ -12 & -6 & 12 & -6 \\ 6 & 2 & -6 & 4 \end{bmatrix} \begin{bmatrix} a_1 \\ a_2 \\ a_5 \\ a_6 \end{bmatrix} = \begin{bmatrix} f_1 \\ f_2 \\ f_5 \\ f_6 \end{bmatrix} + \begin{bmatrix} 0.5 \\ 0.125 \\ 0.5 \\ -0.125 \end{bmatrix}$$

5.7 $k = 4.8980 \times 10^5$

5.8 (a) beam: $k = 0.66667 \times 10^6$ N/m, bar: $k = 1.12291 \times 10^6$ N/m
(b) $A_2 = 1 \times 10^{-5}$: $M_{\text{mid-point}} = 12.6551$ kNm, $A_2 = 1 \times 10^{-4}$: $M_{\text{mid-point}} = 5.5879$ kNm, $A_2 = 1 \times 10^{-3}$: $M_{\text{mid-point}} = 1.5274$ kNm.

5.9 (a)

(b) $K_{1,1} = 3k$, $K_{2,1} = -2k$, $K_{1,2} = -2k$, $K_{2,2} = 3k$

5.10 (a)

(b) $K_{4,4} = 210.896$ MN/m, $K_{5,4} = 0$, $K_{6,4} = 1.344$ MN, $K_{6,6} = 6.048$ MNm

5.11   Horizontal displacement of BDE 0.107 m, rotation at B -0.0166, rotation at D -0.0050.

6.1 $a_3 = -6.11$ mm $N = -24.42$ kN

6.2 $u_A = u_B = 0.20$ mm, $v_A = 1.12$ mm, $v_B = -5.12$ mm, $\theta_A = \theta_B = -6.24 \times 10^{-3}$

6.3   (a) $\delta_C = 8.37$ mm

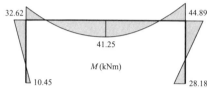

(b) $\delta_C = 15.41$ mm

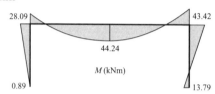

6.4   (a) $k_x u(0) = 1.313 \dfrac{P}{L}$, $k_x u(L) = 0.851 \dfrac{P}{L}$

      (b) $k_x u(0) = 5.00 \dfrac{P}{L}$, $k_x u(L) = 0.067 \dfrac{P}{L}$

      (c) $u(0) = 6.46$ mm, $u(10) = 0.134$ mm, $u(20) = 5.58 \times 10^{-3}$ mm, $N(0) = -100$ kN, $N(10) = -2.079$ kN, $N(20) = 0$

      (d) $u(0) = 2.92$ mm, $u(20) = -1.25$ mm

      (e) $u(0) = 4.31$ mm, $u(10) = -0.556$ mm, $u(20) = 0.139$ mm. If the number of elements is increased, the solution approaches the exact one.

      (f) $u(0) = 5.63$ mm, $u(10) = 0.040$ mm, $u(20) = 0.559 \times 10^{-3}$ mm, $N(10) = -0.70$ kN

6.5   (a) $\theta_B = 0.719 \times 10^{-3}$

      (b) Two elements give $\theta_B = 1.849 \times 10^{-3}$

6.6   (b)

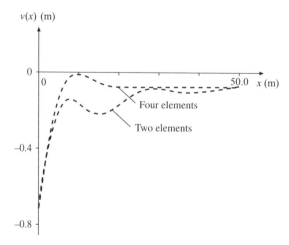

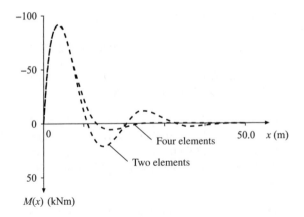

7.1
$$G = \begin{bmatrix} 0.81111 & 0.32444 & -0.48666 & 0 & 0 & 0 \\ 0 & 0 & 0 & 0.81111 & 0.32444 & -0.48666 \end{bmatrix}$$

7.2 (a) $\bar{u}_1^{(5)} = 4.286$ mm, $\bar{u}_2^{(5)} = 0$
(b) change in length $= -4.286$ mm, $N^{(5)} = -173.2$ kN, $\sigma^{(5)} = -173.2$ MPa

7.3 $u_{A,x} = -1.115$ mm, $u_{A,y} = -6.233$ mm, $u_{A,z} = 0.874$ mm, $N^{AB} = 63.74$ kN,
$N^{AC} = -19.76$ kN, $N^{AD} = -26.52$ kN

7.4 (a) $$\bar{f}_l^e = \frac{q_\omega L}{2} \begin{bmatrix} 1 \\ 1 \end{bmatrix}$$

(b) $u_B = \dfrac{q_0 L^2}{2 D_{GK}}$

7.5
$$G = \begin{bmatrix} \mathbf{C} & 0 & 0 & 0 \\ 0 & \mathbf{C} & 0 & 0 \\ 0 & 0 & \mathbf{C} & 0 \\ 0 & 0 & 0 & \mathbf{C} \end{bmatrix}; \quad C = \begin{bmatrix} 0.89443 & 0.44722 & 0 \\ -0.44722 & 0.89443 & 0 \\ 0 & 0 & 1.00000 \end{bmatrix}$$

7.6 (a) $a_{13} = 0$, $a_{17} = 0$, $a_{18} = 0$
(b) $a_{14} = 0$, $a_{15} = 0$, $a_{16} = 0$

7.7 Largest vertical displacement at E $= 4.541$ mm (downwards); Section forces at B: shear force $= 7.014$ kN, torque $= 0.966$ kNm, bending moment $= 4.799$ kNm (tension at upper edge); Section forces at C: shear force $= 12.986$ kN, torque $= 0.923$ kNm, bending moment $M = 7.308$ kNm (tension at upper edge)

8.1 (a) $K = 8000$ MN/m
(b) $K = 420$ MN/m
(c) $K = 2.5$ W/m²K
(d) $K = 25$ W/m²K
(e) $K = 0.0833$ Ω^{-1}

8.2 (a) Temperature distribution: -10.0, -9.6, -9.1, 18.0, 18.6, $20.0\,°C$, Heat flux: $10.8\,W/m^2$

    (b) Temperature distribution: -10.0, -9.5, -9.0, 21.0, 21.7, $23.2\,°C$

8.3 $a_B = 80.5$ V, $a_C = 57.5$ V, $I^{e,(AB)} = -1.95$ A, $I^{e,(BC)} = -1.15$ A, $I^{e,(BD)} = -0.80$ A, $I^{e,(CD)} = -1.15$ A, $I_A = 1.95$ A, $I_D = -1.95$ A

8.4 $p_E = 187.8$ kPa, $p_F = 136.1$ kPa, $p_G = 140.8$ kPa, $p_H = 128.2$ kPa, $H_{AE} = 149.2$ cm^3/s, $H_{EF} = 52.9$ cm^3/s, $H_{FB} = 36.9$ cm^3/s, $H_{EG} = 96.2$ cm^3/s, $H_{FH} = 16.0$ cm^3/s, $H_{GH} = 12.8$ cm^3/s, $H_{HD} = 28.9$ cm^3/s, $H_{GC} = 83.4$ cm^3/s. For $D = 0.005$ m the flow is laminar if $H < 106.3$ cm^3/s, and for $D = 0.010$ m if $H < 212.5$ cm^3/s, that is laminar in all pipes.

8.5 (a) Temperature between brick work and mineral wool $= -7.64$ °C.

    (b) Vapour concentration between brick work and mineral wool $= 4.60 \times 10^{-3}$ $kg/m^3 > 2.61 \times 10^{-3}$ kg/m^3.

    (c) Amount of water condensed $= 0.11$ $kg/m^2/week$.

9.1 Force in spring $= 2H = 0.3$, Horizontal displacement $= 0.1$. The horizontal force H results in a slight misalignment of the hinged columns. In the deformed state this misalignment causes a tensile force in the spring of $2H$. This means that the stairwell needs to be designed for a horizontal force that includes both the horizontal load H and an equally large additional force due to the misalignment of the hinged columns.

9.2 (a) $a_5 = -21.07$ mm, $a_6 = -39.20$ mm, $Q_{\bar{x}}^{(1)} = -5.268$ MN, $Q_{\bar{x}}^{(2)} = 0.333$ MN

    (b) $a_5 = -21.87$ mm, $a_6 = -48.81$ mm, $Q_{\bar{x}}^{(1)} = -5.468$ MN, $Q_{\bar{x}}^{(2)} = 0.590$ MN

9.3 $\alpha_{cr} = 2.48$

9.5 (a) $\theta_B = 0.250\frac{M_0 L}{EI}$, $M_A = -0.500 M_0$, $M_B = M_0$, $M\left(\frac{L}{2}\right) = 0.250 M_0$

    (b) $\theta_B = 0.288\frac{M_0 L}{EI}$, $M_A = -0.614 M_0$, $M_B = M_0$, $M\left(\frac{L}{2}\right) = 0.335 M_0$

    (c) $\alpha_{cr} = 7.6$, Euler's third buckling case $\alpha_{cr} = 5.1$

9.6 Horizontal displacement at B (mm)

	Clamped at C	Roller support at C
Linear	0.20	135.04
Geometrically nonlinear	0.20	162.72

Support moment at A (kNm)

	Clamped at C	Roller support at C
Linear	−42.23	60.50
Geometrically nonlinear	−43.68	80.14

9.7 (a) $2.2 < \alpha_{cr} < 8.9$
(b) $\alpha_{cr} = 5.90$

9.8 (a) $4.4 < \alpha_{cr} < 8.7$
(b) $\alpha_{cr} = 11.08$
(c) $\alpha_{cr} = 6.28$

10.1 $P_u = 77.3$ kN

P (kN)	δ_D (mm)
0	0
69.5	2.10
77.3	4.97

10.2 (a) Proposed model:

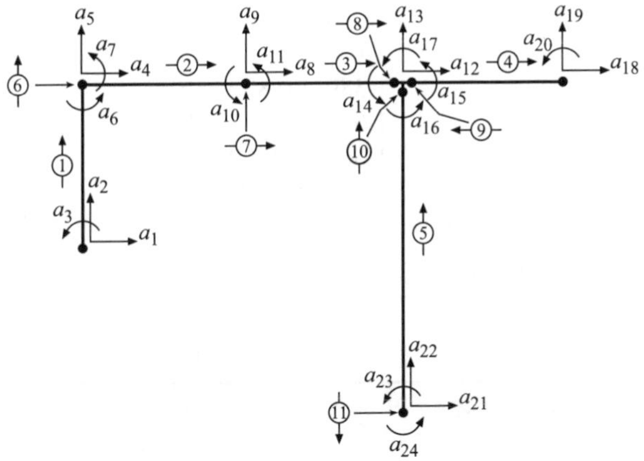

(b)

$$
\text{topology} =
\begin{bmatrix}
1 & 1 & 2 & 3 & 4 & 5 & 6 \\
2 & 4 & 5 & 7 & 8 & 9 & 10 \\
3 & 8 & 9 & 11 & 12 & 13 & 14 \\
4 & 12 & 13 & 15 & 18 & 19 & 20 \\
5 & 21 & 22 & 23 & 12 & 13 & 16
\end{bmatrix};
\quad
\text{topology} =
\begin{bmatrix}
6 & 6 & 7 \\
7 & 10 & 11 \\
8 & 14 & 17 \\
9 & 15 & 17 \\
10 & 16 & 17 \\
11 & 23 & 24
\end{bmatrix}
$$

(c) $a_1 = a_2 = a_{19} = a_{21} = a_{22} = a_{24} = 0$
(d) Three plastic hinges, for example

(e) Four plastic hinges (threefold statically indeterminate structure), for example

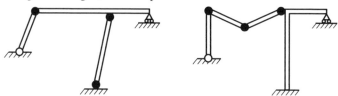

10.3 $\alpha_{cr}P = 1.667, \theta_C = 4.17 \times 10^{-5}$.

αP	$-a_6$
0	0
1.389	0.800×10^{-4}
1.620	1.083×10^{-4}
1.667	1.500×10^{-4}

10.4   (a) $\alpha = 2.01, \delta_B = 4.34$ mm

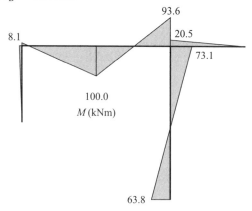

(b) $\alpha_{cr} = 2.50$, $\delta_B = 15.42$ mm

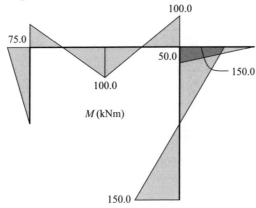

(c)

αP (kN)	δ_B (mm)
0	0
201.1	4.34
206.8	4.54
246.9	12.78
250.0	15.42

(d) At the mid-point of beam BD: $\phi = 22.78 \times 10^{-3} < \phi_{cr}$

(e) $M_Y = 124.3$ kNm

Index

Structural Mechanics: Modelling and Analysis of Frames and Trusses, First Edition.
Karl-Gunnar Olsson and Ola Dahlblom.
© 2016 John Wiley & Sons, Ltd. Published 2016 by John Wiley & Sons, Ltd.